Reliability Technology for Complex Software Systems

Yichen Wang • Yikun Wang • Fuqun Huang

Reliability Technology for Complex Software Systems

Springer

Yichen Wang
School of Reliability and Systems Engineering
Beihang University
Beijing, China

Yikun Wang
Antares Ltd.
Beijing, China

Fuqun Huang
Department of Computer Science
Western Washington University
Seattle, USA

ISBN 978-981-95-3888-1   ISBN 978-981-95-3889-8 (eBook)
https://doi.org/10.1007/978-981-95-3889-8

Jointly published with National Defense Industry Press

Translation from the English language edition: “复杂软件系统可靠性技术” by Yichen Wang et al.,

This Springer imprint is published by the registered company Springer Nature Singapore Pte Ltd.
The registered company address is: 152 Beach Road, #21-01/04 Gateway East, Singapore 189721, Singapore

Preface

For clarifying the concept of complex software systems, it is necessary to clarify two key points regarding complex software systems. First, the term "complex software" is not a newly introduced concept. As early as the 1970s, researchers used the term to describe certain types of software systems. Over time, with rapid advancements in computer and network technologies, the connotation of "complex software" has continued to evolve and expand. This evolution has led to the emergence of various typical complex systems, such as ultra-large-scale systems (ULSS), cyber-physical systems (CPS), and the Internet of Things (IoT). These systems exhibit far greater complexity than earlier software systems in terms of scale, structure, and data, making them a representative of modern complex software systems. Second, the "complexity" of complex software systems extends far beyond mere scale. While factors such as the number of lines of code or the quantity of files within a system are indicators of complexity, contemporary complex software systems exhibit a multidimensional complexity. This complexity arises primarily from the integration of numerous advanced technologies and methodologies, including intelligent systems, network technologies, and self-adaptive mechanisms. The incorporation of these cutting-edge technologies is the fundamental driver of the multidimensional complexity observed in modern complex software systems.

This book summarizes the recent research achievements of the authors' team over the past few years. Anchored in the field of software reliability engineering, the book systematically reviews and analyzes the concept of complex software systems. It introduces the latest research findings on fault mechanisms and reliability models for such systems. Furthermore, it explores two distinct approaches to reliability engineering: architecture-based and data-driven reliability technologies. The book also addresses the increasingly prominent topic of model-based testing techniques. Finally, it presents state-of-the-art advancements in human factors engineering within complex software systems, reflecting the forefront of research in this domain.

This book serves as a valuable reference for professionals engaged in software reliability engineering within the field of equipment development, as well as for graduate students specializing in related disciplines.

The content of this book represents the culmination of years of research conducted by the authors and their team. We extend our heartfelt gratitude to all the teachers and students who contributed to this effort. In particular, we would like to express our special thanks to Shu Shaoguang, Wang Kun, Wang Xun, and Yan Xiaobo for their contributions to Chaps. 2–4; Zong Pengyang and Song Zekun for their work on Chap. 5; and Lan Xinsheng, Yue Ming, Lei Haishen, and Cao Zhiqin for their contributions to Chap. 6.

The authors also consulted a wide range of references during the writing of this book and would like to extend their sincere gratitude to the authors of these works for their invaluable contribution. Given the limitations of the authors' knowledge and experience, errors or inaccuracies may be unavoidable. We sincerely welcome readers' feedback and constructive suggestions for improvement.

Beijing, China — Yichen Wang
Beijing, China — Yikun Wang
Seattle, USA — Fuqun Huang
2025

Contents

Chapter 1
Introduction of Complex Software

From the inception of computer software programs—commonly recognized as starting with the program designed by Ada Lovelace, the daughter of the renowned British poet Lord Byron, for Charles Babbage's Analytical Engine to solve Bernoulli equations—software practitioners have continuously been engaged in a challenging struggle to achieve high software reliability while managing the inherent complexity of software. The human brain is considered the most intricate structure in the universe, and software is one of its most complex creations, making complexity an intrinsic characteristic of software. Throughout the processes of software design, development, and verification, managing complexity has remained a crucial factor.

Following the software crisis of the 1960s, the complexity of software had reached a level that was beyond human control, prompting the emergence of software engineering as a formal discipline. Stephen Hawking once remarked that complexity is the science of the twenty-first century. In the domain of software, the rapid advancements in network technologies and intelligent systems have led to an unprecedented increase in software complexity. Complexity has evolved from being an inherent property of software to becoming a defining descriptor for an entire category of software systems, giving rise to the concept of complex software systems. The development of complex software systems is poised to have profound impacts on the future. Consequently, theories and technologies related to software engineering must undergo corresponding transformations to address the challenges posed by complex software systems.

1.1 About the Complexity of Software

The development of software has greatly benefited from advancements in computer hardware. The increasing speed of computation and the expanding storage capacity of hardware have laid a strong foundation for software growth. Conversely, software development is gradually shifting the focus away from hardware products, while

Y. Wang et al., *Reliability Technology for Complex Software Systems*,
https://doi.org/10.1007/978-981-95-3889-8_1

also driving an increasing demand for software reliability and security. On one hand, people's daily lives are becoming increasingly intertwined with software, while on the other, their expectations for software quality continue to rise. The software revolution has become an essential part of the ongoing technological transformation.

To better understand the complexity of modern software, we can explore it from three key dimensions: scale, behavior, and development process.

1.1.1 Infinite Expansion of Software Scale

As a result of the rapid development of networking and information technology, the complexity of software systems has primarily manifested in the exponential expansion of their scale. The Internet, widely recognized today, originated from ARPANET, which was established by the Advanced Research Projects Agency (ARPA) of the United States Department of Defense in 1969. Initially, ARPANET consisted of only four nodes: the University of California, Los Angeles (UCLA), the Stanford Research Institute (SRI), the University of California, Santa Barbara (UC Santa Barbara), and the University of Utah. These nodes were all located in the western United States. By 1990, ARPANET had completed its mission, having grown from just four nodes to approximately 300,000, and played a pivotal role in the development of foundational network technologies, including email, Telnet, FTP, and TCP/IP. By 2014, QQ, a distributed application supporting instant messaging, had achieved the capacity to support over 200 million concurrent online nodes.

However, the challenges posed by increasing complexity cannot be overlooked. For instance, the UK government initially planned to invest £11 billion in a project to install smart meters in households and small businesses by 2020. Following reassessment and revision, the project has been postponed to 2024, with the total cost expected to exceed an additional £2 billion. This example underscores that while software systems can enhance efficiency and convenience, their development and maintenance entail significant costs, requiring substantial time and effort to ensure quality and reliability.

In the realm of military equipment, the U.S. military is currently developing the "Global Information Grid" (GIG), which aims to establish seamless information links—ranging from sensors to shooters and from the White House to individual soldier positions—through advanced networking and distributed computing software. This ambitious project is expected to encompass more than two million nodes. Simultaneously, the U.S. military is planning the development of equipment software systems comprising over 1 billion lines of code. These efforts underscore the dramatic evolution in the scale of equipment software systems over time. Software has become an integral component of modern military equipment, with its complexity and scale continuously expanding to meet the demands of evolving technologies and operational requirements (Table 1.1).

Table 1.1 Software scale statistics

Approximate time	Software name	Lines of code (LOC)
1990s	Space-Based Infrared System (SBRIS) satellite software	Two hundred fifty thousand
The year 2009	Boeing 787 Aircraft Flight Control System	Six million five hundred thousand
The year 2010	General Motors (GM) Volt in-vehicle computer software	Ten million
The year 2015	F-35 onboard software	Twelve million
The year 2020	Linux Kernel Code	Twenty-seven million

1.1.2 Complexity of Software Failure Behaviors

The behavior of complex software is characterized by significant uncertainties and emergent properties, which increase the likelihood of software failures. The consequences of such failures are becoming progressively more severe, occasionally reaching catastrophic levels. The following three cases demonstrate the complexity of software failure behavior and the gravity of its consequences.

1.1.2.1 F-35 Fighter Jet

An article in The Daily Telegraph on June 28, 2014, revealed that the F-35 fighter jets, purchased from the United States and flown back to the UK by British pilots, "encountered problems." All ten aircraft were affected. In an interview, the UK Air Force Commander Bertche admitted that the F-35s had "faced a series of issues" upon their return to the UK, with software problems being particularly prominent. These software failures rendered all F-35s non-operational, temporarily disabling them from participating in training or combat readiness missions.

The UK had originally planned to designate the F-35 as the next-generation mainstay fighter jet for its air force and a carrier-based aircraft for its navy.

How could a fighter jet purchased at such a high cost encounter significant issues after a single flight back from the United States? An article in The Daily Telegraph quoted an internal source from the UK Air Force, stating, "There were some minor hardware issues, which can be resolved quickly, but the software failures are much more problematic." In modern fifth-generation fighter jets, while hardware components such as air-to-air missiles, engines, and aerodynamic design remain critical, electronic systems and software related to combat operations, command, and communication have become equally vital. These advanced fighter jets are akin to computers—hardware alone is insufficient without robust and reliable software.

According to media reports, the Japanese F-35A fighter jet that crashed into the sea in 2009 has yet to be recovered. However, the United States has issued a warning to Japan, suggesting that the downed F-35A may have been hacked. The

U.S. investigation concluded that the crash was caused by the pilot losing consciousness due to hypoxia, resulting in the aircraft becoming uncontrollable and plunging into the sea. However, public opinion widely suspects that the F-35 has significant software vulnerabilities, leaving it susceptible to hacking, which could potentially be one of the real causes of the crash.

1.1.2.2 F-22 Fighter Jet Malfunction

In February 2007, a US F-22 fighter jet was flying from Hawaii to Japan, but the software was not designed to consider geographic boundaries. When it crossed the International Date Line, it triggered a software defect, causing the global positioning system on the aircraft to fail and the computer system to crash. The pilot could not determine the position of the fighter jet and could only return to the Hickam Air Force Base in Hawaii. Lockheed Martin maintained the software and provided a new version within 48 h.

1.1.2.3 Malfunction of Beijing International Airport Information System

At 13:28 on October 10, 2007, the departure system of China National Aviation Information Network Co., Ltd., located at Beijing Capital International Airport, experienced a sudden malfunction. Within just 50 min, at least 84 departing flights from airports in Beijing, Guangzhou, Shenzhen, and Changsha were delayed. The incident impacted numerous cities, including Shanghai, Changchun, Nanjing, Nanning, Wenzhou, Chengdu, Zhengzhou, Taiyuan, Hohhot, Chongqing, Lanzhou, Hong Kong, Tokyo, and others. Post-incident analysis revealed that the system had been developed by a U.S. company, and software defects were identified as the root cause, raising concerns about the security of the information system.

1.1.3 Complexity of Software Engineering Process

Complex software systems are typically composed of numerous locally autonomous software systems that are intricately coupled and interdependent. They exhibit characteristics of "systems of systems," "cyber-physical systems," and "socio-technical systems," and demonstrate new properties such as heterogeneity of components, open boundaries, emergent behaviors, and continuous evolution. These features challenge the fundamental assumptions of traditional software engineering theories and methodologies, which are rooted in reductionist principles, rendering them insufficient for constructing complex software systems.

From the perspective of software engineering participants, the development of complex software involves a diverse array of stakeholders, including investors with differing objectives, developers, maintainers, and end-users. From the perspective of system components, this process entails the continuous integration of highly heterogeneous and autonomous systems. Traditional software development assumptions—such as the ability to precisely define requirements in advance, strictly control development processes, and predict outcomes—are no longer applicable. Consequently, classic development approaches, such as "top-down decomposition" and "bottom-up assembly," struggle to effectively support the construction of these highly complex software systems.

Moreover, due to the immense scale and intricate internal structure of complex software systems, it is impractical to design, develop, and deploy them in a single iteration. Over time, the process of continuous development, deployment, updates, and adjustments may span several years or even decades. While the general public may perceive the increasing complexity of software systems through their usage, for professionals engaged in software engineering, this growing complexity poses even greater challenges and difficulties.

1.2 The Basic Concept of Complex Software Systems

The rapid escalation of software complexity has garnered significant attention and spurred extensive research within both industry and academia. Relevant studies have been conducted across various domains of software engineering, leading to the formalization of the concept of complex software systems.

1.2.1 Definition of Complex Software Systems

A complex software system refers to a large-scale software system formed through the continuous integration of numerous locally autonomous software systems that are inter-coupled and closely intertwined with the social and physical systems in which they operate. The coupling and interaction among the elements of such systems are dynamically evolving and increasingly intricate, making it difficult to describe the behavior of the entire system as a mere aggregation of the characteristics of its autonomous components. A defining characteristic of complex software systems is emergence, whereby the system's behavior cannot be reduced to the simple sum of its constituent parts. Examples include typical internet software systems, energy internet software systems, national and global financial information systems, large-scale joint command and control software systems in the field of equipment, and complex avionics integrated systems.

Although the complexity of software has been a topic of discussion since its inception, systematic research into complex software only began with the integration of complex systems theory into software engineering. A complex system is defined as a system consisting of a moderate number of intelligent, adaptive agents that make decisions based on local information. In contrast to simple systems, which have been the focal point of scientific inquiry since the Newtonian era, complex systems exhibit fundamental differences. The components of a simple system interact weakly, allowing the system's behavior to be described through classical reductionist approaches in physics. In contrast, complex systems are composed of multiple interacting units, and their behavior is nonlinear, often giving rise to complex structures with numerous hierarchical levels. In these systems, the individual agents typically possess a degree of intelligence. Examples include the cells in an organism, stock traders in a financial market, or drivers in an urban transportation network. These agents are capable of making intelligent judgments or decisions based on their own set of rules, shaped by the environment they operate in.

According to complex systems theory, the formation mechanisms of complex software systems are characterized by three primary features:

1. **Correlation**: This refers to the fact that the formation of a complex software system is not driven by the properties of individual agents, but rather by the long-range interactions between them. These interactions often result in nonlinearity and emergent synergistic effects within the system.
2. **Temporality**: Complex software systems are typically linked to external physical time, necessitating the characterization of continuous time-varying processes. However, computer systems themselves are inherently discrete, creating a distinction between the two. Moreover, the study of time-varying processes often requires consideration of how the current state influences the system's output in the next time step, i.e., feedback. Feedback can be either positive or negative. Negative feedback tends to drive the software system toward a stable equilibrium, while positive feedback can result in instability within the system.
3. **Usability:** The components of a complex software system are autonomous entities, each potentially exhibiting distinct usage profiles—such as varying users, times, and contexts. The inherent complexity of usability can lead to a substantial increase in the system's state space during operation, which may, in turn, result in instability.

1.2.2 Types of Complex Software Systems

Based on differences in software configuration, complex software systems are typically classified into two categories: integrated complex software and distributed complex software.

Integrated complex software has significantly advanced in terms of code size, the number of modules, and the complexity of inter-module interactions when compared to conventional software. However, it remains an integrated single-machine system. The modules communicate with each other through mechanisms such as method calls, inter-process communication, onboard fiber-optic networks, and others, forming a cohesive and unified whole. Each software module is highly complex and autonomous, yet they collaborate towards a common mission and objective.

The flight control software of the F-35 fighter jet, which utilizes the Integrated Modular Avionics (IMA) architecture, and the Aircraft Information Management System (AIMS) used in the Boeing 777 are representative examples of complex integrated software systems. The F-35's flight control software consists of approximately five million lines of code, while the AIMS, based on the IMA architecture, exceeds 14 million lines of code.

Compared to integrated complex software, distributed complex software is characterized by spatial or geographical distribution. Each software system executes its own independent functions, collaborating and competing autonomously, while interconnecting through network technologies to form a larger-scale complex system. The primary feature of such systems is the complexity of interactions, with system failures primarily arising from abnormal interactions between the various subsystems.

Typical examples of distributed complex networks include Internet of Things (IoT) systems and Cyber-Physical Systems (CPS).

1.2.3 The Characteristics of Complex Software Systems

Both integrated and distributed complex software systems exhibit characteristics that are not found in simpler software systems.

The scale of complex software systems is one of their key characteristics. These systems typically feature vast amounts of code, often comprising millions of lines. Furthermore, the number of users can be exceptionally large, ranging from dozens to hundreds or even thousands, with many users potentially interacting with the system simultaneously. Finally, the volume of data accessed and exchanged within these systems is enormous, with extensive data interactions occurring among the large user base. The continuous increase in system scale is a fundamental cause of the other defining characteristics of complex software systems.

From a structural perspective, complex software systems exhibit the typical characteristics of a "System of Systems" (SoS). Four prominent features of such systems are: "decomposability of structure," "heterogeneity of components," "networked architecture," and "decentralization."

The AI provided a more detailed explanation:

In terms of their structure, complex software systems exhibit several key characteristics, often referred to as "system-of-systems" characteristics. These characteristics include:

1. Decomposability: Complex software systems are composed of a large number of subsystems or modules, which can be decomposed or analyzed independently. This allows the system to be designed, implemented, and maintained in a modular way.
2. Heterogeneity: Complex software systems are often composed of heterogeneous subsystems or modules, which may have different architectures, programming

languages, and data formats. This requires the use of interfaces and protocols that enable these subsystems to communicate and interact with each other.
3. Networked: Complex software systems are often built on top of networked infrastructures, which allow for distributed computation, communication, and data storage. This enables the system to scale and to support large numbers of users and data interactions.
4. Decentralized: Complex software systems are often designed to be decentralized, with no central point of control or coordination. This allows for greater autonomy and flexibility in the system, but also increases the complexity of interactions and coordination among subsystems.

These characteristics pose significant challenges for the design, implementation, and maintenance of complex software systems, requiring the use of specialized techniques and methodologies such as software architecture, interface design, network protocols, and distributed computing.

1. Complex software systems consist of various subsystems, and their structure is both decomposable and describable. Typically, the subsystems of a complex software system can be categorized into three types: physical systems, software systems, and human systems, all of which can be modeled. A common approach to modeling such systems involves first creating an abstract model of each subsystem based on its input-output behavior. The modeling techniques for these individual subsystems may vary, including discrete systems, continuous systems, hybrid systems, and the modeling of non-functional attributes. Subsequently, an abstract model of the composite system is defined based on the individual subsystem models. Throughout this process, the modeling framework progresses from abstraction to specification and, ultimately, to the full complex system, starting with individual subsystems and advancing to system compositions, with continuous refinement of abstract definitions.
2. The components of a complex software system not only exhibit structural differences but also vary in terms of their development teams and processes. Each subsystem may either be an existing system or a newly developed one. Consequently, the development processes, architectures, software and hardware platforms, management strategies, and other aspects of these subsystems are often heterogeneous. Moreover, these subsystems typically involve a large number of stakeholders across different management domains, which can give rise to conflicts of interest. Existing systems are often developed by different teams, each with distinct schedules, processes, and objectives, serving a variety of stakeholder groups. Even newly developed systems may be designed using different methodologies and approaches, leading to discrepancies in environmental assumptions, data consistency, and other factors.

   These differences can lead to various challenges in the development, integration, and maintenance of complex software systems, such as communication barriers, technical incompatibilities, and performance issues. Therefore, it is crucial to adopt appropriate methodologies and tools to manage the development and integration of complex software systems, including software architecture frameworks, software engineering processes, system testing strategies, and quality assurance practices.

3. The degree of networked structure in complex software systems has increased significantly. The vast majority of complex software systems are built upon network-based architectures, characterized by strong interconnectivity and complex topologies.

   The networked nature of complex software systems enables distributed computing, communication, and data storage, which can help the system to scale, to support large numbers of users and data interactions, and to operate more efficiently. However, this also increases complexity in the system's architecture and communication infrastructure.

   The network topology of complex software systems may include various structures, such as centralized, decentralized, peer-to-peer, hierarchical, and hybrid structures. Moreover, the topology may change dynamically due to system evolution, user behavior, or environmental factors.

   As a result, managing the network topology of complex software systems is an important challenge for software engineers and system administrators. It requires specialized techniques and methodologies such as network analysis, network design, network optimization, and network security.

4. Complex software systems typically lack a central control unit to manage all components. The system's components may exist across different spatial and temporal dimensions. In other words, each component may be operated by different users in various locations and at different times, and may possess its own processes for data control, development, usage, and evolution.

   This decentralization of control can bring many benefits to complex software systems, such as increased system autonomy and flexibility, the ability to support distributed computing and data storage, and the ability to tolerate faults and failures. However, it also poses significant challenges for system management and maintenance, such as ensuring data consistency, maintaining system integrity and security, and managing system evolution and integration.

   To manage complex software systems with a decentralized structure, various techniques and methodologies have been developed, such as distributed system architecture, data synchronization and replication, version control, and software configuration management. These techniques aim to ensure the consistency, reliability, and security of the system, while also facilitating system evolution and adaptation to changing requirements and environments.

The complex interconnections between components in complex software systems give rise to emergent behavior, which is one of the fundamental causes of the emergence of complexity in such systems. The interdependencies between components are inherently intricate, and the structure of a complex software system can be formally described using a recursive approach that characterizes the relationships between subsystems (or components). However, even if the relationships between components can be described, this does not imply that the overall behavior of the system can be fully captured by a reductionist approach based solely on the individual components. Emergent behavior at the system level, arising from nonlinear interactions between subsystems, may be either beneficial or detrimental to the system.

In fact, emergent behavior is a common and important phenomenon in complex systems, including complex software systems. The behavior of such systems often arises from interactions between components that cannot be predicted by analyzing the behavior of individual components in isolation. As a result, it is essential to understand the emergent behavior of complex software systems in order to manage and control them effectively. This requires the use of advanced techniques such as system simulation, system testing, and system monitoring to analyze, measure, and manage emergent behavior in complex software systems.

Emergent behavior is a fundamental characteristic of complex software systems, and it is one of the key reasons why software systems are classified as complex. While the structure of a complex software system can often be described in detail, such descriptions alone cannot fully determine the system's behavior. This presents one of the most significant challenges of complex software systems. Furthermore, complex software systems must be capable of autonomously adapting to dynamic changes in their physical environment, and they must possess self-adaptive and reconfigurable capabilities.

This means that the system must be able to detect changes in its environment and adjust its behavior and configuration accordingly. This requires sophisticated techniques such as machine learning, data analytics, and artificial intelligence, which enable the system to learn from its environment and adapt to changing conditions.

Overall, the ability of complex software systems to exhibit emergent behavior, coupled with the need for self-adaptation and self-configuration, make the development and management of such systems a challenging task, requiring the use of advanced software engineering techniques and methodologies as well as interdisciplinary collaboration.

Complex software systems exhibit new characteristics in their construction process. Typically, the formation of complex software systems occurs through two approaches: one-time construction and continuous construction.

One-time construction involves building a large and complex software system from the ground up, adhering to software engineering principles. This process includes design, coding, and verification before the system is deployed for use. While ongoing maintenance, updates, and requirement changes may occur post-deployment, the construction process is largely completed prior to the system's use. Such systems are typically found in fields such as aerospace, industrial control, and command and control.

In contrast, continuous construction involves the expansion of a simple software system over time. Initially, a basic software system is developed and deployed, but as it is used, its scope continuously grows, with new requirements added. This results in the incremental evolution of the system, adding components and improving structure, which gradually transforms it into a more complex system. These types of systems are commonly employed in fields such as communication networks, network systems, and social services.

Additionally, from a complex systems perspective, complex software systems exhibit a phenomenon known as "boundary erosion," which refers to the blurring of boundaries between the system and its human users. In traditional software systems, humans occupy distinct roles as either developers or users. In the development phase, different technical personnel handle various tasks at different stages, with clear role definitions. Users, positioned outside the system, interact with the software via a well-defined human-machine interface.

However, in complex software systems, these roles become less distinct. Developers may also assume the role of users, and users may participate in the software's maintenance and evolution. In some cases, humans may become integral parts of the system itself, leading to the dissolution of the boundary between humans and software. Consequently, the roles of the system, developers, and users become increasingly fluid and less clearly defined.

In terms of software evolution, complex software systems exhibit distinct characteristics. Unlike complex systems in fields such as natural sciences, economics, or sociology, complex software systems are entirely artificial, being deliberately designed and developed by engineers to meet the specific needs of targeted user groups. The formation of complex software systems—whether through one-time construction or continuous construction—always involves a step-by-step process initiated from scratch by the design and development team.

Moreover, complex software systems are characterized by continuous evolution. These systems typically have long lifecycles, during which their usage demands, operating environments, and user interaction patterns may change due to factors such as societal development, hardware advancements, and shifts in user demographics. Such changes tend to persist over time, necessitating ongoing evolution of the software system.

Additionally, due to the autonomy of their constituent components, complex software systems exhibit a certain level of intelligence and possess the capacity for self-evolution.

This means that the system can adapt to changes in its environment and user needs, and continually improve itself. This self-evolving capability is an important characteristic of complex software systems and can be achieved through various techniques such as machine learning, data mining, and evolutionary computation.

Overall, the continuous evolution and self-evolving capability of complex software systems pose significant challenges to software engineers and require the use of advanced software engineering techniques and methodologies to ensure that the system remains reliable, secure, and effective throughout its lifecycle.

1.3 Several Typical Types of Complex Software Systems

This section introduces several terms used to describe different types of complex software systems. These terms have been employed by various researchers to characterize software systems that exhibit complex properties.

1.3.1 Ultra Large Scale Systems (ULS)

Ultra Large Scale Systems (ULS) refer to software systems that are characterized by an exceptionally large number of components, subsystems, or interacting elements. These systems typically involve complex interdependencies and operate in environments where scalability, fault tolerance, and adaptability are critical. Due to their enormous scale, ULS systems often present significant challenges in terms of design, implementation, testing, and maintenance.

The term Ultra Large Scale (ULS) Systems was first introduced in a 2006 research report by Linda Northrop's team at the Carnegie Mellon University Software Engineering Institute. The report aimed to address a question posed by the U.S. Department of Defense regarding how to construct software systems that could potentially contain billions of lines of code, given the current challenges in software engineering. The report emphasized that when software-intensive systems reach unprecedented scale—primarily determined by factors such as the number of lines of code, users and stakeholders, system functions, data storage, data access, data manipulation, refinement efforts, inter-component connections and dependencies, and hardware elements—traditional software engineering and management methods no longer suffice. Once a software system evolves into an ultra-large-scale system, it transcends the scope of conventional systems engineering or system-of-systems engineering and becomes a challenge of socio-technical system engineering.

In 2013, Linda Northrop's team produced a follow-up report, revisiting the 2006 findings and assessing the current state of ULS research. This report argued that rapid technological advancements have accelerated the growth of system scale, leading to the pervasive presence of ULS systems in society and driving significant changes in social and public organizational structures. Consequently, the 2006 research was likely too conservative, and the 2013 report suggests that numerous unresolved issues still require further investigation.

The characteristics of ULS can be summarized as follows:

1. The decentralization of processes in ULS refers to the dispersion of the development process, evolution process, and operational control process.
2. The diversity, conflicts, and unpredictability of requirements are key characteristics of ULS.
3. The continuous evolution of ULS refers to the addition of new functionalities and the removal of existing ones during the system's operation.
4. The heterogeneity and dynamic variability of ULS refer to the presence of heterogeneous, inconsistent, and dynamically evolving components within the system.
5. The boundary between the system and humans becomes blurred in ULS. People are no longer merely users of the system but integral components of it, influencing the emergent behavior of the entire system.
6. The normalcy of failure is a characteristic of ULS, where failure is not considered an exceptional event but rather a routine occurrence. The stability and

reliability of the system cannot depend on stringent assumptions about other member systems.

1.3.2 Large-Scale Complex IT Systems

The term Large-Scale Complex IT System (LSCITS) is primarily used within the academic community in the United Kingdom and carries a meaning similar to that of Ultra-Large-Scale Systems (ULS). This term originated from a research and graduate education project in the UK that focused on the development of large-scale IT systems. The project was funded by the Engineering and Physical Sciences Research Council (EPSRC) and received over £10 million in funding between 2006 and 2013. The initiative was launched following a 2004 research report jointly produced by the Royal Academy of Engineering and the British Computer Society. This report investigated the causes of failure in numerous large-scale software projects and offered several recommendations to address these challenges. The objective of the LSCITS project is to "improve the technical methodologies for complex systems engineering and to develop new socio-technical approaches to better understand the intricate interactions between organizations, processes, and systems."

1.3.3 Cyber-Physical Systems (CPS)

Cyber-Physical Systems (CPS) are a new class of intelligent and complex systems that integrate multidimensional, heterogeneous computing units and physical entities within a networked environment, enabling intricate interactions. CPS are characterized by real-time capabilities, robustness, autonomy, efficiency, and high performance. Below are several commonly cited definitions of CPS.

1. CPS is a tightly integrated system of computational and physical process components, where the operation of physical entities is monitored through a computational core. These physical entities, in turn, leverage networks and computational components to perceive and control their environment.
2. CPS is a highly reliable system in which various computational and physical elements are closely integrated and dynamically coordinated in response to uncertain events.
3. From the perspective of computational science and information storage and processing, CPS is typically regarded as a networked computer system that integrates computing, communication, and storage capabilities, enabling real-time, reliable, secure, stable, and efficient monitoring of physical entities in the physical world.
4. From the perspective of embedded systems and device development, "Cyber" refers to the integration of computing, communication, and control technologies

with physical processes and biological characteristics. The essence of CPS lies in its integration of reliable computing, communication, and control capabilities, forming an intelligent system capable of monitoring and controlling various physical entities.

CPS typically involves embedded systems, sensor technologies, real-time data processing, control algorithms and other technologies. It achieves deep integration and real-time interaction through feedback loops between computing and physical processes to increase or expand new functions in a safe, reliable, efficient and real-time manner for detecting or controlling physical entities.

5. CPS is a controllable, trustworthy, and scalable networked system of physical devices that deeply integrates computing, communication, and control capabilities, based on environmental sensing. Through a feedback loop, where computing processes interact with physical processes, CPS enables deep integration and real-time interaction, allowing for the addition or expansion of new functions in a safe, reliable, efficient, and timely manner, to detect or control physical entities.

Security and reliability are the primary criteria for large-scale complex systems. In a CPS environment, the interaction between information and physical components is more seamless and frequent than in traditional network communication structures. As the network exhibits a "decentralized" nature, users within the network, including both humans and intelligent components, enjoy greater equality and freedom. In such an environment, ensuring the security and privacy of user communications, enhancing the resilience and reliability of CPS components, and achieving continuous monitoring and management of the system in uncertain and complex environments, are all highly challenging and critical issues.

From the perspective of ensuring system security, CPS must be capable of promptly detecting network threats and predicting the potential consequences of attacks. Additionally, it is crucial to recognize the distinct differences between CPS security and that of traditional information systems. As such, it is necessary to consider establishing comprehensive CPS security mechanisms that address multiple levels of defense, including prevention, detection, defensive repair, system recovery, and mitigation of similar future attacks.

Furthermore, current research indicates that during the prediction phase, CPS systems can integrate knowledge from fields such as network science, social science, and dynamics. Technologies such as preference analysis, behavior detection, and percolation prediction can be employed to detect potential threats and issue timely warnings. In terms of improving reliability, this can be achieved by enhancing the system's real-time performance. Additionally, leveraging existing technologies for network resilience and cascade failure prevention can help in the study of sudden abnormal event prevention, as well as the development of real-time warning and repair technologies for system recovery. Furthermore, ensuring the sustained maintenance of system energy is essential. This can be accomplished through coordination and scheduling among components to ensure uninterrupted power supply for production systems, or by developing new energy storage devices with extended lifespans.

1.3.4 *Software-Intensive Systems*

The concept of Software-Intensive Systems (SIS) was first introduced in the 1960s. SIS refers to systems in which software interacts with other software, systems, devices, sensors, and humans. Examples of SIS include embedded systems used in the automotive and aerospace industries, as well as dedicated systems in wireless communications.

Early research distinguished SIS as a separate category of systems due to the fact that their development and operation require expertise from a broad range of domains, including both domain-specific knowledge and software engineering. Furthermore, those involved in the development of SIS often come from diverse professional backgrounds, which can give rise to knowledge gaps. These gaps may stem from differences in how individuals perceive or understand certain technical aspects, which can impede knowledge sharing and reuse, ultimately having a significant impact on the success and efficiency of SIS development.

As the application domains of Software-Intensive Systems (SIS) continue to expand and the number of applications increases, SIS are exhibiting several new characteristics. First, the environments in which SIS operate have become increasingly open and uncertain. Second, the complexity of interactions—both among system components and between the system and its environment—has risen significantly. Lastly, the incompleteness of system requirements and the frequent changes in these requirements have further exacerbated the system's complexity.

In 2005, a report by TNO/IDATE titled "Future Software-Intensive Systems" offered a clearer definition of SIS from a different perspective: "Software-intensive systems are systems in which software plays a dominant role in system functionality, development cost, development risk, and development time." This definition places particular emphasis on the central role of software, which is the primary reason SIS are often categorized as complex software systems.

During the development of Software-Intensive Systems (SIS), the evolution of complexity has consistently been a major driving force behind research in this field. The escalating complexity of SIS has introduced new characteristics, presenting unprecedented challenges to system development, verification, and reliability. These challenges can be summarized as follows:

The following are features provided by AI.

Interdisciplinary Complexity SIS involves various disciplines such as software engineering, hardware engineering, and domain knowledge, which makes it difficult to integrate and coordinate different disciplines.

Heterogeneity Complexity SIS comprises of heterogeneous components, which have different characteristics, functions, and interfaces to communicate with each other, leading to challenges in the effective interaction and collaboration of these components.

Dynamic Complexity The dynamic behavior of SIS is highly complex, and it changes rapidly over time due to various factors, such as user input, environmental changes, and system errors. This makes it difficult to predict and control system behavior.

Uncertainty Complexity The environment in which SIS operates is highly unpredictable, which leads to challenges in dealing with incomplete and uncertain system requirements.

Safety and Reliability Complexity As SIS is increasingly used in mission-critical applications, ensuring the safety and reliability of the system has become a significant concern, which requires the development of novel methods for fault tolerance, error detection, and self-recovery.

1. As software technologies and computer hardware continue to evolve, the complexity of the internal logic of Software-Intensive Systems (SIS) has reached an unprecedented level

   In these systems, the size and complexity of the software have reached new heights, with millions or even tens of millions of lines of code. The coupling between different functional modules within the system is becoming higher and the logical relationships are becoming more complex. This makes the development, maintenance, and testing of SIS extremely challenging, requiring the implementation of more rigorous software engineering methods and standards to ensure the correctness and reliability of the system.

   Traditional software tends to have a smaller scale, with fewer lines of code and less modularization. Additionally, there is less interaction between software components, which makes it possible to use traditional process models for software development, testing, and integration.

   However, the code volume of complex software has increased dramatically, with modularity becoming more pronounced and both the number of modules and their interactions rising significantly. Software components have become more autonomous, and network-based complex software now exhibits a trend of geographical distribution, as seen in systems like command and control. This shift requires more sophisticated methods to manage the transfer of data and workflows, as well as to facilitate collaboration among components. As a result, controlling and predicting the interactions between internal modules has become increasingly challenging. Errors are more likely to occur, are harder to detect, and can spread more easily, making them difficult to diagnose and pinpoint. Traditional methods of reliability analysis, design, and evaluation are ill-equipped to address these new challenges.
2. The complexity of the external environment with which complex software interacts has increased rapidly, causing the system's boundaries to become increasingly blurred.

   Firstly, the complexity of the hardware and environmental resources that complex software-intensive systems depend on for operation is continuously increasing. The frequency of interactions between software and the external environment is rising, requiring real-time adjustments based on environmental factors collected by sensors. Hardware and the environment are no longer merely external elements; they have become integral components of complex software systems. The anomalies arising from interactions between software and hardware or the environment have increasingly become a significant cause of failures in complex software systems.

Furthermore, complex software-intensive systems are increasingly interacting with people, including not only the buyers and developers of the software but also its users. Different stakeholders have varying expectations regarding the functional and non-functional requirements of the software, which leads to uncertainty and even conflicts in the system's requirements. With the rise of ubiquitous computing and the advent of wearable devices, the integration between users and complex software is becoming increasingly seamless. As a result, complex software must dynamically adjust its resources or functionalities based on user needs to accommodate the constantly evolving demands of users.

However, traditional software engineering views software as a standalone product and seldom considers external factors such as the environment and human elements, making it difficult to address failures arising from interactions.

3. As complex software-intensive systems are required to handle increasingly complex tasks, they are unlikely to "retire" in the short term, implying a longer lifecycle.

   Throughout their long lifecycle, complex software-intensive systems must continuously endure internal errors, resist external attacks, and experience failures becoming increasingly normalized. However, such systems must possess the capability to maintain "quasi-stable" operation. "Quasi-stable" refers to high fault tolerance, resilience, graceful degradation in the face of attacks, and a high level of operational health, rather than the traditional notion of absolute "reliability."

   Furthermore, a long lifecycle also implies that requirements will continually evolve, necessitating the constant evolution of complex software to eliminate errors, enhance robustness, meet new demands, and adapt to changing environments.

Various modern weapon systems and information systems, especially computer-controlled systems such as flight control, fire control, guidance, and command-control systems, are typical examples of software-intensive systems. The application of software-intensive systems has expanded across various military domains, including combat command, weapon control, military training, intelligence, communications, reconnaissance, electronic warfare, as well as research, design, testing, simulation, and support. These systems have become a critical technological foundation and pillar for the development of modern defense technology and weapon systems. They not only serve as key enablers for modern warfare, operational command, communications, logistical support, and other decisive factors in warfare, but also have a profound impact on traditional military theory and military concepts.

1.3.5 Internet of Things (IoT) System

The concept of the Internet of Things (IoT) originated from the networked wireless Radio Frequency Identification (RFID) system proposed by the Auto-ID Labs at the Massachusetts Institute of Technology (MIT) in 1999. This system connects all objects to the Internet through information sensing devices, such as RFID, enabling intelligent identification and management. The concept was subsequently widely adopted across various application fields, leading to the formal definition of IoT. IoT refers to an interconnected network of physical devices, vehicles (also known as "connected devices" and "smart devices"), buildings, and other objects embedded with electronics, software, sensors, actuators, and network connectivity, allowing these objects to collect and exchange data. By 2013, the IoT Global Standards Initiative (IoT-GSI) defined IoT as "the infrastructure of the information society." IoT enables objects to be remotely sensed and/or controlled via existing network infrastructure, thereby integrating the physical world more directly into computer-based systems, which improves efficiency, accuracy, and economic benefits. When IoT incorporates sensors and actuators, it becomes an instance of a cyber-physical system, which may include technologies such as smart grids, smart homes, smart transportation, and smart cities. Each object can be uniquely identified through its embedded computing system and can operate seamlessly within the existing Internet infrastructure.

According to reports, by 2021, the number of objects connected by the Internet of Things in China had reached 4.53 billion, and it is expected to exceed 10 billion by 2025.

From the perspective of communication entities and processes, the core of the Internet of Things lies in the exchange of information between objects and between people and objects. The key characteristics of the Internet of Things can be summarized as comprehensive sensing, reliable transmission, and intelligent processing.

1. Comprehensive sensing: By utilizing technologies such as radio frequency identification (RFID), QR codes, and sensors for sensing, capturing, and measuring, information about objects can be collected and obtained anytime and anywhere.
2. Reliable transmission: By connecting objects to the information network and leveraging various communication networks, reliable information exchange and sharing can be conducted anytime and anywhere.
3. Intelligent processing: By leveraging various intelligent computing technologies, vast amounts of sensory data and information can be analyzed and processed to enable intelligent decision-making and control.

To provide a clearer description of the key components of the Internet of Things, an information functional model has been abstracted from the perspective of information science, focusing on the flow of information, as shown in Fig. 1.1.

1. Information acquisition function: This includes both information perception and information recognition. Information perception refers to the sensitivity and

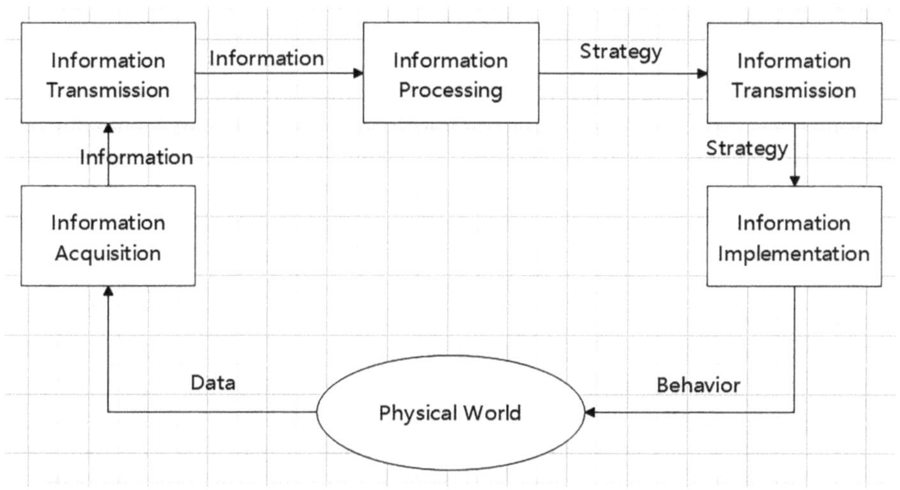

Fig. 1.1 Information functional model of the Internet of Things (IoT)

awareness of the state of objects and their changes. Information recognition refers to the ability to represent the motion state of perceived objects and their changes.

2. Information transmission function: This includes the processes of information sending, transmission, and reception, ultimately completing the task of transmitting the state of objects and their changes from one point in space (or time) to another. This constitutes the general communication process.
3. Information processing function: This refers to the process of processing information, with the aim of acquiring knowledge, understanding the nature of objects, and generating new information from existing data, i.e., the decision-making process.
4. Information application function: This refers to the process in which information is ultimately utilized, manifesting in various forms. The most important aspect is the adjustment of an object's state and its changes through information, in order to achieve the desired motion state of the object.

1.4 Research Areas Related to Complex Software Systems

With the dramatic increase in software complexity, the research field of software science has been continuously expanding. Traditional software engineering techniques are increasingly integrating research outcomes and methodologies from various other fields, leading to cross-disciplinary interactions between software engineering and numerous other disciplines. This section introduces several research areas related to complex software systems.

1.4.1 System Engineering

Systems engineering is a discipline that studies the dynamic trends of complex information feedback systems. Based on information processing and computer simulation technologies, systems engineering analyzes the behavior patterns that emerge in complex systems over time. It views the behavior of a system as being determined by the internal information feedback mechanisms. By developing systems engineering models, the dynamic relationships between the system's structure, function, and behavior can be examined to identify optimal system configurations and functionalities.

The techniques and methods used in systems engineering to address complex problems offer valuable insights and references for the study of complex software systems.

In particular, the understanding of the behavior patterns of complex systems, and the analysis of the dynamic relationships between structure, function, and behavior can be applied to the research of complex software systems, to help us better understand and optimize software systems. Additionally, the application of information feedback mechanisms can improve the reliability and stability of complex software systems. It is thus beneficial for software engineers to learn and apply the principles and methods of systems engineering in order to improve the design and development of complex software systems.

The theory of complex systems is a cutting-edge field in systems science and serves as the main research focus of complexity science. Often referred to as the science of the twenty-first century, its primary goal is to uncover the dynamic behaviors of complex systems that are difficult to explain using existing scientific methods. In contrast to traditional reductionist approaches, the theory of complex systems emphasizes a combined methodology that integrates both holism and reductionism to analyze systems. Although the theory of complex systems is still in its infancy, it holds the potential to spark a revolution in system science and even challenge traditional scientific methods. Life systems and social systems are both examples of complex systems, and the application of complex systems theory is of significant importance in fields such as systems biology and the mathematical modeling of biological systems. The theory of complex systems mainly encompasses the following aspects:

1. Complexity and complexity science;
2. Complex networks and graph theory;
3. Self-organizing systems and emergence;
4. Chaos theory and nonlinear dynamics;
5. Fractal theory and scaling laws.

Models. The system models are typically described using entities and their interactions, or by depicting structural changes that occur through evolution.

Objectives. The overall behavior of the system, such as emergence, serves as the primary research objective and focus of description.

Regularities. The aim is to explore general laws of evolutionary dynamics, such as power laws, genetic rules, self-organized criticality, and others.

The theory of complex systems emphasizes the integration of mathematical theory with computer science. Tools such as cellular automata, artificial life, artificial neural networks, and genetic algorithms can be seen as virtual experimental methods for this field.

The modeling techniques used in complex systems research, along with the understanding and study of emergence, hold significant implications for complex software systems.

By utilizing modeling techniques, we can better understand the behavior patterns of complex software systems and predict their long-term evolution. Additionally, the study of emergence can help us identify and analyze the underlying factors that contribute to the emergence of certain behaviors in software systems, such as scalability or adaptability. Understanding and utilizing these factors can greatly enhance the performance and effectiveness of complex software systems. Therefore, the application of complex system research in software engineering has the potential to improve the development and design of complex software systems.

1.4.2 Human Factors Engineering

Human factors engineering is an emerging and rapidly developing interdisciplinary field that spans a variety of disciplines, including physiology, psychology, anatomy, management, engineering, systems science, ergonomics, safety science, environmental science, and more, with a wide range of applications. As a result, throughout the formation and development of this field, scholars from different disciplines, regions, and countries have provided definitions and names for the field from their respective perspectives, reflecting diverse research emphases and application areas. To date, a universally accepted definition of human factors engineering has yet to be established.

The research methodology of human factors engineering is grounded in epidemiology, physiology, psychology, and psychobiology. It involves systematically applying human-related information, including abilities, inherent limitations, behaviors, and motivations, to design objects, processes, and environments. The collection of such information typically requires ongoing experimentation and statistical analysis. Therefore, human factors engineering necessitates a high level of statistical and experimental design expertise. In simple terms, human factors engineering is a discipline that explores and applies knowledge of human behavior, abilities, inherent limitations, and other characteristics to the design of tools, machines, equipment, systems, tasks, work, and their surrounding environments, with the goal of enhancing productivity, safety, comfort, and efficiency, ultimately improving the quality of human life.

There are many areas involving human factors in the study of complex software systems, such as the human factors mechanisms behind software failures, human-computer interaction technologies, the continuous evolution of requirements, and the complexity of software usage profiles.

1.4.3 Socio-Technical Engineering

The socio-technical systems approach is a relatively new management theory that emerged after World War II, evolving from the social systems approach. Most of the literature on socio-technical systems focuses on the impact of science and technology on individual and group behaviors, organizational structures, and management practices. As a result, it particularly emphasizes issues related to industrial engineering and human-machine engineering. The socio-technical systems approach views an organization as both a social system and a technical system, with a strong emphasis on the importance of the technical system, which is seen as the medium through which the organization connects with its environment.

The socio-technical systems approach advocates for the effective coordination between social and technical systems in order to enhance organizational productivity and management effectiveness. In situations where conflicts arise between these two systems, the approach suggests that modifications should be made to the technical system to better accommodate the social system. This perspective is primarily focused on industrial engineering, which closely examines the interactions between technological systems and the workforce in industrial contexts, such as transportation, product assembly, and chemical processing.

The development and construction process of complex software systems is a typical research domain of socio-technical engineering.

1.4.4 Software Engineering

Software engineering is a discipline that focuses on using engineering methods to construct and maintain effective, practical, and high-quality software. It involves aspects such as programming languages, databases, software development tools, system platforms, standards, design patterns, and more.

Ultimately, complex software systems are still a problem within the field of software engineering. Therefore, the research on the various new features of complex software systems is combined with the traditional research fields of software engineering, as listed in Table 1.2.

As software engineering continues to evolve, new research directions related to the study of complex software systems have emerged, including high-integrity

Table 1.2 The relationship between complex software systems and software engineering

Research field	Design and evolution	Planning and control	Monitoring and evaluation
Human interaction	√	√	–
Computation emergence	√	√	–
Design	√	–	–
Computing engineering	√	–	–
Adaptable system architecture	√	√	–
Adaptable and predictable system quality	√	√	√
Policy, procurement, and management	√	√	–

software engineering, predictable software systems, adaptable software systems, and the mathematical foundations of software engineering, among others.

1.5 Challenges of Complex Software Systems

Software reliability, as a crucial branch of software engineering, faces significant challenges posed by complex software systems. Both the foundational principles of software engineering and the novel characteristics of complex software systems have the potential to influence, and even expand, the scope of research in software reliability engineering.

1.5.1 Limitations of Traditional Reductionism

Traditional software engineering theories and classical software technologies are grounded in reductionism. The fundamental approach to addressing software complexity is the 'divide and conquer' strategy, employing a top-down, plan-driven methodology. Once precise user requirements are obtained, these requirements are decomposed in the design phase into manageable, easily developed functional components, adhering to the principle that the whole is the sum of its parts. The system is then constructed through a static, bottom-up assembly process.

While current software engineering techniques are adequate for building standalone systems, they lack both a scientific understanding and the technical capabilities necessary for large-scale, complex software systems. Given the vast scale of such systems, the most pressing challenge is not to adjust tactics, but to revise the overarching strategy.

In recent years, the field of software engineering has increasingly acknowledged the limitations of reductionism and has introduced new software development paradigms and technologies, including aspect-oriented programming, service-oriented architecture and technologies, agent-oriented software engineering, open-source software development, and trustworthy software engineering. More recently, the focus within the software engineering community has shifted towards supporting the effective development, flexible deployment, and continuous evolution of large-scale software systems. As a result, emerging software development methods and technologies have become research hotspots, such as approaches for constructing adaptable software, software engineering for ultra-large-scale systems, enabling technologies for online software evolution, and 'system alliances' aimed at addressing complex systems.

1.5.2 Heterogeneity of Systems

Components in complex software systems are typically distributed across different locations, developed by various organizations using distinct methods, and often adhere to different development standards, which inevitably leads to variations in the reliability levels of the components. Moreover, it is widely acknowledged in the industry that software reliability is an order of magnitude lower than hardware reliability, and human reliability is even lower. Consequently, for complex software systems that integrate hardware, software, and human components, the reliability levels can vary significantly. This creates considerable challenges for reliability assessment, as constructing reliable models for these heterogeneous components with large discrepancies in reliability levels is highly complex.

1.5.3 Impact of the Constructive Development Process

According to the constructive principle of complex software systems—'complex software systems are constructed through the dynamic interconnection of numerous autonomous systems'—the reductionist approach of traditional software engineering is increasingly inadequate for the development of such systems. The principle of evolutionary construction addresses the following two fundamental issues:

How to Support the Dynamic Evolution of Autonomous Systems: Autonomous systems are the fundamental building blocks of complex software systems. Supporting the dynamic evolution of autonomous systems is a core issue in the evolutionary construction of complex software systems. This involves addressing problems such as the adaptability of autonomous systems and the collaboration and interaction between autonomous systems.

How to Support the Dynamic Evolution of the System as a Whole: The dynamic evolution of the system as a whole is another fundamental issue in the evolutionary construction of complex software systems. This includes challenges related to the overall evolution, improvement, and optimization of the system.

1. Software Modeling Problem:
   How can an appropriate model be selected to support the construction of complex software systems, and, in turn, facilitate their evolution and growth? How can software models capture the growth and metabolic characteristics inherent in the construction of complex software systems?
2. Development Methodology Issues:
   The development methodology for complex software systems is another critical issue. What process model should be adopted for the construction of complex software systems? What techniques should be utilized to support their construction? What methods should be employed for the procurement and management of such systems? These all fall within the scope of the development methodology problem.

Reliability evaluation for complex software systems presents a dilemma. On one hand, it is impossible to clearly define the reliability level of the software system at the requirements stage, or even to fully understand the software requirements at that point. As a result, it is not feasible to develop a comprehensive and detailed reliability design and implementation plan based on reliability requirements. Consequently, it is not possible to ensure that the software system meets its required reliability through effective testing, evaluation, and enhancement measures during the development process.

On the other hand, complex software systems require high reliability due to their large user base, complex usage scenarios, and long service life. Only with high reliability and low failure rates can the system ensure normal operation. Furthermore, when a failure occurs in complex software systems, the consequences are often significant, making it essential to adopt reliability measures to minimize the occurrence of failures.

1.5.4 Impact of Continuous Evolution

According to the principle of continuous evolution in complex software systems—'complex software systems evolve continuously by adapting to changes in their environment and requirements'—'adaptive evolution' is a key characteristic of this process. The evolution of complex software systems in terms of adaptability exhibits the following distinctive features:

1. The System Exhibits Strong Adaptive Capabilities:
   Complex software systems possess significant adaptive capabilities, enabling them to adjust their structure and functionality in response to changes in the environment and requirements. This adaptability ensures the stability and reliability of the system over time.
2. Long Evolutionary Cycle:
   The evolutionary cycle of complex software systems is typically long, requiring multiple iterations and evolutions before the system's functionality and performance can be gradually refined and optimized.

3. Incremental Evolution of the System:
   The evolution of complex software systems is often incremental, involving the continuous addition, modification, and removal of functionalities and modules. This incremental approach facilitates the system's evolution and updates, as opposed to a complete overhaul or full-scale reengineering.
4. Support for Interactive Development Process in System Evolution:
   The evolution of complex software systems necessitates an interactive development process, wherein developers engage with the system continuously, utilizing feedback to optimize and evolve the system iteratively.

In summary, complex software systems are inherently characterized by strong adaptability and continuous evolutionary capabilities. To support their evolution and growth effectively, appropriate development methodologies and techniques must be employed.

1. Traditional software evolution typically takes place during the operational phase, after the software system has been deployed and delivered. In contrast, the adaptive evolution of complex software systems is often intertwined with the system's construction, spanning the entire software lifecycle.
   The adaptability evolution of complex software systems is characterized by their strong adaptability, long evolution cycle, incremental evolution, and interactive development process. Therefore, appropriate development methods and technologies must be used to support their evolution and growth.
2. Traditional software evolution is typically carried out in a centralized manner, with periodic or phased updates. In contrast, the adaptive evolution of complex software systems is a gradual, continuous, and long-term process.
3. Traditional software evolution primarily refers to the improvements made to software systems during the operational phase in response to changing requirements or issues. In contrast, the adaptive evolution of complex software systems is a metabolic process that builds upon existing systems and responds to various changes. It emphasizes the coexistence of new and old elements, adaptation to evolving conditions, and continuous growth.
4. Traditional software system evolution typically employs an offline approach, whereas the evolution of complex software systems requires an online approach. This means achieving dynamic adaptation while ensuring the software system remains continuously operational and providing consistent services.

In traditional software systems, most of the reliability work is completed during the design and development phases, which is why reliability is often regarded as 'designed-in.' Once the software system is delivered to the user, the user experiences the reliability level that was built into the system by the developer. In contrast, the evolution of complex software systems is continuous; the system continues to change during its usage by the user. Consequently, the reliability level of the software system is also a dynamic, ongoing process. As a result, reliability-related technical tasks have become a continuous, normalized process. Traditional methods such as reliability prediction, assessment, and verification based on a fixed reliability model are difficult to apply, as it is challenging to establish a relatively stable reliability model for a system that is constantly evolving. Therefore, new scientific and technological approaches are required to develop such a reliability model.

1.5.5 The Impact of the Normalization of Failures

Software reliability is a technical field concerned with the failure and malfunction of software systems. As noted earlier, failures and malfunctions have become commonplace in complex software systems. This suggests that software reliability technologies may evolve in two distinct directions: either reliability technologies will cease to exist in the future, or they will undergo a fundamental transformation in terms of their underlying principles.

1.5.6 System Boundary Erosion

The fuzzy boundaries of complex software systems integrate humans as part of the system. In contrast, in traditional software reliability engineering, humans, as users, are considered external to the system. As a result, software usage models and operational profiles developed using traditional methods do not account for human involvement or modeling.

1.6 Typical Complex Software Systems in Equipment

This section will introduce a typical complex software system: the integrated avionics system.

In the field of avionics, as systems evolve towards digitization, modularization, and integration, avionics systems have transformed into highly integrated modular systems. Traditional standalone electronic equipment has been replaced by the unified design of multiple independent electronic systems with various functions, offering unmatched advantages in performance, size, weight, cost, and other aspects compared to conventional methods. However, as the scale of the systems continues to grow, their complexity increases exponentially, and the highly integrated modular avionics system has become an exceedingly complex system.

1.6.1 Complex Avionics Architecture: IMA Architecture

This section begins with the development history of avionics architecture. It analyzes the differences between the federated digital architecture and IMA (Integrated Modular Avionics) from the perspective of the increasing complexity of demands for avionics systems, and explores the reasons behind the emergence of the IMA structure. Additionally, it compares the characteristics of the IMA structure with those of complex systems, analyzing the complexity of the IMA structure and the solutions it offers to address this complexity.

1.6.1.1 Development History of Avionics Architecture

From the 1940s to the early 1960s, avionics equipment on fighter jets consisted of dedicated sensors, controllers, displays, and analog computers. The equipment was minimally interconnected, operating largely independently without a central control computer. This represents the first generation of avionics architecture, known as independent/analog avionics, or alternatively referred to as federated, discrete, or analog architecture. An example of this structure is the F-4. Its characteristics include strong specialization, low flexibility, and difficulties in information exchange.

In the mid-1960s, digital computers began to be widely deployed for on-board navigation and fire control calculations, establishing a central control system. Other analog subsystems, such as the atmospheric data system, interacted with the digital computer through A/D and D/A conversion. With the introduction of a central control computer, the avionics architecture of this period is referred to as centralized architecture, with the F-111D serving as a representative example of this technology.

In the 1970s, analog computers in centralized architectures were gradually replaced by digital computers, resulting in functionally independent subsystems or avionics equipment interconnected via the 1553B multiplex data bus and communicating with the central computer. This centralized-distributed architecture is a product of the digitization of avionics, facilitating resource sharing between the back-end control and display components of the information chain. The adoption of modular software design techniques not only reduced development costs and shortened development cycles, but also enhanced system maintainability and scalability. Representative aircraft models for this technology include the F-15 and F-16.

The centralized and centralized distributed architectures both represent stages in the full transition of avionics computers from analog to digital. As a result, most researchers tend to group them together under the term 'federated avionics,' classifying them as part of the second generation of avionics architecture.

In the 1980s, the 'Gem Pillar' program outlined a new integrated avionics architecture, proposing requirements such as modularity, openness, high fault tolerance, and high flexibility. Based on VLSI technology, digital signal processing (DSP) technology, and image processing technology, it achieved highly integrated processing of sensor signals and data for various avionics subsystems (such as radar,

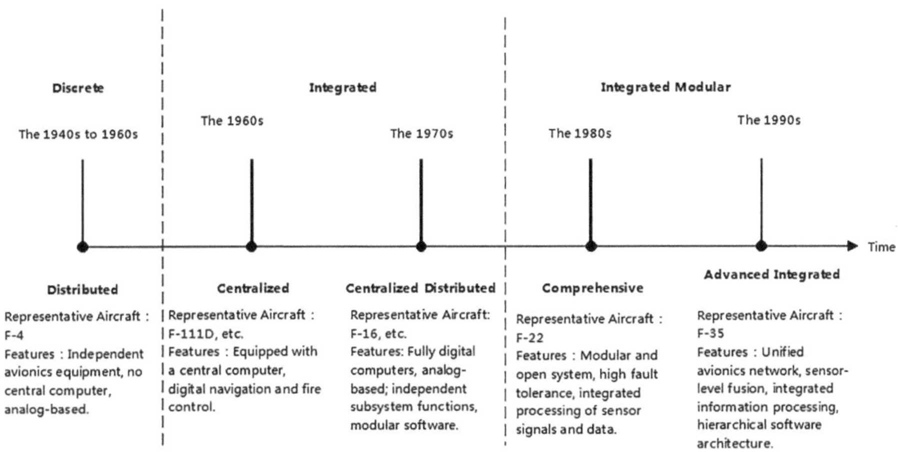

Fig. 1.2 Evolution of avionics architecture

electronic warfare, etc.) through extensive sharing of RF components and antenna apertures. The F-22 is an example of this technology.

Since 1990, integrated avionics has been further advanced with the development of the 'Gemstone' program. It adopts an open architecture, making extensive use of commercial off-the-shelf (COTS) products to implement software and hardware functional units. The system connects all functional areas via a unified fiber-optic network and promotes the integration of radar, electronic warfare, CNI, and other RF components. The overall system integration capability has been significantly enhanced compared to the 'Gem Pillar' program, earning it the designation of advanced integrated avionics. The F-35 fighter is the most advanced example of this technology.

Generally speaking, integrated and advanced integrated avionics are classified as the third and fourth generations of avionics architecture. However, internationally, the term IMA is commonly used to refer to the structures defined by the 'Gem Pillar' and 'Gemstone' programs.

Unlike the classification mentioned above, some researchers have attempted to categorize the development of avionics architecture into four stages based on the evolution of buses and unit modules: distributed analog structure, distributed digital structure, federated digital structure, and IMA. Overall, avionics architecture has evolved through three generations (independent, federated, and integrated modular) and five stages (discrete, centralized, centralized distributed, integrated, and advanced integrated). Figure 1.2 illustrates the evolution of avionics architecture.

1.6.1.2 Comparison Between Fedhitecture and IMA Architecture

A federated avionics system consists of avionics units provided by different equipment vendors. The embedded software in these systems is typically tailored to the specific hardware of each unit. Furthermore, communication between units is

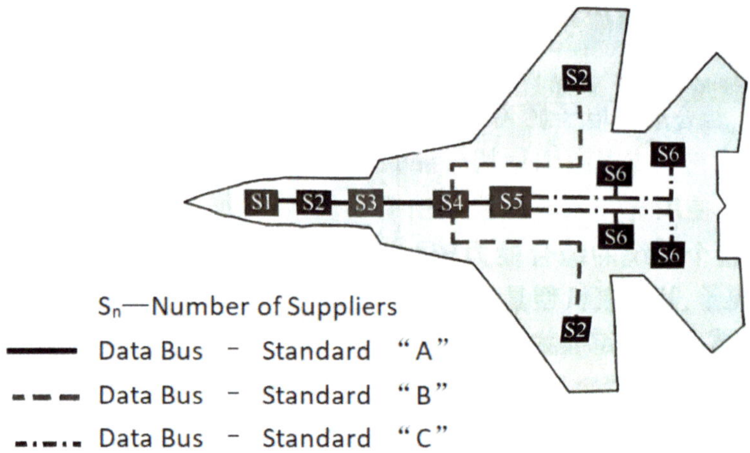

Fig. 1.3 A typical federated avionics system

predominantly facilitated through various data buses, with two or three commonly used bus standards. Figure 1.3 illustrates the typical structure of a federated avionics system.

The aviation industry generally agrees that developing aircraft using a federated avionics system leads to several challenges: frequent maintenance, low availability of individual components, limited hardware and software reuse, and the need for a large inventory of spare parts. These issues significantly increase both the initial production and ongoing maintenance costs of avionics systems. Additionally, as user requirements for mission capabilities and operational complexity continue to grow, avionics systems have become increasingly complex. However, the availability of suitable components in the market is steadily declining.

It is evident that, with the advancement of aircraft technology, the federated avionics architecture can no longer meet the growing demands for mission execution, operational efficiency, and life cycle cost (LCC) in avionics systems. Consequently, a new system architecture is required to address these needs. The characteristics of this new architecture and its corresponding requirements are outlined in Table 1.3.

Once the requirements at these three levels are translated into architectural characteristics, a new avionics system architecture—the IMA architecture—emerges. The structure of the IMA system is shown in Fig. 1.4. The IMA core system consists of one or more integrated architectures, which are made up of standardized, limited types of modules, hardware-independent reusable functional applications, operating systems, and system management software. These modules communicate through a unified network. IMA can be regarded as an independent entity containing multiple integrated processing resources, which can be used to build processor platforms of varying specifications and complexities, thus addressing the trade-off between mission execution, operational efficiency, and life cycle cost requirements.

Table 1.3 Correspondence between structural characteristics and requirements

Structural features	Task execution	Operational effectiveness	LCC
Define a widely applicable set of modules	–	√	√
Design modules to be replaceable	–	√	√
Maximize interchangeability and operability between modules	–	√	√
Adopt an open system architecture	–	–	√
Maximize the use of COTS (Commercial Off-The-Shelf) technology	–	√	√
Maximize technical transparency between software and hardware	–	–	√
Minimize the impact of software and hardware upgrades	–	–	√
Maximize software reuse and portability	–	√	√
Define integrated BIT (Built-In Test) and fault tolerance techniques to delay repairs	√	√	√
Provide support for highly integrated functions and physical designs	√	–	√
Reduce re-certification to ensure growth capability	√	–	√

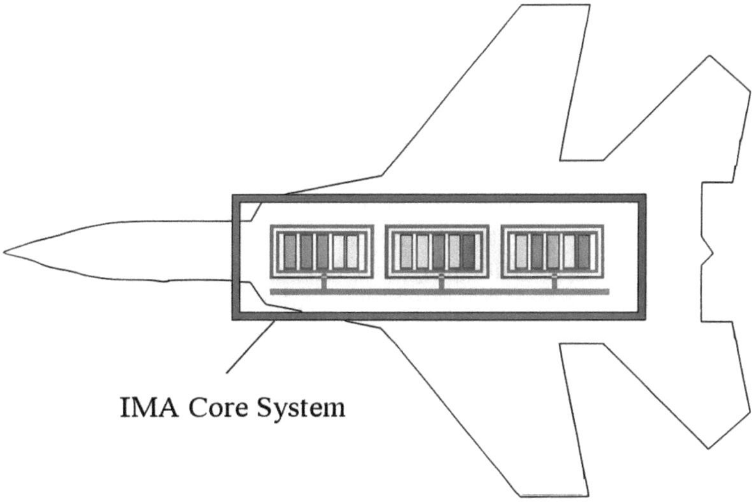

Fig. 1.4 IMA system architecture

1.6.1.3 IMA Architectural Features

IMA represents the highest level of development in avionics architecture, with the objectives of reducing aircraft life cycle costs (LCC), enhancing avionics functionality and performance, and addressing challenges such as software upgrades and hardware obsolescence. In contrast to the federated avionics approach, where each subsystem's software and hardware are specialized and function independently,

IMA is fundamentally a highly open, distributed real-time computing system capable of supporting avionics mission programs across various criticality levels. The key characteristics of IMA are summarized as follows:

AI Suggested:

- IMA is a highly integrated, modular, and open architecture that can support various avionics functions and applications.
- IMA uses a unified network for data communication, which allows for efficient and flexible information sharing among different modules and systems.
- IMA supports real-time processing of multiple applications with different criticality levels, which enhances the system's overall functionality and performance.
- IMA features a high degree of redundancy and fault-tolerance, which ensures system reliability and availability.
- IMA enables software and hardware upgrades to be performed independently, reducing maintenance costs and improving system scalability.

System Integration
IMA maximizes system integration by establishing a core hardware processing platform and enabling shared use of RF sensors. It seamlessly integrates data from various sensors, allowing the processed results to be reused across multiple applications.

The system is capable of unified control, scheduling, and display, while also supporting the pilot in tactical decision-making and system management.

Hierarchical Structure
The IMA system isolates application programs from the underlying hardware through various standard interfaces, ensuring that application programs are only related to aircraft functionality, not the hardware implementation. This modular approach allows new application programs to be loaded without hardware changes, thereby enhancing software portability. Additionally, replacing hardware components does not affect the execution of application programs, addressing issues such as hardware obsolescence. This design also improves system scalability and flexibility, enabling autonomous configuration and upgrades based on evolving requirements.

Functional Softwareization
IMA aims to replace hardware with software to implement avionics functions wherever possible. This approach helps reduce the number of subsystems, minimize aircraft weight, space, and cost, improve resource utilization, and leave more room for future expansion.

By adopting a software-based approach, IMA systems can also provide greater flexibility and adaptability to changing mission requirements and technological advancements, without requiring significant hardware upgrades or replacements.

Network Standardization
IMA has standardized the avionics network, eliminating the coexistence of multiple data buses typical in federated architectures. This approach helps reduce costs, decrease system weight, and enhance data transmission speed.

Product Commercialization
In the IMA architecture, both software and hardware components make extensive use of COTS (Commercial Off-The-Shelf) products. This approach promotes product standardization and modularity, which facilitates product portability and helps reduce system life cycle costs (LCC).

Schedule Flexibility
IMA divides application programs into fine-grained units and employs time-slicing and/or priority preemptive scheduling strategies to ensure that each application or safety-critical program is completed within the specified deadline.

Cumulative Certification
IMA emphasizes affordability by introducing the concept of Safety Cumulative Certification. When replacing or adding a hardware or software component, only the new component needs to undergo safety certification, rather than re-certifying the entire system. This approach helps reduce certification costs.

Maintenance Centralization
IMA introduces the concept of central maintenance for avionics, linking onboard fault prediction and health monitoring systems with the ground maintenance center in real-time to establish a central maintenance system. When a fighter jet is away from maintenance stations, the onboard component of the central maintenance system can dynamically reconfigure the avionics, ensuring continuous functionality and performance. The ground component of critical maintenance systems is also easier to maintain as needed.

1.6.1.4 Analysis of the IMA Architecture from the Perspective of Complex Systems

A "system" refers to an organic combination of elements that are interconnected and interact with each other, consisting of three components: entities, properties, and activities. Entities are the specific object elements that form the system, properties describe the states and parameters of these entities, and activities represent the changes in the states of objects over time. Systems can be classified into three types: simple systems, complex systems, and open systems.

A simple system is one in which the coupling between functional entities is low. A complex system is characterized by relatively high coupling between its functional entities. An open system exchanges matter, energy, and information with its external environment. This exchange allows the system to absorb negentropy from the environment, thereby reducing its total entropy and increasing its level of order.

The key characteristics of a complex system are openness, complexity, emergence, instability, nonlinearity, uncertainty, unpredictability, and pathological structures.

The Integrated Modular Avionics (IMA) system is a highly complex open system that exhibits all the characteristics of a complex system.

1. Openness

   An Integrated Modular Avionics (IMA) system typically has a lifecycle exceeding 30 years, from development to operational deployment. Over such a long period, the external environment undergoes continuous changes, microelectronic technologies evolve in line with Moore's Law, which predicts that performance doubles every 18 months. The performance benchmarks of application systems consistently improve, new systems are continuously integrated, and operational concepts and strategies evolve. To ensure the IMA system can quickly adapt to these dynamic changes, openness must be a key consideration.
2. Complexity

   An Integrated Modular Avionics (IMA) system integrates multiple sensor functions, including radar, electronic warfare, communication/navigation/identification, electro-optics, and other multi-sensor capabilities. These application systems are highly interdependent and exhibit strong coupling. On the horizontal level, the modular IMA system encompasses intelligent control management, intelligent information fusion, application software, foundational framework software, network management software, hardware platforms, hardware modules, and more. A typical IMA system currently includes over 50 application function threads, more than 150 defined system classes, over 40,000 reconfigurable items, and more than five million lines of code. The design and development team consists of more than 300 individuals, making the system exceptionally complex both technologically and managerially.
3. Other Characteristics

   A review of the development of avionics system architecture reveals that in the third-generation Joint Service Digital Architecture (JSS), each functional device operates independently, with interconnections between devices achieved through a data bus. These independently functioning devices exhibit high internal cohesion, with related components grouped together, achieving maximal localization. Furthermore, the devices operate in a loosely coupled manner. The lower the degree of coupling between the devices, the easier it is to limit system changes to the individual devices without affecting others.

In contrast, the fourth-generation Integrated Modular Avionics (IMA) architecture adopts a fundamentally different approach. At the physical level, functional entities are highly integrated, resulting in high interdependency and minimal internal cohesion. Localization is minimized, and the degree of coupling between components is maximized. Consequently, any modification to one component may potentially impact other functional entities, leading to reduced system reliability and scalability, as well as a significant increase in system complexity.

1.6.1.5 Managing the Complexity of IMA

From a system integration perspective, the third-generation Joint Service Digital Architecture (JSS) is better suited for system integration. Each functional device is physically independent, with high internal cohesion and loose coupling between devices. Upgrades or expansions of any individual device are confined within that device, without impacting other parts of the system. Therefore, the key to addressing the complexity of the Integrated Modular Avionics (IMA) system lies in how to transform a physically highly coupled system into a logically loosely coupled one, thereby enhancing the internal cohesion of system components and reducing the interdependence between functional entities.

Adopting a layered structural model is an effective approach to addressing the complexity challenges in electronic information systems. Through the process of layering, a complex system is divided into multiple hierarchical levels, each composed of different functional components. Each component performs specific functional tasks and has well-defined interfaces. Communication between components is facilitated through standardized protocols. After applying a layered logical model to the complex system, the three key elements of the system—components (entities), component attributes (properties), and the interconnections between components (activities)—are established.

By applying a layered structural model to abstractly decompose the physically highly cohesive fourth-generation Integrated Modular Avionics (IMA) system, each layer is composed of functional entities that perform specific tasks. These functional entities exhibit high internal cohesion, while the coupling between them remains loose. This approach endows the system with the structural characteristics of an integrated system, similar to those found in the third-generation Joint Service Digital Structure (JSS) avionics system.

Although the layered processing model offers an effective approach to addressing the complexity of electronic information systems, the issue of system complexity persists. However, the nature of this complexity has shifted from the previously chaotic and intricate relationships between functional entities and factors within the system to a more structured issue of coordination among the various layers and functional entities. Figure 1.5 illustrates the increasingly complex interrelationships between the system's layers (application layer, system layer, module support layer, COTS software layer, COTS hardware layer) after the highly integrated consolidation of the functional threads.

Through effective layering, the highly integrated avionics system at the module level is designed around components, their attributes, and the interconnections between them. Each component is designed to adhere to standardized external interfaces, ensuring that system updates are confined to the relevant components without impacting others. This approach forms the foundation of an integrated modular system capable of adapting to external environmental changes, ensuring its openness. Furthermore, maintaining and managing the interconnections between components through a "framework" technology enables the seamless addition or removal of

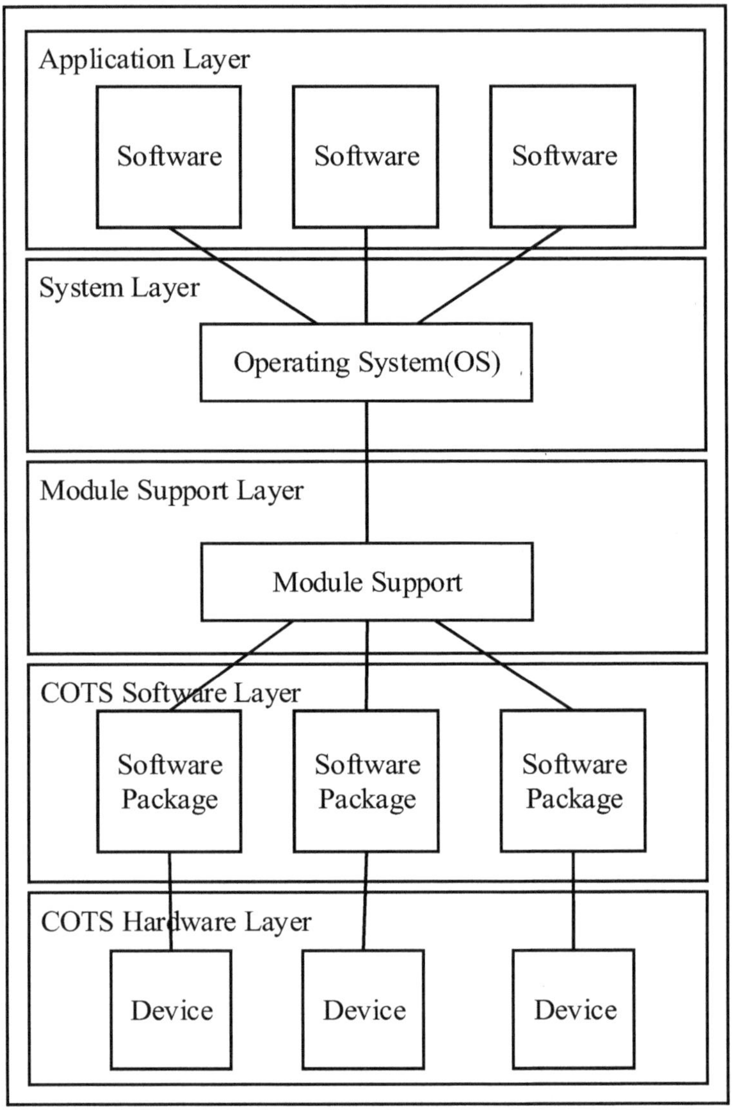

Fig. 1.5 The complex interrelationships between the layers of the system

components, which serves as the basis for achieving the system's scalability in Integrated Modular Avionics (IMA) systems.

In summary, the emergence of the Integrated Modular Avionics (IMA) architecture reflects both the inherent complexity of integrated avionics systems and the use of a layered structure to mitigate this complexity. The IMA architecture can be regarded as a quintessential example of complex systems in the field of avionics.

1.6.2 IMA Software Architecture

In the development of avionics system architecture, avionics systems have increasingly moved toward integration and modularization. The proportion of avionics software in airborne systems and equipment has gradually increased, becoming one of the key factors in achieving the aircraft's mission objectives. Modern avionics systems have transitioned from being electromechanical-intensive to software-intensive, with the scale of avionics software growing significantly. The concept of software-based avionics has become increasingly prominent. On the F-22, 80% of avionics functionality is implemented through software, with the software code totaling 1.7 million lines. In the F-35, this number has risen to over five million lines. This demonstrates that software has become a critical means of avionics development and the realization of modernization.

Compared to general application software, avionics software has the following characteristics:

1. The majority of avionics software is subject to real-time requirements, necessitating timely and accurate responses to external, unpredictable events.
2. Avionics is a safety-critical system, where the security and reliability of avionics software directly impact the overall safety of the aircraft
3. Avionics software is a complex system, where mission programs such as electronic warfare, CNI, electro-optics, and others share hardware resources and information, making it difficult to isolate them from each other without mutual interference.

The ideal IMA software architecture should both implement the principles of IMA and reflect the characteristics of avionics software. In practical implementation, IMA adopts a layered software strategy, with communication between layers occurring through standard interfaces. This approach aims to achieve mutual isolation between application software and hardware, facilitating the upgrade and evolution of both software and hardware products. At the same time, application programs are designed based on functional requirements and support partitioning strategies. Currently, there are four typical IMA software architectures in the aerospace field: ARINC 653, ASAAC, GOA, and the software architecture on the F-22 Universal Integrated Processor. Of these, ARINC 653 and ASAAC are the most representative. The following sections provide a detailed explanation of these two architectures.

1.6.2.1 ARINC 653 Software Architecture

Strictly speaking, the ARINC 653 specification primarily defines a standard interface (APEX) between the operating system layer and the application software layer and does not fully address system architecture at a broader level. However, it introduces the concept of program partitioning by dividing application programs into

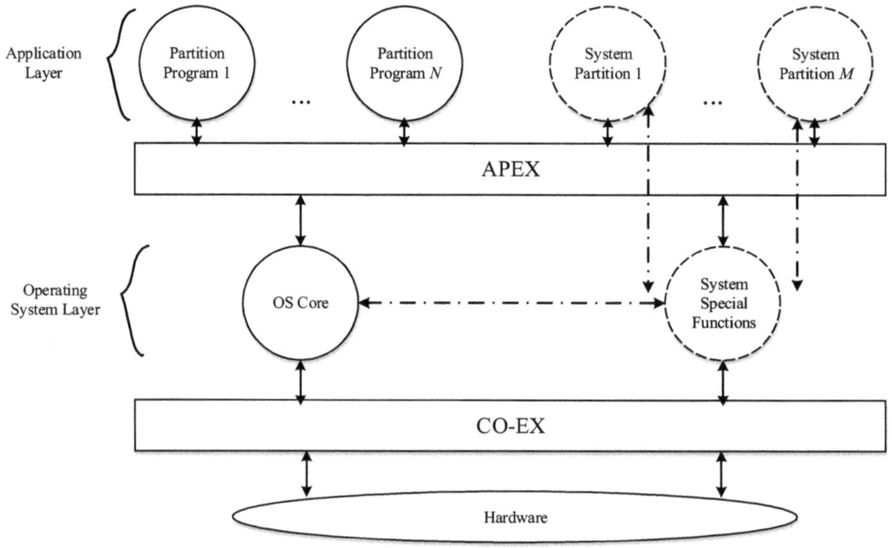

Fig. 1.6 ARINC 653 software architecture

multiple partitions. Each partition is allocated a specific amount of memory space and a designated CPU time slot, confining failures within individual partitions. This achieves mutual isolation among partitions and enhances system security and predictability to some extent.

Building upon program partitioning, the revised ARINC 653 specification introduces the concept of system partitioning, as indicated by the dotted lines in Fig. 1.6. This addition is designed to address potential system-level issues, such as external events and system failures.

Avionics operating systems based on the ARINC 653 software architecture, such as Integrity 2178B, Multicomputing OS, LynxOS 2178B, CsLEOS, and VxWorks AE653, are implemented in aircraft including the F-35, B767, C-130, and S-92. These examples demonstrate that ARINC 653 is not only applicable to civil avionics but also to military avionics. In particular, as the openness of Integrated Modular Avionics (IMA) systems continues to increase, the potential for Commercial Off-The-Shelf (COTS) products in military avionics is expected to expand significantly.

1.6.2.2 ASAAC Software Architecture

Compared to the ARINC 653 software architecture, the IMA software architecture proposed by the Allied Standards Avionics Architecture Council (ASAAC) is more aligned with the architectural domain, as shown in Fig. 1.7. ASAAC adopts a hierarchical structure, dividing the software system into the application layer, operating

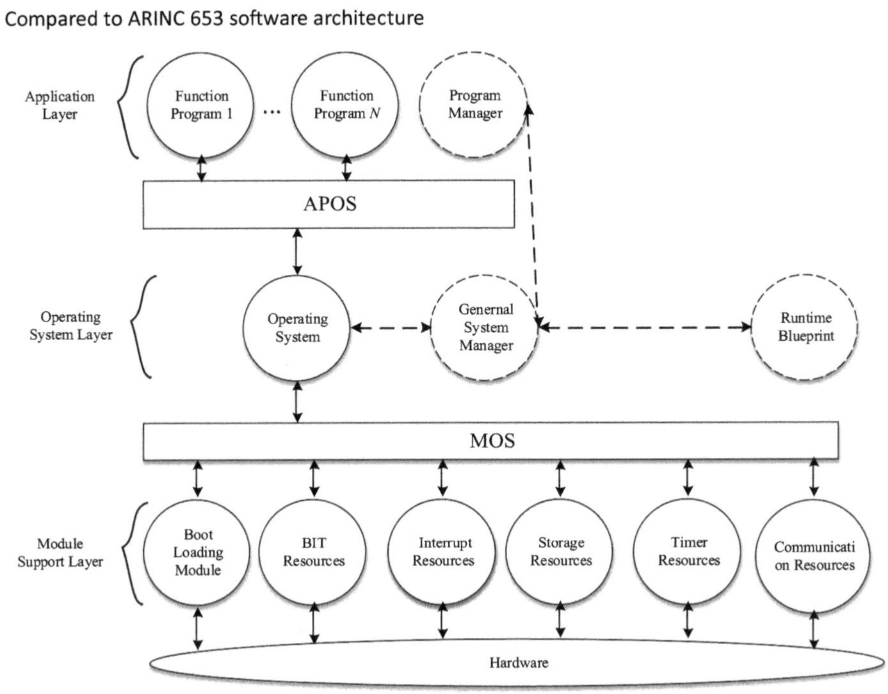

Fig. 1.7 ASAAC IMA software architecture

system layer, and module support layer. Interfaces such as APOS and MOS are used between layers to conceal the specific implementation details.

A comparison of Fig. 1.6 and Fig. 1.7 reveals that, in essence, the two architectures are quite similar. Both implement a real-time operating system (RTOS) and a system manager (SM) responsible for handling system transactions in the operating system layer. The key differences between the two architectures are as follows:

1. In the ASAAC architecture, programs are divided into processes rather than partitions, and a priority-based preemptive scheduling strategy is employed, while ARINC 653 uses partitioning and a time-sharing scheduling strategy.
2. In ASAAC, scheduling control, communication ports, configuration management, and health management are all governed by the runtime blueprint, rather than the APEX API., while in ARINC 653, APEX is used for control.
3. ASAAC further subdivides the operating system in ARINC 653 into the module support layer (equivalent to CO-EX), the general system manager (for health monitoring), the runtime blueprint (which manages the pre-defined system configuration data files), and the operating system layer (which manages the entire system and is responsible for responding to application requests).

In 2005, the UK Ministry of Defence adopted the ASAAC software architecture as a provisional defense standard (DS 00274). However, due to its relatively recent technological advancements, high level of standardization, and the complexity of its implementation, there have been no successful engineering application cases to date.

Based on the above analysis, there are few significant differences between ASAAC and ARINC 653 at the architectural level, as both exhibit the characteristics of IMA software partitioning and layering. However, in software testing, greater emphasis is typically placed on the characteristics of the application layer software (hereafter referred to as applications). The following section provides a comparative analysis of the applications in these two widely used IMA architectures, ASAAC and ARINC 653.

1.6.3 Comparative Analysis of IMA Applications

Table 1.4 lists the characteristics of the software architectures of ASAAC and ARINC 653. This section primarily focuses on ASAAC, with ARINC 653 as a supplementary reference, and compares the similarities and differences in the applications of these two IMA software architectures in terms of the following three aspects: general modules, inter-process communication methods, and program models.

1.6.3.1 General Modules

The core of the IMA system is comprised of chassis, which house a limited and standardized range of modules. These modules communicate through a unified network. The modules can be regarded as the fundamental building blocks of the IMA architecture.

By combining and connecting different modules, complex system functions can be created. Through modular design, system maintenance and upgrades are made easier, while also increasing system flexibility and scalability.

In the ASAAC architecture standard, the Common Functional Module (CFM) is a replaceable module responsible for providing computational capabilities, network support, and power conversion. The IMA core system includes the following five modules:

Table 1.4 Comparison of ASAAC and ARINC 653 characteristics

IMA Software architecture	ASAAC	ARINC 653
Application domain	Military-use	Primarily civil-use, with military integration
Degree of architecture standardization	High	Low
Level of adoption	No engineering application cases	Widely used

These are the five modules identified by AI:

1. **Processing Module (PM)**: Provides processor computing capability and supports multiple processor architectures.
2. **Storage Module (SM)**: Provides system storage and supports various storage media such as RAM, Flash, etc.
3. **Communication Module (CM)**: Provides network support capabilities, supporting multiple network protocols such as MIL-STD-1553B, ARINC 429, Ethernet, etc.
4. **Power Module (POM)**: Provides power conversion capabilities, supporting various power types such as DC, AC, and battery power.
5. **Clock Module (CLKM)**: Provides system clock signals, supporting multiple clock types such as UTC, GPS, etc.

Additional modules:

1. Signal Processing Module (SPM)
2. Data Processing Module (DPM)
3. Mass Memory Module (MMM)
4. Network Support Module (NSM)
5. Power Conversion Module (PCM)

The CFM standard defines the functional and interface specifications of the Common Functional Module (CFM) to ensure interoperability among CFMs and provides design guidelines for its implementation. The CFM standard is transparent and open to a wide range of markets, enabling the maximal utilization of COTS (Commercial Off-The-Shelf) technologies.

Within the CFM, the following elements need to be defined to ensure its installation and deployment:

1. The general CFM defines a set of basic functions that are applicable to all CFMs.
2. Processing capability specifies the unique functions associated with each CFM type.
3. Logical and physical interfaces define the interoperability and interchangeability of CFMs.

The internal architecture of each CFM consists of a series of functional units, each of which is applied to the respective CFM. As shown on the left side of Fig. 1.8, all functions, except for the processing unit (PU), are applicable to each CFM type. Additionally, Fig. 1.8 illustrates the relationship between ASAAC software and CFMs, with each CFM incorporating an ASAAC software structure.

In the ARINC 653 standard, this type of general module is referred to as a "core module"; however, the standard does not provide a detailed description of the structure of this module. Additionally, unlike CFMs, the core module in ARINC 653 does not have specialized modules for different data processing methods or applications.

As shown in Fig. 1.9, the ARINC 653 core module supports program partitioning. A partition in an IMA system is a set of functionally related application

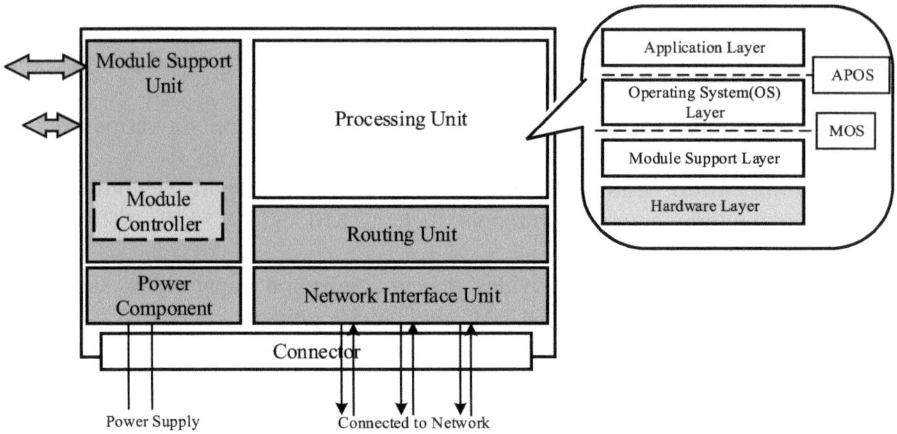

Fig. 1.8 The relationship between CFM and ASAAC software architecture

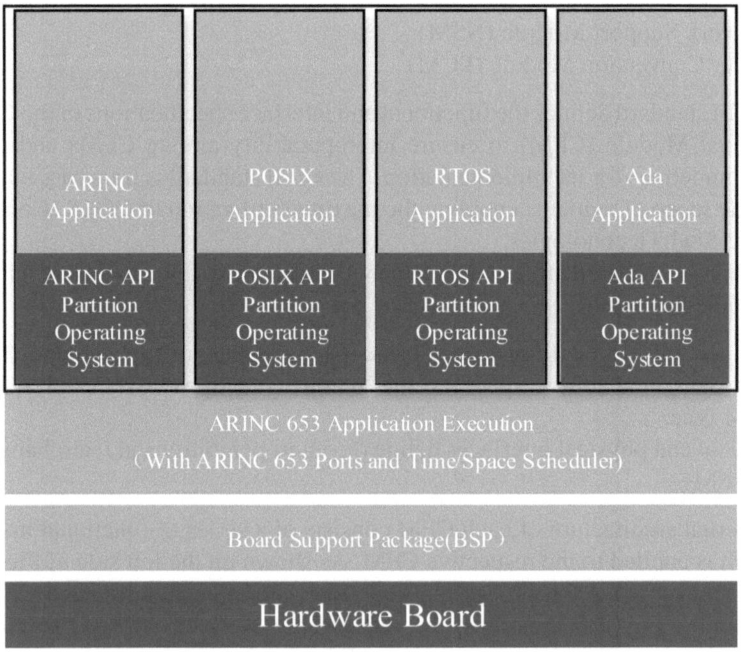

Fig. 1.9 Core modules and their software partitions in ARINC 653

software that is treated as a single entity for configuration and execution. Essentially, partitions are equivalent to processes in a general ASAAC operating system and include their own memory, context, and configuration attributes. Users can configure spatial and temporal partition scheduling information via configuration files, which are then compiled and integrated into the ARINC 653 operating system to enable dynamic configuration of spatial and temporal partition scheduling.

The ARINC 653 operating system ensures spatial isolation of partitions through the Memory Management Unit (MMU) and enforces temporal partition scheduling through strict periodic round-robin methods. Within each partition, priority or round-robin scheduling strategies can be implemented. This feature of program partitioning effectively prevents fault propagation between programs—if one software component fails, the failure does not affect others—thus enhancing the reliability of the system.

The similarity between CFMs and core modules lies in the fact that both are general-purpose modules capable of running standard applications, are interchangeable with each other, and can communicate through a unified network.

1.6.3.2 Communication Between Programs

Communication between programs in ASAAC is based on virtual channels (VCs), which have the following properties:

1. **Non-directionality**—Communication does not require a fixed direction for message flow.
2. **Message-oriented**—A message is defined and routed based on the virtual channel to which it is assigned.
3. **Managed by the operating system layer**—The operating system is responsible for creating, deleting, and routing virtual channels.
4. **Predictability of time and resources**—The system ensures predictable timing and resource usage for communication.

The properties mentioned above allow a process to send data to one or more processes. For the sending process, the receiving process can reside in the same processing unit, the same CFM, or even in different CFMs. The sending process has no knowledge of the receiving processes; it simply sends specific data to a designated VC. Similarly, the receiving process is unaware of the sending processes; it merely receives specific data from a designated VC.

Figure 1.10 illustrates the communication between programs across different general functional modules.

In Fig. 1.10, at the sending end, a communication frame header is added every time the VC drops a layer; at the receiving end, a frame header is removed every time the VC rises a layer, until it reaches the receiving process. The specific description is shown in Fig. 1.11.

In Fig. 1.10, at the sending end, a communication frame header is added each time the VC descends a layer; at the receiving end, a frame header is removed each time the VC ascends a layer, until it reaches the receiving process. A more detailed description is provided in Fig. 1.11.

Due to the unique partitioning characteristics of ARINC 653, communication between programs is referred to as partition communication. This communication

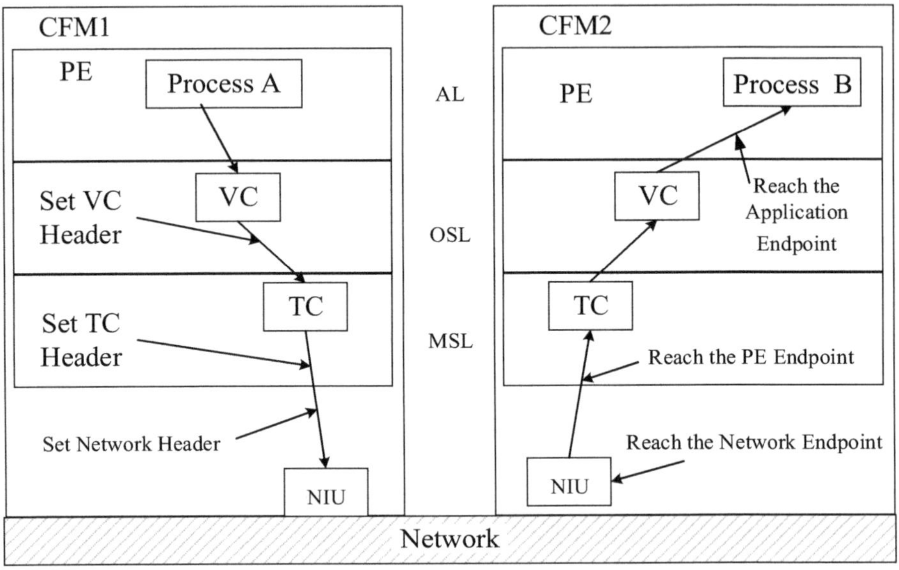

Fig. 1.10 Communication routing in the application layer

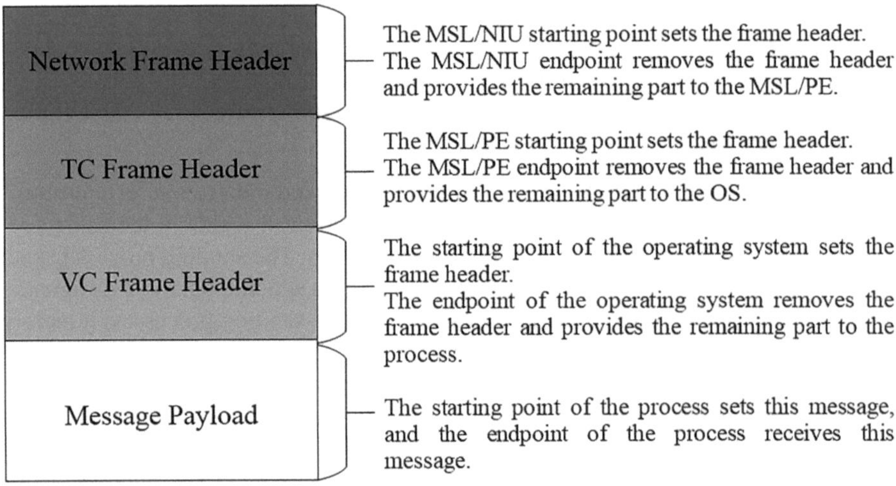

Fig. 1.11 ASAAC message transmission frame header

can occur either within the same processing module or across different processing modules, and can also involve communication between partitions and external devices. The communicating entities are unaware of each other's names or physical locations; instead, they send and receive messages via local ports, with the message being directed to a port rather than a specific process. All communication is message-based, and the fundamental mechanism for linking partitions through

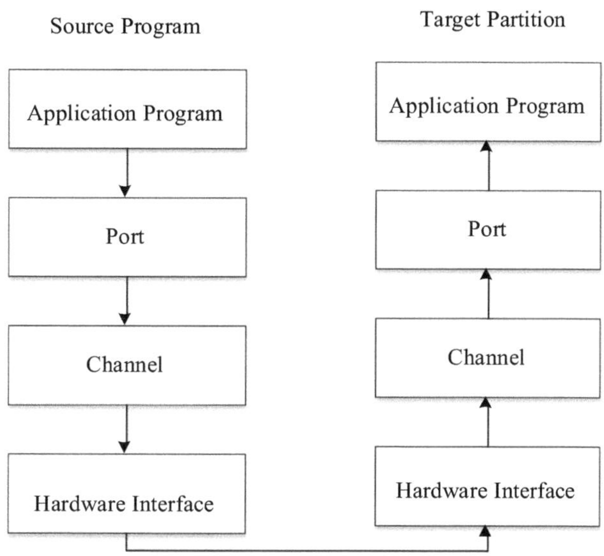

Fig. 1.12 Communication protocol stack structure for inter-partition communication

messages is a channel, which establishes a logical connection between a message source and one or more destinations. Applications interact with these channels through ports.

Partition communication is implemented by the operating system. To facilitate partition communication, ARINC 653 specifies a channel-based communication and synchronization mechanism. The structure of the communication protocol stack for this communication service mechanism is shown in Fig. 1.12. The source partition application program invokes the APEX function defined by ARINC 653 to send data to a port. The port organizes the data according to the port communication protocol and transmits it to the channel, which then forwards it to the physical interface of the target partition through the physical layer interface, ultimately delivering it to the application program in the target partition. An analysis reveals that the communication methods of ASAAC and ARINC 653 are quite similar.

1.6.3.3 Program Model of the Application Layer

The ASAAC software model ensures the system's real-time performance, but this does not imply that the program design methodology is constrained. On the contrary, any appropriate design approach can be adopted in program development, such as data flow and object-oriented design methodologies.

As shown in Fig. 1.13, in the ASAAC software architecture, all programs and system management functions are decomposed into processes. The interactions between these processes are mapped to virtual channels.

In the ARINC 653 standard, the concept of a program model is not explicitly defined. However, it does specify that communication between processes is

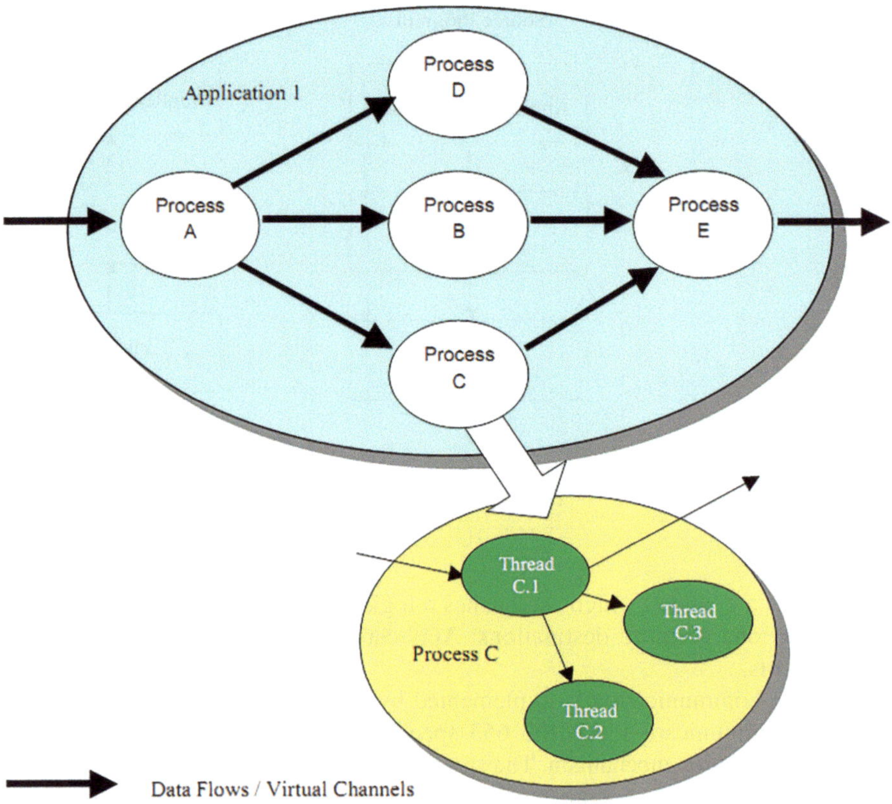

Fig. 1.13 The relationship between programs, threads, and virtual channels

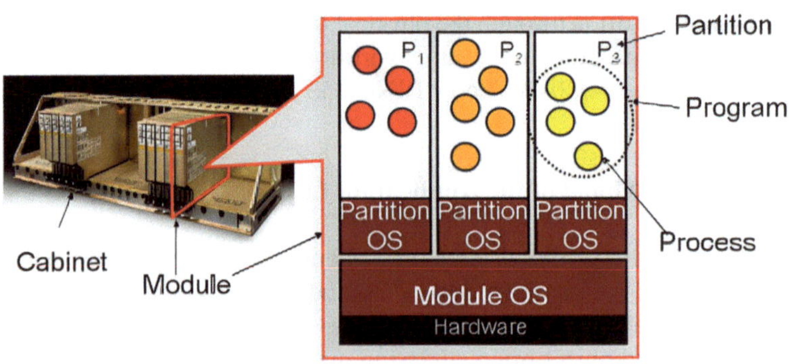

Fig. 1.14 Spatial partitioning in ARINC 653

accomplished through message channels, and services are invoked via threads. Furthermore, when discussing spatial partitioning, it is clearly stated that application programs within each partition are composed of processes. Thus, in terms of program modeling, ARINC 653 bears significant similarities to ASAAC (Fig. 1.14).

1.7 Summary

Complex software systems represent an inevitable trend in the future development of software. However, the corresponding theoretical foundations and technologies are still immature, and there is no established engineering methodology for constructing such systems. Researchers in the field of software reliability have only recently begun to shift their focus toward the challenges posed by complex software systems. Against this backdrop, this book aims to summarize and synthesize the research on complex software systems by scholars both domestically and internationally. Combining the author's years of engineering practice and insights, the book offers a comprehensive overview of the reliability technologies for complex software systems and provides bold prospects for their future development.

Chapter 2
Failure Mechanisms of Complex Software Systems

In her 1993 book Before the Digital: Why We Shouldn't Depend on Software, Ruth Wiener stated, "Even a medium-sized software system is one of the most complex products ever created by human beings, and the software development process is one of our most intricate activities. Regardless of the resources—labor, capital, or time—we invest, the outcome is often only basic reliability, and even after the most thorough and rigorous testing, errors will inevitably persist. We can never test all potential inputs or exhaustively verify all executable paths within a system."
This observation remains as pertinent today as it was over two decades ago. Over the past 20 years, software failures have continued to cause loss of life, economic damage, and mission failures. In the twentieth century, notable incidents included the Ariane 5 rocket disaster, the Therac-25 radiation overdose case, and the failure of the Patriot missile defense system. In the twenty-first century, a 2006 software failure at the U.S. Internal Revenue Service (IRS) led to financial losses of between $200 million and $300 million, alongside an additional $21 million in costs for error correction.

In safety-critical domains, such as the Integrated Modular Avionics (IMA) system discussed in Chap. 1, while the system provides benefits like reduced weight, simplified installation and maintenance, and lower costs, it simultaneously increases complexity exponentially. Functions that were once isolated within physically separate hardware units are now consolidated onto a single platform. This integration means that when a hardware failure occurs, it can simultaneously disable multiple critical functions. Such an increase in complexity complicates safety and reliability assessments.

For example, in 2008, an Airbus A330-303 experienced two rapid and severe downward maneuvers within just 180 s during its ascent to 37,000 feet (11,277.6 m), resulting in injuries to 110 passengers and over half the crew. The Australian Transport Safety Bureau's investigation report concluded that the incident was caused by a failure of the air data inertial reference unit (ADIRU), which transmitted erroneous data to other systems. Consequently, the main flight control

Y. Wang et al., *Reliability Technology for Complex Software Systems*,
https://doi.org/10.1007/978-981-95-3889-8_2

computer (FCPC) used this faulty data to issue dive commands, leading directly to the accident.

The large scale and intricate structure of complex software systems not only render the software development process inherently complex and difficult to ensure, but also introduce a diversity of relationships among the various components within the software. These relationships cause faults to propagate differently across heterogeneous components, resulting in faults of varying intensity and type. Consequently, the fault behaviors in complex software systems are both intricate and multifaceted. Moreover, these systems exhibit non-linear interactions, self-organization, and the emergence of global behaviors, which further complicates the software's behavior and makes it difficult to predict. As a result, fault behaviors and failure causes in complex software systems are characterized by significant uncertainty. Understanding the characteristics of software under abnormal conditions is critical, particularly with regard to how faults propagate and impact the execution of the software. This chapter focuses on the behaviors of "faults" and "failures" in complex software systems, which are central to reliability concerns.

2.1 Software Fault Complexity

Researching the fault characteristics and mechanisms of complex software systems is a fundamental starting point for analyzing the uncertainty and reliability of these systems. In simple software systems (relative to complex software systems), once a defect is triggered, the resulting behavior failure or malfunction is exposed during the system's execution, often referred to as the "single point of failure assumption." However, complex software systems consist of numerous locally autonomous subsystems that are integrated into a larger whole. The coupling relationships between these subsystems, as well as between the systems and human operators, are intricate and dynamic, making it difficult to describe the behavior of the entire system by merely aggregating the characteristics of individual subsystems. As a result, failures within these systems manifest in complex and unpredictable ways.

Faults in complex software systems typically occur in the interconnection points between subsystems. As a result, the failure of one subsystem is highly likely to trigger and propagate failures in other interconnected subsystems. This is particularly true for networked systems, such as power grids, the internet, and transportation networks, where network congestion and large-scale system-level failures are often the result of cascading failures. Currently, there are two primary areas of focus in the study of such faults. The first is to investigate how a fault, once it occurs, propagates from one module (or system) to another. This includes exploring the impact of different software architectures on fault propagation and how to accurately describe the process of this propagation. The second area of focus is to study the propagation patterns of various fault types (including hardware faults, data faults, etc.) in different types of software systems (such as embedded systems, distributed systems, and operating systems). This research aims to uncover these patterns in order to facilitate fault mining, detection, localization, and reliability prediction.

The complexity of software architecture is a primary factor contributing to the complexity of software faults. In recent years, a research framework has gradually emerged to study fault mechanisms, with its foundation being a deep understanding of the structure of complex software systems. This framework seeks to explore the intrinsic relationship between software architectural characteristics and fault propagation behaviors. By developing key quantitative metrics or measurement methods, it aims to reveal the main statistical patterns of fault propagation within complex software systems and assess the potential harm that such propagation may cause.

For example, the Department of Computer Science at the New Jersey Institute of Technology focuses its research on characterizing specific properties of software architectures. This research employs concepts such as error propagation probability, change propagation, and requirement change propagation probability to describe the propagation behaviors of various types of faults within software architectures.

Error propagation probability quantifies the likelihood of an error originating in a component during runtime and propagating to other components within the architecture.

Change propagation captures the impact of maintenance modifications or enhancements made to a specific component (e.g., component A), which necessitate corresponding changes in another component (e.g., component B) to maintain the overall system functionality.

Requirement change propagation probability measures the likelihood that a requirement change in one component will propagate to other components within the system's architecture.

Another relevant example is the work of Sizarta Sarshar, author of Analysis of Error Propagation between Software Processes in Source Code, who focuses on fault propagation at the source code level. In his research, Sarshar developed a methodology for studying the propagation of faults between software processes, aiming to identify distinct characteristics of error propagation mechanisms detectable within program source code. His approach leverages Failure Mode and Effects Analysis (FMEA), applied specifically to shared memory associated with system calls. Unlike traditional methods that analyze software system functionality, this approach focuses on identifying failure modes when programs interact with these services, particularly in the context of parameter transmission and variable return. This analytical technique is capable of uncovering numerous potential failure modes, some of which may impact other processes and thereby lead to cascading error propagation.

Studying the mechanisms of failure propagation provides critical insights into the underlying principles and inherent patterns of software failures. Increasingly, researchers have recognized that failure propagation significantly impacts various branches of research on complex software systems.

There are two prevailing perspectives on the mechanisms of failure propagation. The first posits that system architecture determines failure propagation behavior, asserting that the same failure may evolve into different types or severities of system failures depending on the architecture of the software system. This perspective grounds the study of failure propagation in system structure analysis, leading researchers to focus on identifying the propagation patterns within the system's architectural framework.

The second perspective contends that different types of software failures possess distinct characteristics, which subsequently influence their propagation behaviors and the types of system failures they may induce. According to this view, even within the same software system, failures of different types exhibit varying propagation patterns. Consequently, these researchers prioritize the study of failure characteristics as the key determinant of propagation behavior.

2.2 Research on Fault Propagation

Generally, the propagation mechanisms of faults are examined from two primary dimensions: the software evolution dimension and the software operation dimension. Software systems are progressively developed and modified throughout their entire life cycle, and faults follow a similar trajectory.

The software evolution dimension focuses on modeling and analyzing fault propagation across the software life cycle. This involves examining how faults introduced at a particular stage grow and propagate during subsequent development stages, drawing on theories related to software evolution and system evolution.

This book, however, centers on the reliability of complex software systems, placing greater emphasis on the study of fault propagation within the software operation dimension. This dimension examines how faults propagate during the operational phase of software systems, providing insights that are critical for ensuring system reliability.

2.2.1 Concept of Fault Propagation

In the current literature on software fault propagation, terms such as error propagation, defect propagation, fault propagation, and failure propagation are often used interchangeably. In the context of propagation issues, these terms can be regarded as consistent and need not be strictly differentiated. This is because, based on the definitions of error, defect, fault, and failure, the following points emerge:
Temporal distinctions: These terms are defined at different stages, corresponding to distinct phases along the time dimension of software development and operation.

Granularity of software entities: Since these faults arise at different points in time, they naturally align with different software stages and entities, which vary in granularity (e.g., functions, components, modules, and processes). Defects correspond to the smallest granularity of software entities, while failures are associated with the largest granularity.

Additionally, when describing software entities, it is essential that the fault propagation process corresponds to the characteristics of the described software entities. Consequently, when addressing propagation issues, as long as the terminology aligns with the software development phase, the granularity of software entities, and the propagation characteristics, the research subject can be considered fundamentally consistent.

Given that fault propagation is inherently a dynamic process occurring during software operation, this book adopts the term software fault propagation for clarity and consistency.

Fault propagation describes the process by which faults in components lead to system-level failures. This process can be outlined as follows: if a defect—an internal error in the code implementation of a component's internal activity—exists, its execution will trigger an internal fault within the component. If this fault reaches the component's interface, it will result in the component's failure. However, if the fault is masked by instructions from other internal activities, the component failure will not manifest.

Similarly, within a component-based system, a fault in one component may propagate to the system's interface, leading to a system failure. Conversely, if the fault is masked by instructions or behaviors from other components, it will not cause the system to fail.

From the perspective of system science, software fault propagation is a complex process that occurs between various system entities. These entities may include physical components, processes running on single or multiple CPUs, data objects in databases, functions within programs, or even individual statements in the code.

In complex software systems, the process whereby a failure in one entity, triggered by internal or external factors, causes subsequent failures in other entities is referred to as fault propagation. The representation or characterization of fault propagation using a formalized approach is termed a fault propagation model.

The study of fault propagation models can be broadly categorized into several aspects, as illustrated in Fig. 2.1.

2.2.2 Architecture-Based Fault Propagation Model

The architecture-based fault propagation model posits that the behavior of fault propagation is determined by the system's architecture. Specifically, the same fault may evolve into different types or levels of severity in terms of system failures, depending on the architecture of the software system. Building upon the architecture model, researchers typically define various characteristics of fault propagation to effectively describe its behavior. The following introduces several common concepts associated with fault propagation:

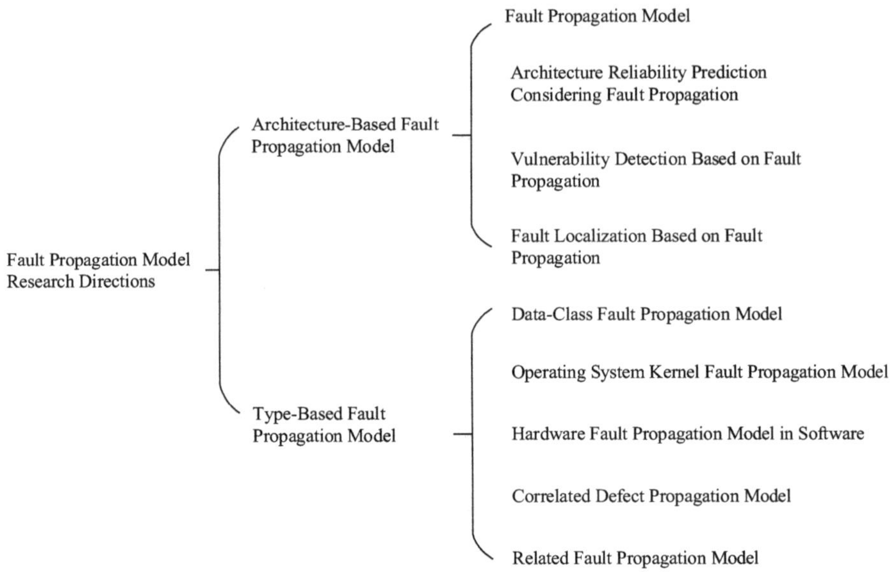

Fig. 2.1 Research directions of fault propagation models

2.2.2.1 Fault Propagation Probability

The fault propagation probability is an attribute of the software architecture that quantifies the likelihood of an error occurring in component A propagating to component B during runtime. This probability can be formally defined as follows:

$$\mathrm{EP}(A,B) = \mathrm{Prob}([B](x) \neq ([B](x')|x \neq x') \tag{2.1}$$

The fault propagation probability describes the probability of error propagation from component A to component B when component A calls component B with a probability of 1. To make the model closer to the actual usage scenario, the unconditional fault propagation probability is proposed, which is expressed as: $\mathrm{EP}(A,B) = \mathrm{EP}(A,B) \times T(A,B)$. It is determined by the error propagation probability $\mathrm{EP}(A,B)$ and the probability transition matrix $T(A, B)$. The elements of the probability transition matrix represent the probability of activation of the connector, that is, the probability that component A calls component B.

2.2.2.2 Error Propagation Rate

The error propagation rate refers to the probability that an error in an input signal propagates to the corresponding output signal. Each pair of input and output signals is associated with a specific error propagation rate, defined as the conditional probability of an error occurring in the output signal given that an error exists in the input signal.

$$0 \le P^{M}_{i,k}, = \Pr\{\text{err in o} / P_k \text{ err in o} / p_i\} \le 1 \tag{2.2}$$

Building on the concept of error penetration rate, the notions of relative penetration rate and unweighted relative penetration rate have subsequently been proposed. The error penetration rate pertains to a specific input-output pair, while the relative penetration rate quantifies the probability of an error propagating through a module from its input to its output.

$$0 \le P^{M} = \left(\frac{1}{m} \cdot \frac{1}{n}\right) \sum_i \sum_k P^{M}_{i,k} \le 1 \tag{2.3}$$

Relative penetration rate may not necessarily reflect the overall error penetration rate of a module, but it can serve as an abstract measure to characterize the relative error permeability among modules. Therefore, the unweighted relative penetration rate has been proposed.

$$0 \le P^{M} = \sum_i \sum_k P^{M}_{i,k} \le m \cdot n \tag{2.4}$$

2.2.2.3 Error Transmission Probability and Error Transparency

The Error Transmission Probability quantifies the likelihood of an error propagating from one component to another. It serves as a critical metric for assessing the stability of a software system. When the error transmission probability is high, proactive measures are required to mitigate it, such as by restructuring the software architecture or optimizing interface design.

Additionally, Error Transparency refers to the system's capability to promptly and accurately report errors when they occur, facilitating debugging and resolution. High error transparency enables developers to quickly identify and address issues, thereby enhancing the maintainability, reliability, and usability of the software.

Therefore, during the design, development, and testing phases of a software system, it is essential to consider both error transmission probability and error transparency to improve the system's quality and performance.

If the software architecture is conceptualized as being composed of Error Control Modules (ECMs), the process of error propagation can be divided into three stages: first, the presence of an error in the source ECM (denoted as ECM_S); second, the propagation of the error from ECM_S; and finally, the resulting occurrence of an error in the target ECM (denoted as ECM_T).

Building upon the concept of error penetration rate, the notions of Error Transmission Probability and Error Transparency are introduced. Error Transmission Probability is defined as the probability that a transient error occurs in the input of ECM_S, propagates through its output channel M_j, and subsequently reaches the input set of ECM_T. The formal definition is expressed as:

$$P^{1}_{j} = \left(\Pr\{I\} / N\right) \cdot \sum_{k=1}^{N} \Pr\{M_j | I_k\} \tag{2.5}$$

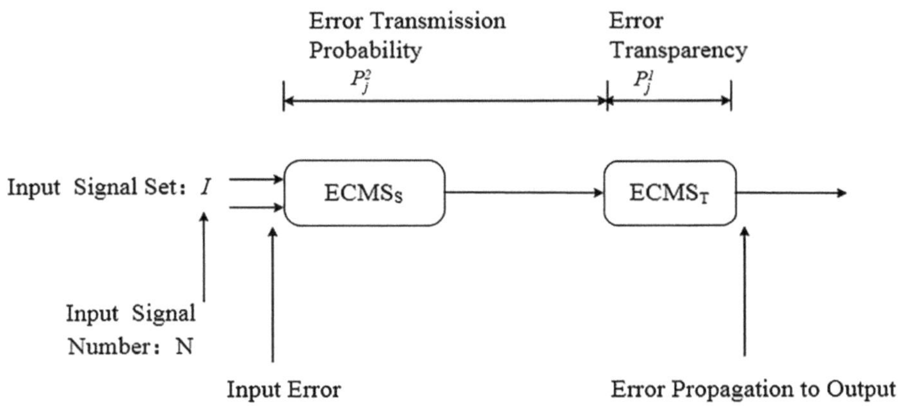

Fig. 2.2 The process illustrating the propagation of errors from the input of ECMS to the output of ECMT

This conceptualization provides a foundation for analyzing and quantifying the reliability characteristics of software architectures.

In the formula, $\Pr\{M_j | I_k\}$ denotes the probability that an error located at the k-th input k passes through M_j. $\Pr\{I\}$ represents the probability of errors occurring in the input set I of an ECM, which is set to $\Pr\{I\} = 1$ during injection experiments. N refers to the number of inputs in the ECM module. Error Transmission Probability characterizes the first two steps of error propagation: the generation of errors in ECM_S and their subsequent propagation. Conversely, Error Transparency defines the likelihood of errors occurring in ECM_T after receiving erroneous input M_j. This parameter encapsulates the final step of error propagation, where the results of the initial two steps culminate in errors within ECM_T.

The process of error propagation from the input of ECM_S to the output of ECM_T is illustrated in Fig. 2.2.

2.2.2.4 Error Propagation Function

By applying statistical methods, the fault propagation model can be represented as $y = f(X + Z)$, where f is the system function or propagation function, X is the input vector of the system, Z is the error vector of the X vector, and y is the system output. This method uses mathematical statistics to expand $f(X)$ with Taylor's formula and derive the mean and variance of y when calculating or estimating the variance distribution of error Z. In addition, some researchers have also applied interval mathematics theory to propose a data flow model based on interval arithmetic analysis, which is used to quantitatively describe fault propagation.

2.2.3 *Type-Based Fault Propagation Model*

According to the type-based fault propagation model, the behavior of fault propagation is determined by the specific type of fault. Different fault types, even within the same software architecture, may evolve into distinct forms or lead to system failures of varying severity levels. In this line of research, faults are typically categorized based on their underlying causes and points of occurrence. Representative fault types are introduced below, along with their corresponding propagation models:

1. **Control-flow faults:**
   These faults are caused by errors in the control flow of the software system, such as loops, branches, and function calls. The propagation model for control-flow faults is based on the control flow graph, which describes the structure of the software system.
2. **Data-flow faults:**
   These faults are caused by errors in the data flow of the software system, such as incorrect data values or data dependencies. The propagation model for data-flow faults is based on the data flow graph, which describes the flow of data through the software system.
3. **Timing faults:**
   These faults are caused by errors in the timing behavior of the software system, such as a delay in processing or an incorrect synchronization of events. The propagation model for timing faults is based on timed automata, which describe the behavior of the system as a sequence of timed transitions between states.
4. **Communication faults:**
   These faults are caused by errors in the communication protocol between different components or systems. The propagation model for communication faults is based on the communication protocol, which describes the rules and procedures for exchanging messages between systems.

By developing propagation models for different types of faults, engineers can better understand the behavior of the system under different failure scenarios and develop more effective strategies for fault detection, prevention, and recovery.

2.2.3.1 Data-Related Fault Propagation Model

Data-related faults primarily propagate through data flows, with shared data serving as the most complex propagation mechanism. This complexity arises because shared data can influence a broad range of components, significantly expanding the scope of the fault's impact. Furthermore, due to the intricate computations and processing performed by software systems, data-related faults often evolve and mutate during propagation, further increasing the complexity of analyzing and mitigating these faults.

Therefore, it is important to develop effective propagation models for data-related faults to analyze their behavior and impact on the system, and to develop strategies for their detection, prevention, and recovery.

In studying the propagation patterns of data-related faults, fault injection techniques are commonly employed. Faults are introduced into the input signals of a module, one fault at a time per input signal. The module's output is then recorded and compared with the output obtained under fault-free conditions. This comparison enables the tracing of the fault's propagation path, facilitating the analysis of how data errors spread within modular software.

It is widely recognized that errors in the original input signals can propagate along multiple pathways, potentially influencing the system's outputs. Consequently, the root nodes of a data-related fault propagation model are not restricted to the system's inputs but typically encompass all signals of interest within the analysis.

Such a model aids in identifying the vulnerable points within the system, where errors are most likely to propagate. By performing vulnerability analysis, the most effective trustworthiness structures and design mechanisms within the system's components can be determined, providing critical insights for improving system reliability and fault tolerance.

The type-based fault propagation model is not inherently in conflict with the architecture-based model. Instead, the concept of error permeability, introduced earlier, serves as a unifying framework for defining a series of related parameters. These parameters facilitate the study of data-related fault propagation at both the signal and module levels. By constructing an error propagation tree, it becomes possible to identify the modules most prone to error propagation.

Furthermore, by analyzing error permeability and other associated parameters, the model provides insights into how and where to implement error detection mechanisms (EDM) and error recovery mechanisms (ERM). This approach enables a systematic strategy for optimizing fault-tolerance measures within the system, ensuring accurate detection and efficient recovery from faults.

This approach enables a comprehensive understanding of the behavior of data-related faults within the software system, providing engineers with the insights needed to develop effective strategies for fault management and control.

2.2.3.2 Error Propagation Model of the OS Kernel

Operating system (OS) faults are categorized separately for three primary reasons:

1. **Complexity of Modern Operating Systems:** Contemporary operating systems are inherently complex systems. For instance, the Windows operating system is reputed to have one of the largest source code bases among all software, making faults within the OS inherently intricate and multifaceted.
2. **Fundamental Role in Complex Systems:** Operating system software constitutes a foundational component of nearly all complex systems. Faults within the OS can propagate broadly, exerting widespread and profound impacts. These

faults are notoriously challenging to detect, trace, and resolve, further underscoring their significance.
3. **Critical Bridging Role:** The operating system functions as the intermediary between application software and hardware. It is highly sensitive to time-related factors and hardware errors and plays a critical role in ensuring system security. Any fault in the OS can compromise this bridge, leading to severe consequences for both software and hardware layers.

The propagation of internal faults within an operating system primarily occurs through the interactions and invocations among its program components. Based on the fundamental architecture of the operating system, fault propagation can be analyzed at two distinct levels: (1) between the operating system kernel and device drivers, and (2) between the kernel and application programs.

1. By conceptualizing device driver programs as atomic units that interact solely with the operating system kernel and not with each other, and by focusing exclusively on the interactions between the driver programs and the kernel when defining parameters, the constructed propagation model is primarily designed to illustrate two key aspects: (1) how hardware-related faults propagate to the software level, and (2) how software faults, in turn, influence hardware states.
2. The operating system kernel offers a variety of services that can be invoked by application programs, with faults in the kernel propagating primarily through call relationships. Constructing a call graph is a fundamental approach to understanding this propagation. However, for complex operating systems that incorporate concurrency mechanisms and support distributed architectures, it is often necessary to utilize more expressive modeling languages, such as state machines or Petri nets, to accurately capture and analyze fault propagation behaviors.

2.2.3.3 The Propagation Model of Hardware Fault Inside Software

A defining characteristic of complex software systems is the increasingly close integration between software and hardware, resulting in a blurred boundary between the two. In this context, hardware faults have emerged as a distinct type of fault within complex software systems. In the field of reliability, the mechanisms of software and hardware faults are fundamentally different. Hardware faults are primarily caused by physical and chemical factors, such as fatigue, aging, fractures, and corrosion. In contrast, software faults ultimately stem from human errors during development or usage.

However, in complex software systems, hardware failures can have significant repercussions on software behavior due to the intricate interactions between software and hardware. This tight coupling allows hardware faults to propagate easily into the software layer, often leading to unexpected consequences. For software designers, mitigating the impact of hardware faults presents a substantial challenge, as addressing such faults requires a deep understanding of both hardware behavior and the software-hardware interaction dynamics.

Hardware faults are generally categorized into two main types based on their "distance" from the software: on-board hardware faults and device-level hardware faults. In some complex systems, cross-linked device-level hardware faults are also considered. This discussion, however, focuses solely on research related to on-board hardware faults. On-board hardware faults typically refer to hardware components directly accessed and operated by the software during runtime, such as memory and registers. Based on their occurrence frequency, these faults can be further classified into two types: transient faults and intermittent hardware faults.

The English translation reflects the original meaning fairly well, but it can be refined to improve clarity, readability, and academic precision. Here's a polished version:

Transient faults typically occur in buffer memories or processor registers, with a high probability of occurrence due to the high utilization efficiency of buffer memories. When a processor performs read/write operations exclusively through buffer memory, the overall system performance can be significantly enhanced. Additionally, higher processor clock frequencies reduce cycle times, which shortens the time allocated for buffer memory read/write operations to meet bandwidth requirements. However, this reduction in read/write time substantially increases the likelihood of transient faults.

To address such faults, some operating systems offer protective mechanisms. For example, the MVS/SP operating system provides efficient error recovery features, automatic error detection and correction, and real-time recording of the system state during error occurrences. Nonetheless, studies on the handling of hardware-related software errors and the efficiency of fault recovery mechanisms reveal that operating systems are generally ineffective at diagnosing software errors linked to hardware faults. Furthermore, system recovery mechanisms tend to be less efficient at resolving hardware-related software errors compared to purely software-related faults.

Interestingly, research has also demonstrated that hardware-related software faults exhibit specific patterns. Investigating these patterns holds promise for enabling fault prediction and recovery, paving the way for more effective management of this type of error.

Intermittent hardware faults are typically addressed at the hardware level through fault avoidance techniques, such as reducing process variations, limiting voltage ranges, or implementing temperature management. However, these hardware-level approaches cannot completely eliminate the occurrence of such faults. Similar faults may still arise and propagate to the software, with most intermittent hardware faults ultimately leading to system crashes. Consequently, it is essential to mitigate the impact of these faults on detection, diagnosis, and recovery processes at the software level.

Research on intermittent hardware faults often involves constructing a crash model to identify which types of intermittent faults lead to program crashes and how these faults propagate within the software prior to a crash. Dynamic Dependency Graphs (DDG) are commonly utilized in such research to model the propagation of intermittent faults at the instruction level. Through fault injection experiments conducted via simulators, researchers have concluded that the primary impact of intermittent faults on software is program crashes. Among all injected faults, 62–79% result in program crashes, 20–34% are benign, and only 4% lead to silent data corruption (SDC).

The DDG model is capable of predicting all crashes identified in fault injection experiments, with a missed detection rate of approximately 0.06% and a false positive rate of 20%. The crash locations and fault propagation predicted by the DDG model are accurate within 100 instructions in 93% of cases, with the remaining inaccuracies likely due to the program being unsuited for DDG modeling. Analyzing a benchmark program using the DDG model typically takes 4–89 s, whereas performing the same analysis using a simulator requires 6–70 h.

2.2.3.4 Defect Correlation

In complex software systems, the various types of faults mentioned above rarely occur in isolation but rather exhibit intricate interrelations, commonly referred to as defect correlation. Researchers regard defect correlation as a prevalent phenomenon in software systems. When certain defects in software are activated, they may influence the detection rate of other defects or even determine whether those defects are activated. If the presence of defect Db affects the defect detection capability of defect Da, the two defects are termed correlated defects.

Correlated defects can be classified into two categories: positively correlated defects and negatively correlated defects. At its core, software defect correlation is the fundamental cause of cascading failures and correlated failures within software systems.

The emergence of correlated defects has significantly influenced the field of software reliability research. Most traditional reliability models and reliability growth models share a common assumption: they do not account for differences among defects, assuming all defects have the same severity and identical detection probabilities. Furthermore, these models assume that the removal of a defect does not introduce new defects. However, the existence of correlated defects renders such assumptions unrealistic.

As a result, in practical software testing, many researchers treat software failures as correlated rather than entirely independent. This perspective has led to the development of numerous new software reliability models. In these models, defects are considered interdependent, exhibiting various correlation relationships. Additionally, defect removal may inadvertently introduce new defects, thereby increasing the overall number of defects. Markov stochastic processes and queuing theory serve as the primary foundations for modeling such phenomena.

Furthermore, correlated defects have introduced new requirements for software testing techniques. Research in this field has focused on optimizing static testing methods to account for correlated defects, selecting regression test case sets that consider these correlations, and evaluating the adequacy of testing processes. These topics have become key areas of interest in the study of software testing.

2.2.4 Reliability Prediction Considering Fault Propagation

As previously mentioned, faults in complex software are influenced by the system architecture, resulting in variations in propagation paths. These differences in propagation also affect the impact on software reliability. Based on this understanding, the system architecture can be utilized for reliability prediction or evaluation.
In general, the process by which a component failure leads to system failure can be described as follows: if there exists an internal error in the implementation of a component's internal code, this error may result in an internal fault within the component. Once the fault propagates to the component's interface, it can cause the failure of the component. However, if the fault is masked by other internal activity instructions, the component failure may be prevented.

Similarly, in the context of a component-based system, if a component failure propagates to the system's interface, it may lead to a system failure. Conversely, if the fault is masked by instructions from other components, the system failure may not occur.

Throughout this process, the reliability of the software system is influenced by three primary factors:

1. The internal failure probability of each component.
2. The fault propagation probability of each component, which is the probability that the component will propagate the erroneous input it receives to its output interface.
3. The fault propagation path probability, which is the probability of each possible fault propagation path from a component to the system output.

To ensure software reliability, it is necessary to analyze and evaluate these factors and optimize the architecture and design of the software accordingly.

The first two factors are intrinsic properties of the components, whereas the third factor is determined by the system's structure. It is assumed that the probability of internal failure and the probability of fault propagation are mutually independent. In some studies, the fault propagation probability is used as a substitute for the fault propagation path probability. Regardless of the terminology, both metrics are defined to characterize the properties of software architecture.

Since there are various methods for representing software architecture—such as employing UML use case diagrams, sequence diagrams, state diagrams, and deployment diagrams in object-oriented design—the calculation of reliability-related metrics must be tailored to the specific architectural representation method used.

Using propagation theory to study the reliability of software architecture can result in various models depending on the underlying assumptions. For instance, it can be assumed that an internal error within a component of the architecture not only directly causes system failure, thereby impacting the overall system reliability, but also propagates to other components, inducing further errors that compound the effects on system reliability. In other words, the impact of internal component errors on system reliability is dual-faceted.

Some researchers also believe that the probability of internal failure and error propagation in each component is independent of each other. That is, when a component fails, it often transfers the error to the next component, regardless of whether the component has received erroneous input from the previous component. Of course, if the fault tolerance mechanism is taken into account, the situation will be more complicated. The failure of a component spreads to another component, and because the faulty output will be overwritten by instructions from other components (fault-tolerant), the component can still generate correct output. The existence of fault tolerance mechanisms changes the path of fault propagation, which in turn affects the reliability of the system.

2.2.5 Fault Localization Leveraging Fault Propagation Analysis

In the field of reliability research, fault localization is typically based on program dependency analysis, which is broadly categorized into two types: control dependencies and data dependencies. Fault localization methods commonly compute a metric known as suspiciousness, where a higher value indicates a greater likelihood of the fault residing within the corresponding basic block. However, as software systems grow increasingly complex, faults often exhibit a "prolonged propagation process." If this propagation process is not taken into account, identifying the module with the highest suspiciousness may still leave uncertainty as to whether the fault originated in that module or was merely propagated from a preceding module. Furthermore, the propagation characteristics of faults can introduce inaccuracies in the calculation of suspiciousness, thereby compromising the effectiveness of fault localization.

In complex software systems, accurate fault localization can be achieved by analyzing fault propagation to refine the calculation of suspiciousness. Specifically, if fault propagation is assumed to occur through control dependency relationships within the program, the calculation of propagation suspiciousness can be formulated based on the suspiciousness of dependency pairs, denoted as $\theta^{\Delta}(\varepsilon)$The method for calculating the propagation rate of dependency pairs is as follows:

$$\mathrm{W}\left(b_j,b_k\right)=\frac{\theta^{\Delta}\left(b_j,b_k\right)}{\sum_{\forall\left(b_*,b_k\right)}\theta^{\Delta}\left(b_*,b_k\right)} \tag{2.6}$$

In the equation, the numerator represents the suspiciousness of dependency pairs between B_j and B_k, where B_k is one of B_j's successor blocks. The fault state of the program may propagate from B_j to B_k through the dependency pairs (B_j, B_k). The denominator represents the sum of all dependencies where B_k is the successor block. The ratio between the numerator and the denominator is used as an indicator to measure the propagation rate of B_j to B_k.

In complex software systems, fault propagation occurs not only through control dependency relationships but also through data dependency relationships. A more effective approach is to integrate both types of dependencies by constructing a program dependency graph. A program dependency graph is a directed graph where nodes represent the states of the program and edges represent control and data dependencies. Control dependency edges are assigned weights that reflect branching conditions, while data dependency edges are assigned weights that represent the transmission of changing variables. By transforming the program dependency graph, a joint dependency graph can be derived, providing the basis for a refined method to calculate suspiciousness.

$$\begin{aligned}\text{final}_{\text{susp}(n_i)} &= \text{suspiciousness}(n_i) \\ &+ \sum\nolimits_{\forall \text{edg}(n_{i,n_j}) \in \text{edg}(n_{i,*})} \left[k_{\text{edg}(n_1, n_j)} \cdot \text{suspiciousness}(n_j) \right]\end{aligned} \tag{2.7}$$

The error in statement N_j has two sources: one is produced by itself, and the other is propagated from its predecessor node. In Eq. (2.7), the bracketed part represents the error from parent node N_j to statement N_i. The variable k represents the probability of propagating error for each data dependency edge.

Through the calculation of suspiciousness, the node with the highest suspiciousness, $V_{\max}$, is first identified. Subsequently, an analysis is performed to determine whether there is a fault propagation trend from its direct predecessor nodes. If such a trend exists, a recursive process is applied to the predecessor node until a node V_0 is reached where no further fault propagation trend is observed. At this point, a virtual edge is established from V_0 (as the starting node) to $V_{\max}$ (as the ending node) to adjust the suspiciousness of V_0. By accounting for the influence of fault propagation, this approach enables accurate differentiation between the following two scenarios:

1. The fault originates in the predecessor nodes of $V_{\max}$ and is propagated to $V_{\max}$ through the edge (V_0, $V_{\max}$).
2. The fault is directly located at $V_{\max}$.

2.3 Research on Fault Propagation Models Based on System Architecture

Section 2.2.2 briefly introduced several fundamental concepts related to fault propagation research based on software architecture. This section will present a detailed explanation of a fault propagation model grounded in the definition of software architecture.

Software architecture represents a high-level abstraction of a software system, encompassing not only the structure but also the principles and documentation for its creation. Once established, the software architecture should serve as a reliable foundation for accurately deriving the software system. A comprehensive software architecture includes software elements, the relationships between these elements, and the properties of both the elements and their relationships. Similar to the importance of building architecture in construction, software architecture is critical to the success of a system. Due to its high level of abstraction, selecting and implementing a specific architecture is a pivotal decision. Changes to the architecture after implementation should be avoided, as they can lead to substantial cost inefficiencies.

Architectural patterns are general and reusable approaches designed to address recurring problems within specific software environments. Analogous to architectural styles in building design, software architecture also has its own distinct styles. It views software systems from the perspective of their organizational structure, comprising a series of components, connectors, and constraints that define their interactions. The Garlan-Shaw model, represented as SA = {components, connectors, constraints}, is a classic example of a software architecture model.

Researchers propose that software architecture consists of three essential elements: components, connectors, and necessary constraints. Components can vary significantly in scale, from a small segment of code to an entire system or subsystem. Connectors represent the interactions and relationships between these components. Furthermore, a software architecture incorporates constraints that regulate the behavior and interactions of its elements.

This book defines the software architecture model as: SA = {components, signals}.

Software architecture comprises a set of elements, specifically components and signals. Components can be identified as inputs, parts, sets or subsets, systems, or subsystems. They perform activities, tasks, or jobs and can execute functions distinct from those of other components within the system. Components can be interconnected to enable a more comprehensive functionality or to form a complete system. Additionally, components are often designed for reuse. Signals, on the other hand, represent the input and output information of components, including control signals, call signals, data signals, and others.

Definition 2.1: Fault Signal: a signal that falls outside the normal range (as specified by requirements) of its values. Signal that falls within the normal range is defined as a normal signal.

Definition 2.2: Relevant Components of Signal *i*: the components affected by signal *i*, either in terms of their behavior or the signal values they produce.

Definition 2.3: Reachable Signal: a signal *j* that can be influenced by signal *i* through the signals produced by the related components of *i* during a single run of the software. Signal *j* is considered a reachable signal of signal *i*.

Definition 2.4: Propagation Path Between Signal *i* and Signal *j*: a path r_{ij} represented by a directed path P(C, S, E) if signal *i* and signal *j* are reachable, where C represents the related component of signal *i* between signal *i* and *j*, S represents the signals transmitted between components, and E is a set of directed edges indicating the direction of signal transmission. In complex software systems, there may be more than one propagation path between signals *i* and *j*. It is defined that if a propagation path r_{ij} is different from other paths in terms of the signals or components it includes, it is considered as a different path.

Definition 2.5: Propagation Path Set R_{ijn}: R_{ijn} which represents the set of different propagation paths between signals *i* and *j*, where *n* represents the number of propagation paths is defined as propagation path set.

Based on the above definitions, a given software system can be represented as a network G(S, C, P), where S denotes the set of all signals in the network, C denotes the set of all components in the network, and P denotes the propagation paths within the software.

The network can be conceptualized as a directed graph, where nodes represent signals or components, and edges denote propagation paths between these nodes. The direction of each edge indicates the flow of signal transmission between components.

Within this framework, the related components of a signal can be determined by identifying the components associated with its input and output. Similarly, the propagation paths between two signals can be established by identifying their related components and tracing the direction of signal transmission along these components.

By analyzing the network, it is possible to identify the set of reachable signals from a given signal, as well as the propagation paths contributing to fault propagation within the software system. This analysis provides valuable insights for diagnosing and resolving faults in software systems.

The definition of the software architecture can be interpreted as illustrated in Fig. 2.3.

In Fig. 2.3, signal *i* is reachable to signal *j*, and components *A*, *B*, and *C* are the related components of signal i. The propagation path from signal i to signal j is represented by P(C, S, E), where C = {component *A*, component *B*, component *C*}, and S = {signal *a*, signal *b*}.

Based on the understanding of software architecture, the process of fault propagation can be described as shown in Fig. 2.4. Suppose there is a defect in the code implementation of a component, which leads to an internal fault of the component when this part of the code is executed. Once the fault reaches the interface of the component, it will cause the component to fail and propagate the fault through signals. However, if the fault is masked by other internal activity instructions, the component will not fail. Similarly, it can be inferred that, within a component system, if a component fault propagates to the system interface and is propagated through signals, the component fault will cause the system to fail. If the fault is masked by the instructions of other components, it will not cause the system to fail.

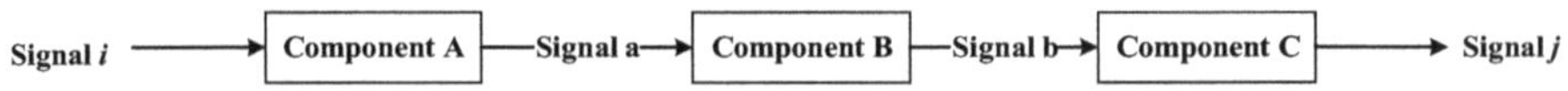

Fig. 2.3 Interpretation of architecture definition

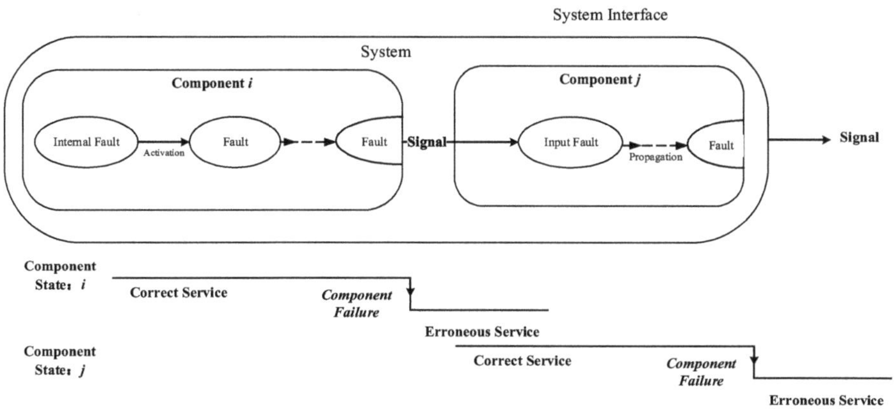

Fig. 2.4 Process of fault propagation

2.3.1 Cellular Automata and Fault Propagation

A cellular automaton is a discrete model that can be used to describe phenomena in computational theory, mathematics, physics, complexity science, theoretical biology, and microscopic structural modeling. It consists of a network of cells governed by specific rules. Each cell within the grid can exist in one of a finite number of states, and at any given moment, a cell can occupy only one state. The grid itself can have any finite dimensionality.

The set of cells adjacent to a given cell is referred to as its neighborhood, with each cell having its own defined neighborhood. At the initial time (t=0t = 0t=0), each cell is assigned an initial state. As time advances (t = t + 1t = t + 1t = t + 1), the cells transition to a new generation based on predefined rules. These rules determine the state of each cell by considering its current state and the states of its neighboring cells.

Typically, the rules remain constant over time and apply simultaneously to all cells in the grid. However, exceptions exist, such as in random cellular automata or asynchronous cellular automata, where the rules or their application may vary.

The concept of cellular automata was first introduced by Stanislaw Ulam and John von Neumann in the 1940s during their tenure at the Los Alamos National Laboratory. Although cellular automata were explored by other researchers in subsequent years, the concept did not garner widespread scholarly attention until the 1970s. This shift occurred with the introduction of the *Game of Life*, a two-dimensional cellular automaton developed by British mathematician John Horton Conway in 1970. The *Game of Life* was first presented in Martin Gardner's "Mathematical Games" column in the October 1970 issue of *Scientific American.*

In the 1980s, Stephen Wolfram advanced the study of one-dimensional cellular automata, which are often referred to as elementary cellular automata. Further research by Matthew Cook demonstrated that these rules possess Turing-completeness, highlighting their computational power.

Cellular automata are an effective tool for modeling complex systems, particularly excelling in the representation of synchronous parallel processes. They are widely employed to simulate nonlinear systems and possess robust capabilities for addressing dynamic problems caused by coupling interactions between system components. Examples include the spread of infectious diseases, network viruses, the evolution of traffic congestion, cascading failures in power grids, rumor propagation, forest fire dynamics, and more. These applications demonstrate the significant role cellular automata play in the study of propagation phenomena across various disciplines, including transportation, networks, sociology, and ecology.

Given this versatility, cellular automata can be effectively utilized to model the process of software fault propagation. They can simulate how local fault information spreads, leading to state changes across all components of a system. Additionally, the structural diversity of software systems allows for the formulation of different fault propagation rules tailored to specific system architectures. These diverse rules can be integrated to collectively determine fault propagation paths. This flexibility makes cellular automata a powerful framework for describing fault propagation processes in software systems.

2.3.2 *Fault Propagation Model Based on Architecture*

2.3.2.1 Fault Propagation Between Two Signals

Based on the above definitions, a representation method for fault propagation between two signals is proposed. Since the state of a signal can be classified as either faulty or normal, the set $S_i = \{0,1\}$ is defined to represent the state of signal i.

$$S_i = \begin{cases} 0 & (\text{Signal } i \text{ is a normal signal.}) \\ 1 & (\text{Signal } i \text{ is a fault signal.}) \end{cases} \tag{2.8}$$

For Component A, if signal i serves as its input signal and signal j as its output signal, the input-output relationship of the component can result in the following four possible scenarios.

1. $S_i = 0$, $S_j = 0$: This indicates that no fault occurs in component A, or any fault within A does not affect its output signal.
2. $S_i = 0$, $S_j = 1$: This indicates that signal i causes a fault in component A, resulting in a faulty output signal j.
3. $S_i = 1$, $S_j = 0$: This indicates that the fault tolerance mechanism within component A successfully handles the faulty input signal i, producing a correct output signal.
4. $S_i = 1$, $S_j = 1$: This indicates that component A fails to handle the faulty input signal i, and the output signal j is also faulty.

The model assumes that signal faults are solely caused by component faults. To quantify the four scenarios described above, the concepts of component failure rate and fault tolerance rate are introduced, defined as follows:

Definition 2.6: Failure Rate (*f*) The failure rate of component *A* is defined as the probability of obtaining a faulty output from a normal input. The formula for calculating the failure rate of component *A* is:

$$f_A = \frac{m}{n} \tag{2.9}$$

Where *nnn* normal signals are input into component *A*, and *m* faulty signals are observed in the output. If component *A* is hardware, its failure rate can be calculated using Eq. (2.10), where $R_A(t)$ represents the reliability function of hardware *A*, defined as:

$$f_A = 1 - R_A(t) \tag{2.10}$$

Definition 2.7: Fault Tolerance Rate The probability of obtaining a normal output when a component A is provided with fault inputs is defined as the fault tolerance rate βA. The formula for calculating the fault tolerance rate is:

$$g_A = \frac{m}{n} \tag{2.11}$$

Definition 2.8: Component State Transition Matrix M_A The state transition of a component can be represented using a 2 × 2 matrix, which is called the state transition matrix M_A.

$$\mathbf{M}_A = \begin{pmatrix} S_{00} & S_{01} \\ S_{10} & S_{11} \end{pmatrix} \tag{2.12}$$

The input signal of component *A* is *i* while the output signal is *j*, then: $S_{00} = \text{Prob}(S_j = 0 \mid S_i = 0)$; $S_{01} = \text{Prob}(S_j = 1 \mid S_i = 0)$; $S_{10} = \text{Prob}(S_j = 0 \mid S_i = 1)$; $S_{11} = \text{Prob}(S_j = 1 \mid S_i = 1)$.

The calculation formula of state transition matrix is:

$$\mathbf{M}_A = \begin{pmatrix} S_{00} & S_{01} \\ S_{10} & S_{11} \end{pmatrix} = \begin{pmatrix} 1 - f_A & f_A \\ g_A & 1 - g_A \end{pmatrix} \tag{2.13}$$

Definition 2.9: Signal State Transition Matrix $\mathbf{M}_{ij}$ If signal *i* and signal *j* are accessible to each other, the state transition matrix $\mathbf{M}_{ij}$ for the signals is defined as follows:

$$\mathbf{M}_{ij} = \begin{pmatrix} \text{Prob}(S_j = 0 | S_i = 0) & \text{Prob}(S_j = 1 | S_i = 0) \\ \text{Prob}(S_j = 0 | S_i = 1) & \text{Prob}(S_j = 1 | S_i = 1) \end{pmatrix} \tag{2.14}$$

According to the definition, the fault propagation path r_0 is an element of the set of propagation paths. Assuming there are n components $(1, 2, \cdots, n)$ in a certain execution path, the state transition matrix M_{s_i} for the path from signal i to signal j can be obtained as follows:

$$\mathbf{M}_{\underline{R}_{ij}} = M_1 M_2 M_3 \cdots M_n \tag{2.15}$$

For example, when $n = 2$:

$$\mathbf{M}_{\underline{R}_{ij}} = \mathbf{M}_A \cdot \mathbf{M}_B = \begin{pmatrix} S_{00} & S_{01} \\ S_{10} & S_{11} \end{pmatrix} \begin{pmatrix} S'_{00} & S'_{01} \\ S'_{10} & S'_{11} \end{pmatrix} = \begin{pmatrix} S_{00}S'_{00} + S_{01}S'_{10} & S_{00}S'_{01} + S_{01}S'_{11} \\ S_{10}S'_{00} + S'_{11}S'_{10} & S_{10}S'_{01} + S_{11}S'_{11} \end{pmatrix} \tag{2.16}$$

where:

$$S_{00}S'_{00} + S_{01}S'_{10} = \text{Prob}\left(S_j = 0 \middle| S_i = 0\right)$$

$$S_{00}S'_{01} + S_{01}S'_{11} = \text{Prob}\left(S_j = 1 \middle| S_i = 0\right)$$

$$S_{10}S'_{00} + S_{11}S'_{10} = \text{Prob}\left(S_j = 0 \middle| S_i = 1\right)$$

$$S_{10}S'_{01} + S_{11}S'_{11} = \text{Prob}\left(S_j = 1 \middle| S_i = 1\right)$$

Thus, the state transition matrix for each propagation path can be obtained. The state transition matrix for the k-th path, $\mathbf{M}_{R_{ijk}}$, can be represented as:

$$\mathbf{M}_{R_{ijk}} = \begin{pmatrix} 1 - f_k & f_k \\ g_k & 1 - g_k \end{pmatrix} \tag{2.17}$$

Then, the state transition matrix from signal i to signal j can be obtained using the following formula:

$$\mathbf{M}_{ij} = \begin{pmatrix} (1-f_1)(1-f_2)\cdots(1-f_n) & 1-(1-f_1)(1-f_2)\cdots(1-f_n) \\ g_1 g_2 \cdots g_n & 1 - g_1 g_2 \cdots g_n \end{pmatrix} \tag{2.18}$$

If signal i and signal j are not reachable from each other, the state transition matrix is defined as:

$$\mathbf{M}_{ij} = \begin{pmatrix} 1 & 1 \\ 1 & 1 \end{pmatrix} \tag{2.19}$$

Based on the definitions outlined above, the necessary concepts for the propagation model, including signals, failure rate, and fault tolerance rate, have been established. These foundational concepts provide the basis for constructing the propagation model.

2.3.2.2 Fault Propagation Model

Assuming that a complex software system has N signals, and $M(t)$ is an $N \times N$ matrix, where the element in the i-th row and j-th column represents the state transition matrix M_{ij} from signal i to signal j at time t. As the MT_{ij} of the software system changes dynamically during runtime, in order to obtain the state of the signals after one cycle of the system running, $d_i(t)$ is defined as the state matrix of signal i at time t.

$$d_i(T) = \begin{pmatrix} \text{Prob}(S_i = 0 | t = T) \\ \text{Prob}(S_i = 1 | t = T) \end{pmatrix} \tag{2.20}$$

The matrix $D(t)$ is defined as the matrix whose diagonal elements are the state matrices $d_i(t)$ of each signal.

$$\mathbf{D}(t) = \begin{pmatrix} \mathbf{d}_1(t) & \cdots & 0 \\ \vdots & & \vdots \\ 0 & \cdots & \mathbf{d}_n(t) \end{pmatrix} \tag{2.21}$$

the state matrix of signal i at time $t + 1$ is defined as:

$$\mathbf{D}(t+1) = \mathbf{M}^{\mathrm{T}}(t) \cdot \mathbf{D}(t) = \begin{pmatrix} \mathbf{m}_{11}^{\mathrm{T}}(t)\mathbf{d}_1(t) & \cdots & \mathbf{m}_{n1}^{\mathrm{T}}(t)\mathbf{d}_n(t) \\ \vdots & & \vdots \\ \mathbf{m}_{1n}^{\mathrm{T}}(t)\mathbf{d}_1(t) & \cdots & \mathbf{m}_{nn}^{\mathrm{T}}(t)\mathbf{d}_n(t) \end{pmatrix} \tag{2.22}$$

In this matrix, the element in the i-th row and j-th column represents the state transition matrix from signal i to signal j at time $t + 1$. It is assumed that if signal i can affect signal j at time t and cause signal j to fail at time $t + 1$, then the state of signal j at time $t + 1$, denoted as $S_j(t + 1)$, is equal to 1.

We can assume that, at time $t + 1$, the state matrix of signal j is:

$$\mathbf{m}_{ij}(t)\mathbf{d}_i(t) = \begin{pmatrix} P_i \\ 1 - P_i \end{pmatrix} \tag{2.23}$$

In the equation, P_i represents the probability that signal j is in state 0 under the influence of signal i, and $1 - P_i$ represents the probability that signal j is in state 1. Assuming that the effects of other signals on signal j are independent of each other, the state matrix of signal j at time $t + 1$ under the influence of other signals can be represented as:

$$\mathbf{d}_j(t+1) = \begin{pmatrix} P_1P_2\cdots P_n \\ 1-P_1P_2\cdots P_n \end{pmatrix} \tag{2.24}$$

2.3.2.3 Using Cellular Automata to Represent the Fault Propagation Model

The complex software system G(S,C,P) has n signals, and the cellular automaton model is defined as CA = (C,Q,V,f).

Cellular Space C: An n-dimensional space where each cell represents a signal within the software.

The neighbors V: The neighbors of a signal j are defined as the set of accessible signals to signal j.

Finite state set Q: $S_i(t) = \{1,0\}$ represents the state of signal i at time t, where 1 represents a fault signal and 0 represents a normal signal.

State transition rule f: The state transition rule is the core of the cellular automaton. The state transition rule followed by the cellular automaton propagation model is:

$$S_i(t+1) = f_i(P_1P_2\cdots P_n) \tag{2.25}$$

where P_i represents the probability that signal j is in state 0 under the influence of signal i. Let $p = P_1P_2...P_n$, the state transition rule can be represented as:

$$f_i(p) = \begin{cases} 0 & (x \le p) \\ 1 & (x > p) \end{cases} \tag{2.26}$$

Where: x is a random number between 0 and 1.

The signals can be classified into the following three types:

1. Signals generated only by hardware during each software running cycle.
2. Input signals for the next software running cycle.
3. Signals that will not affect any other signals in the next software running cycle.

Different types of signals correspond to different transition functions. Therefore, to take into account different types of signals, the state transition rule is improved, where $R_i(t)$ is the reliability function of the hardware that generates signal i.

$$S_i(t+1) = \begin{cases} f_i(1-R_i(t)) & (i \in \text{class}1) \\ f_i(P_1P_2\cdots P_n) & (i \in \text{class}2) \\ 0 & (i \in \text{class}3) \end{cases} \tag{2.27}$$

Definition 2.10: Average Propagation Strength E_n The average number of fault signals between the initial time $t0$ and the end time t_n. The larger the value of E_n, the greater the scope of the impact of the fault propagation.

2.4 Research on Fault Propagation Models Incorporating Correlated Faults

2.4.1 Correlated Faults

The concept of correlated faults was briefly introduced in Sect. 2.2.3. In this section, we will develop a propagation model that accounts for correlated faults.

Due to the influence of factors such as software structure, data flow, and control flow, faults are not entirely independent of each other. The occurrence of one fault may subsequently trigger another fault.

In a narrow sense, correlated faults are defined as follows: In program Pr, given two known faults p and q, p and q are said to be correlated if and only if it can be concluded that if fault p does not occur, fault q will also not occur. In this case, p and q form a pair of correlated faults.

In a broad sense, correlated faults are defined as follows: For a program Pr with m components, the component failure rate is represented by f, and the following properties hold:

1. f represents component fault rate, and $f_i(t) \in (0, 1)$, where 1 represents component fails and 0 represents component is normal.
2. If there are two components $(i,j) \in C$, let $f_i(t) \neq f_{ij}(t)$, which means whether component i fails or not depends on whether component j fails.
3. If $f_i(t) = 0$ and $f_{ij}(t) = 0$, or $f_i(t) = 1$ and $f_{ij}(t) = 1$, component i and j are positive correlated.
4. If $f_i(t) = 0$ and $f_{ij}(t) = 1$, or $f_i(t) = 1$ and $f_{ij}(t) = 0$, component i and j are negative correlated.

In complex software systems, components that interact with each other can be categorized into two types: influencing components and influenced components. Influencing components are those that exert an effect on other components, while influenced components are those affected by other components. Based on the number of influencing and influenced components and the nature of their interactions, correlated faults can be classified into five distinct types, as illustrated in Fig. 2.5.

Figure 2.5a shows a unidirectional correlated relationship, which is the fundamental type of correlated faults, where component i influences component j unidirectionally. Figure 2.5b shows a bidirectional correlated relationship, where component i and component j influence each other, which is the fundamental type of bidirectional correlated faults. Figure 2.5c shows a single component influencing multiple components simultaneously, where the number of influenced components

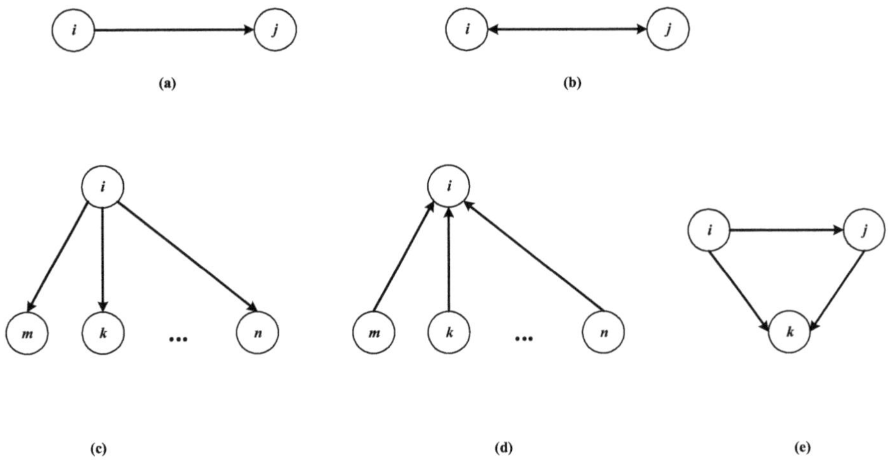

Fig. 2.5 Five types of correlated faults

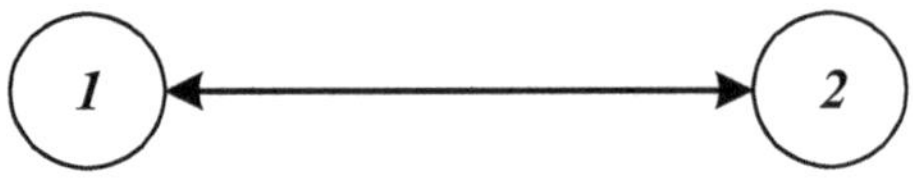

Fig. 2.6 Bidirectional correlation relationship

is p ($p > 1$), known as $1 - n$ correlated faults. Figure 2.5d shows multiple components influencing the same component, where the number of influencing components is q ($q > 1$), known as $n - 1$ correlated faults. Figure 2.5e shows the correlated relationship between components in a complex software system. The correlated relationships between components in a complex software system can be formed by combining the above five types.

2.4.2 *Fault Rate Considering Correlated Faults*

According to the definition of correlated faults, the failure rate of components involved in fault correlation should account for two factors: the intrinsic failure rate of the component when considered independently, and the additional failure rate caused by the influence of other components. Therefore, the failure rate of influenced components can be expressed as a function of their own failure rate and the failure rate of the influencing components.

Suppose that components 1 and 2 in the software system influence each other's failure rates, exhibiting a bidirectional correlation as shown in Fig. 2.6. The failure rates of these two components under correlated faults can be expressed as follows:

$$f_1(t) = \varphi_1\left[f_{I_1}(t), f_2(t)_A, t\right], f_2(t) = \varphi_2\left[f_{I_2}(t), f_1(t)_A, t\right] \tag{2.28}$$

In the equation, $f_1(t)$ and $f_2(t)$ represent the failure rates of components 1 and 2, respectively, when faults are independent; is the fault rate generated by component 2 affecting component 1, while $f_1(t)$ is the fault rate generated by component 1 affecting component 2.

$f_1(t)$ and $f_2(t)$ represent the failure rates of components 1 and 2, respectively, when faults are independent. If we extend this definition to a complex software system with N components, the correlated failure rate of each component can be expressed as:

$$\begin{cases} f_1(t) = \varphi_1\left[f_{I_1}(t), \hat{f}_{j_1}(t)_A, t\right] \\ f_2(t) = \varphi_2\left[f_{I_2}(t), \hat{f}_{j_2}(t)_A, t\right] \\ \vdots \\ f_i(t) = \varphi_i\left[f_{I_i}(t), \hat{f}_{j_i}(t)_A, t\right] \\ \vdots \\ f_N(t) = \varphi_N\left[f_{I_N}(t), \hat{f}_{j_N}(t)_A, t\right] \end{cases} \tag{2.29}$$

In the equation, $f_i(t)$ and $f_{I_5}(t)$ represent the correlated failure rate and the independent failure rate of component i, respectively. $\hat{f}_{j_i}$ denotes the failure rate generated by all components that affect component i. The subscript ji indicates the impact of component j on component i. If there are six components in a system, and component 5 is affected by components 2, 4, and 6, then $j_5 = 2, 4, 6$. We can have the following equations:

$$f_5(t) = \varphi\left[f_{I_5}(t), \hat{f}_{j_5}(t)_A, t\right] = \varphi\left[f_{I_5}(t), f_2(t)_A, f_4(t)_A, f_6(t)_A, t\right] \tag{2.30}$$

According to the Taylor formula, the correlated failure rate in Eq. (2.29) can be expressed as:

$$\begin{aligned} f_i(t) &= \varphi_i\left[f_{Ii}(t), \hat{f}_{j_i}(t)_A, t\right] \\ &= \varphi_i\big|_{f_{j_i}(t)_A=0} + \sum_{j_i} \frac{\partial \varphi_i}{\partial f_{j_i}}\bigg|_{f_{j_i}(t)_A=0} f_{j_i}(t)_A \\ &\quad + \sum_{j_i,k_i} \frac{\partial^2 \varphi_i}{2\partial f_{j_i} f_{k_i}}\bigg|_{f_{j_i}(t)A=0} f_{j_i}(t)_A f_{k_i}(t)_A \\ &\quad + \sum_{j_i} \frac{\partial^2 \varphi_i}{4\partial f_{j_i}^2}\bigg|_{f_{j_i}(t)_A=0} f_{j_i}^2(t)_A + o\left(f_{j_i}^3(t)_A\right) \end{aligned} \tag{2.31}$$

In the equation, $f_{j_i}(t)$ and $f_{k_i}(t)$ represent the failure rates generated by components j and k, respectively, that affect component i. To emphasize the failure rates generated by the mutual influence between components, Eq. (2.31) can be rewritten as:

$$f_i(t) = \varphi_i\big|_{f_{j_i}(t)_A=0} + \left[\sum_{j_i} \frac{\partial \varphi_i}{\partial f_{j_i}}\bigg|_{f_{j_i}(t)_A=0} \sum_{j_i,k_i} \frac{\partial^2 \varphi_i}{2\partial f_{j_i} f_{k_i}}\bigg|_{f_{j_i}(t)_A=0} f_{k_i}(t)_A \right.$$

$$\left. + \sum_{j_i} \frac{\partial^2 \varphi_i}{4\partial f_{j_i}^2}\bigg|_{f_{j_i}(t)_A=0} f_{j_i}(t)_A + \frac{o\left(f_{j_i}^3(t)_A\right)}{f_{j_i}(t)_A} \right] \times f_{j_i}(t)_A \tag{2.32}$$

If component i is not influenced by other components, i.e., $f_{j_i}(t) = 0$, indicating that component i operates independently with respect to faults, the failure rate of component i then simplifies, in accordance with Eq. (2.32), to:

$$f_i(t) = \varphi_i\big|_{f_{j_i}(t)_A=0} = f_{I_i}(t) \tag{2.33}$$

Let:

$$\theta_{ij} = \sum_{j_i} \frac{\partial \varphi_i}{\partial f_{j_i}}\bigg|_{f_{j_i}^{(t)}{}_A=0} + \sum_{j_i,k_i} \frac{\partial^2 \varphi_i}{2\partial f_{j_i} f_{k_i}}\bigg|_{f_{j_i}(t)_A=0} f_{k_i}(t)_A$$

$$+ \sum_{j_i} \frac{\partial^2 \varphi_i}{4\partial f_{j_i}^2}\bigg|_{f_{j_i}(t)_A=0} f_{j_i}(t)_A + \frac{o\left(f_{j_i}^3(t)_A\right)}{f_{j_i}(t)_A} \tag{2.34}$$

θ_{ij} is the correlation coefficient between components with correlated faults. Substituting this into Eq. (2.33), we can obtain the failure rate of component i as:

$$f_i(t) = f_{I_i}(t) + \sum_j \theta_{ij}(t) f_{j_i}(t)_A \quad (i = 1, 2, \cdots, N) \tag{2.35}$$

Equation (2.35) shows that the failure rate of a component under correlated faults can be expressed as the sum of its own independent failure rate and the failure rates generated by other components that affect it. Rewriting Eq. (2.35) in matrix form, we have:

$$\{f(t)\} = [I]\{f_I(t)\} + [\theta(t)]\{f(t)_A\} \tag{2.36}$$

In the formula: $\{f(t)\}$ is $\mathbf{M} \times 1$ the vector matrix represents the corresponding failure rate; $\{f(t)_A\}$ is the $\mathbf{M} \times 1$ vector matrix representing the failure rates related to the influence of other components; $\{f_I(t)\}$ is the $\mathbf{M} \times 1$ vector matrix representing the independent failure rates; $[I]$ is the $\mathbf{M} \times \mathbf{M}$ identity matrix; $\boldsymbol{\theta}(t)$ is the correlation coefficient matrix. It has the following properties:

1. $\{\boldsymbol{\theta}(t)\}$ is a non-negative matrix, meaning $\theta_{ij}(t) \geq 0(i,j = 1,2,\cdots,N)$; If $\theta_{ij}(t) = 0$, 1, it indicates that the failure or non-failure of component j has no influence on component i; If the failure of component j guarantees the failure of component i, then $\theta_{ij}(t) = 1$.
2. The trace of $\boldsymbol{\theta}(t)$ is 0, $\mathrm{tr}([\boldsymbol{\theta}(t)]) \equiv 0$, Given $\theta_{ij}(t) \geq 0$, this implies $\theta_{ij}(t) = 0$, meaning the correlation coefficient of a component with itself is 0. This reflects that a component's failure or non-failure has no influence on itself.
3. In most complex systems, components are rarely independent but instead interact with other components in the system.

Specifically, if the components in a system fail independently, all elements of the correlation coefficient matrix are 0, i.e.,

$$\theta_{ij}(t) = 0(i,j = 1,\ 2,\ \cdots,\ N) \tag{2.37}$$

Substituting this into Eq. (2.36) yields:

$$\{f_i(t)\} = \{f_{I_i}(t)\}(i = 1,\ 2,\ \cdots,\ N) \tag{2.38}$$

Equation (2.38) indicates that the correlated failure rate of component i depends only on the independent failure rate of that component.

The independent failure rate of component C is $f_{I_C}(t)$, and its failure is independent of other components. Suppose that when component C fails at any time, components 1, 2, ⋯, N fail simultaneously, and the failures of components 1, 2, ⋯, N have no impact on each other. This phenomenon is called common cause failure. In this case, component C is the only factor that affects components 1, 2, ⋯, N, and $\theta_{ij}(t)$ can be expressed as:

$$\theta_{ij}(t) = \begin{cases} 1 & (i = 1,2,\cdots,N,j = C) \\ 0 & (\text{Other}) \end{cases} \tag{2.39}$$

Substitute this to Eq. (2.36), we have:

$$f_i(t) = \begin{cases} f_{I_i}(t) + f_{I_C}(t) & (i = 1,2,\cdots,N) \\ f_{I_C}(t) & (i = C) \end{cases} \tag{2.40}$$

Since $f_{I_C}(t) > 0$, the correlated failure rate of component i is greater than its independent failure rate. If $f_i(t) = \lambda_{I_i}$, and $f_{I_C}(t) = \beta\lambda$, then β is the common cause factor.

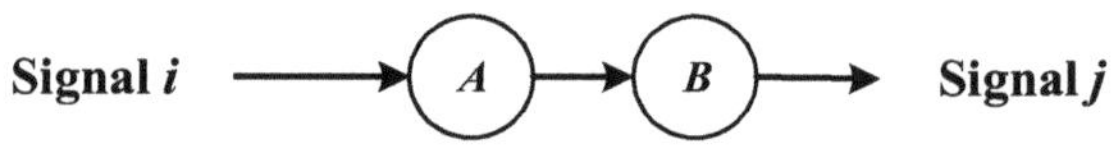

Fig. 2.7 Unidirectional correlated faults

2.4.3 Fault Propagation Model Taking Correlated Faults into Account

2.4.3.1 Fault Propagation Considering Unidirectional Correlated Faults

Considering the impact of component *j* on component *i*, as shown in Fig. 2.7, the state transition matrix of component *A* can be expressed as:

$$\mathbf{M}_A = \begin{pmatrix} S_{00} & S_{01} \\ S_{10} & S_{11} \end{pmatrix} = \begin{pmatrix} 1-f_A & f_A \\ g_A & 1-g_A \end{pmatrix}$$

$$= \begin{pmatrix} 1-\left[f_{IA}(t)+\theta_{AB}(t) f_{B_A}(t)_A\right] & f_{IA}(t)+\theta_{AB}(t) f_{B_A}(t)_A \\ g_A & 1-g_A \end{pmatrix} \tag{2.41}$$

The state transition matrix for the path from signal *i* to signal *j*, $\mathbf{M}_{R_{ij}}$, is given as:

$$\mathbf{M}_{R_{ij}} = \mathbf{M}_A \cdot \mathbf{M}_B = \begin{pmatrix} 1-\left[f_{IA}(t)+\theta_{AB}(t) f_{B_A}(t)_A\right] & f_{IA}(t)+\theta_{AB}(t) f_{B_A}(t)_A \\ g_A & 1-g_A \end{pmatrix} \begin{pmatrix} 1-f_B & f_B \\ g_B & 1-g_B \end{pmatrix} \tag{2.42}$$

Especially, when:

$$\theta_{AB} = 0, \mathbf{M}_{R_{ij}} = \mathbf{M}_A \bullet \mathbf{M}_B = \begin{pmatrix} 1-f_{IA}(t) & f_{IA}(t) \\ g_i & 1-g_i \end{pmatrix} \begin{pmatrix} 1-f_B & f_B \\ g_B & 1-g_B \end{pmatrix}$$

2.4.3.2 Fault Propagation Considering $n-1$ Correlated Faults

Taking into account the impact of component *P, L*,..., *Q* on component *A* (Fig. 2.8), the state transition matrix $\mathbf{M}_A$ of component *A* can be expressed as:

$$\mathbf{M}_A = \begin{pmatrix} S_{00} & S_{01} \\ S_{10} & S_{11} \end{pmatrix} = \begin{pmatrix} 1-f_A & f_A \\ g_A & 1-g_A \end{pmatrix}$$

Fig. 2.8 $n-1$ Correlated faults

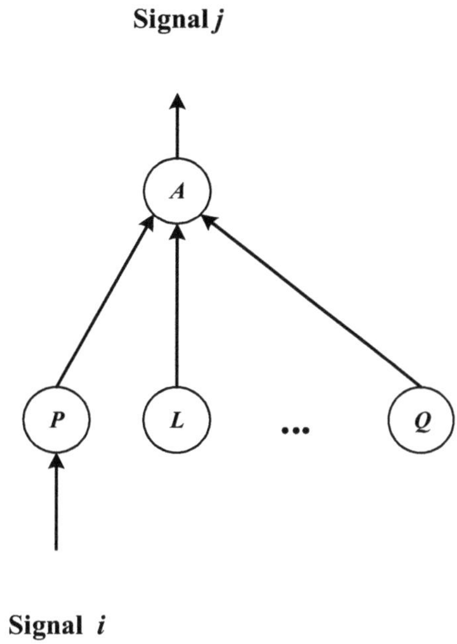

$$= \begin{pmatrix} 1-\left[f_{IA}(t)+\sum_{j}\theta_{Aj}(t)f_{jA}(t)_{A}\right] & \left[f_{IA}(t)+\sum_{j}\theta_{Aj}(t)f_{jA}(t)_{A}\right] \\ g_A & 1-g_A \end{pmatrix} \begin{pmatrix} 1-f_B & f_B \\ g_B & 1-g_B \end{pmatrix} \tag{2.43}$$

The state transition matrix of the path from signal i to signal j is:

$$\mathbf{M}_{R_{ij}} = \mathbf{M}_A \mathbf{M}_B$$

$$\mathbf{M}_{R_{ij}} = \mathbf{M}_A \mathbf{M}_B$$

$$= \begin{pmatrix} 1-\left[f_{IA}(t)+\sum_{j}\theta_{Aj}(t)f_{jA}(t)_{A}\right] & \left[f_{IA}(t)+\sum_{j}\theta_{Aj}(t)f_{jA}(t)_{A}\right] \\ g_A & 1-g_A \end{pmatrix} \begin{pmatrix} 1-f_B & f_B \\ g_B & 1-g_B \end{pmatrix} \tag{2.44}$$

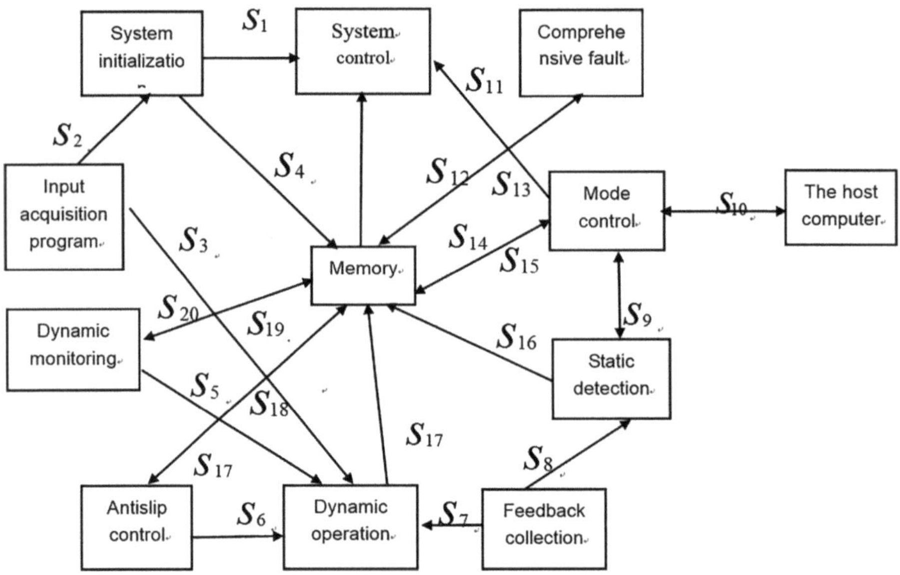

Fig. 2.9 Architecture diagram of the airplane brake system

2.5 A Case Study: System Analysis

In this section, two examples are presented to demonstrate the architecture-based fault propagation model and the fault propagation model that accounts for correlated faults.

2.5.1 Case Study: Analysis of Fault Propagation Model Based on Architecture

In this section, the airplane brake system is used as an example to construct a fault propagation model based on system architecture. The architecture model of the system is illustrated in Fig. 2.9.

To estimate the failure rate and fault tolerance of each software and hardware component (as shown in Table 2.1), 1000 sets of normal signals and 1000 sets of fault signals are used. It is assumed that the fault tolerance rate of hardware components is 0.

Figures 2.10, 2.11, 2.12, 2.13, 2.14, 2.15, and 2.16 show the result of MATLAB running and analysis

1. The X-axis represents the number of execution cycles n, and the Y-axis represents the average propagation strength E_n. From Fig. 2.10, we can see that the propagation strength fluctuates greatly in the initial cycles, but tends to stabilize as the number of execution cycles increases.

Table 2.1 Parameter list of airplane brake system

	Parameters	
Component	failure rate	fault tolerance rate
System initialization	0.003	0.991
System control	0.001	0.916
Integrated fault	0.011	0.972
Mode control	0.007	0.993
Static detection	0.01	0.936
Feedback collection	$\exp(-0.000000023t)$	0
Dynamic operation	0.012	0.994
Anti-slip control	0.005	0.914
Dynamic monitoring	0.013	0.943
Input acquisition	$\exp(-0.000000001t)$	0
Memory	$\exp(-0.000000004t)$	0
Host	$\exp(-0.000000093t)$	0

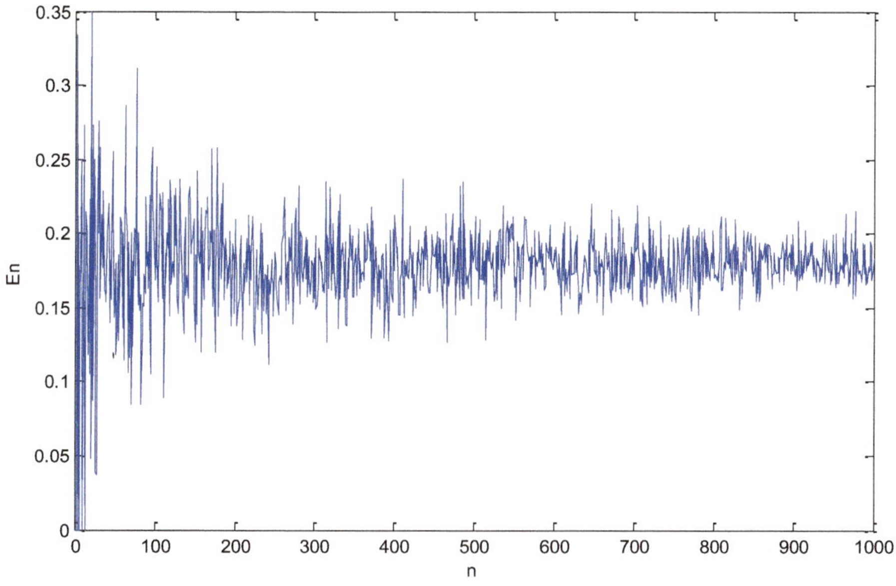

Fig. 2.10 Relationship between n and E_n

2. As shown in Fig. 2.11, the horizontal axis represents the number of execution cycles n, while the vertical axis represents the slope of E_n as it changes with the execution cycles.

$$P_n = \frac{|E_{n+1} - E_n|}{E_n} \tag{2.45}$$

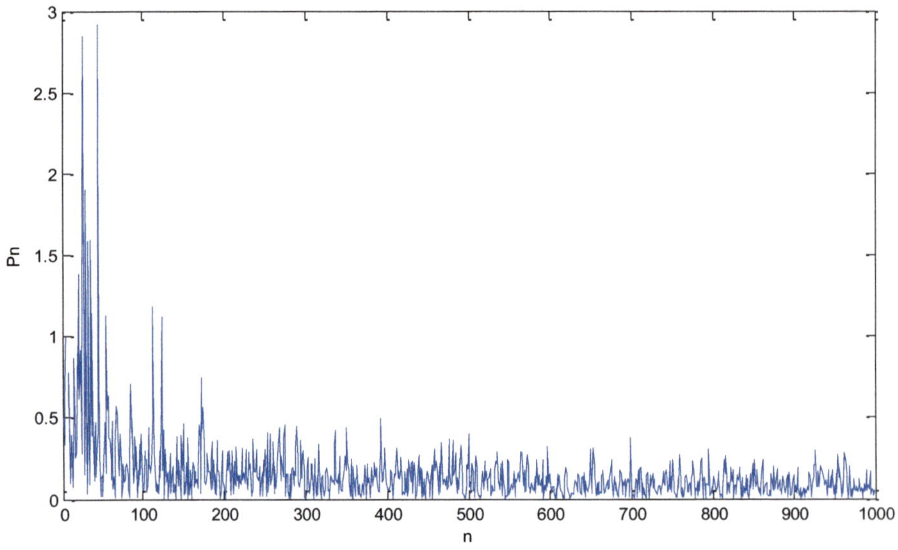

Fig. 2.11 Relationship between n and the rate of change of E_n

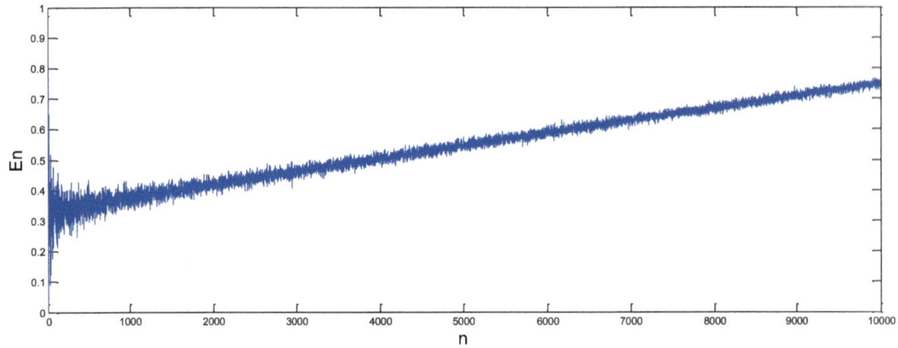

Fig. 2.12 Relationship between n and E_n with increased execution cycles

From Fig. 2.11, it can be observed that as the number of execution cycles increases, the rate of change P_n of En gradually decreases, eventually reaching a steady state close to 0.

3. To investigate the effect of hardware on the system, we set the reliability function of the hardware to $\exp(-0.0003t)$, and then increase the number of execution cycles n to 10000. From Fig. 2.12, we can see that as the number of execution cycles increases significantly, the number of fault signals also tends to increase.
4. By selecting a segment of the 10,000 execution cycles, it can be observed that E_n stabilizes during this period (Fig. 2.13). This behavior is attributed to the hardware failure rate, as an increase in the hardware failure rate results in a higher

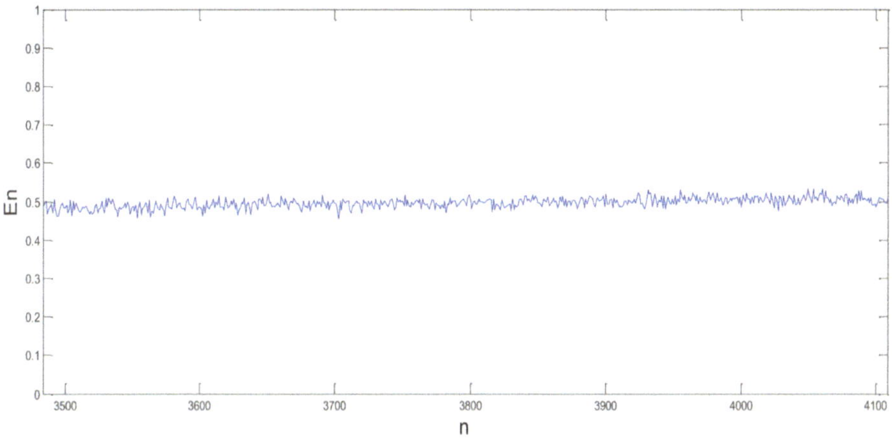

Fig. 2.13 Relationship between n and E_n in a selected segment (magnified view)

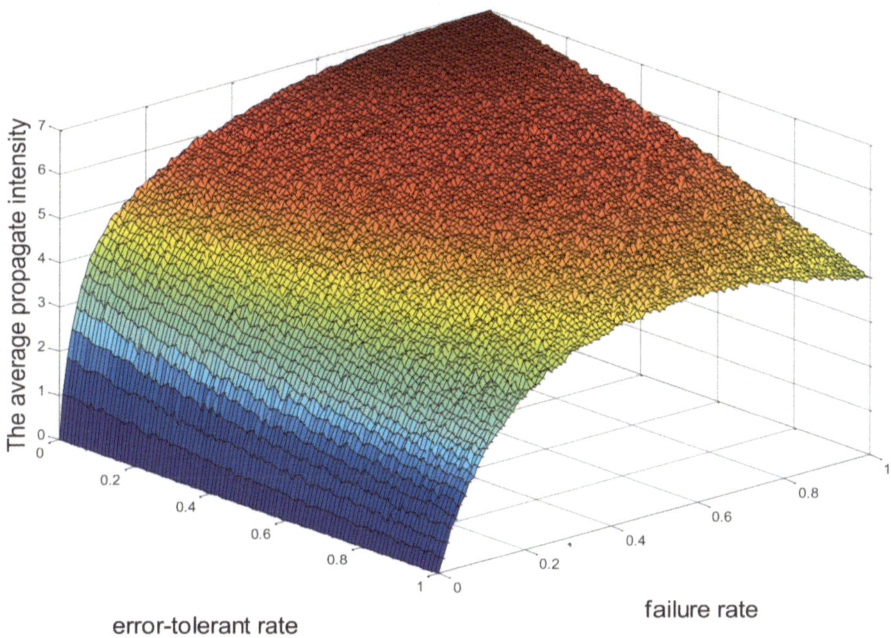

Fig. 2.14 Relationship between fault rate, fault tolerance rate, and E_n

steady-state value of E_n Thus, it is evident that the failure rates of software and hardware components have different impacts on the system.

5. To illustrate the relationship between fault tolerance and failure rate in software, the fault tolerance rate and failure rate of all components are set to the same value. The variation of E_n as the fault tolerance and failure rate range from 0 to 1 is shown in Fig. 2.14.

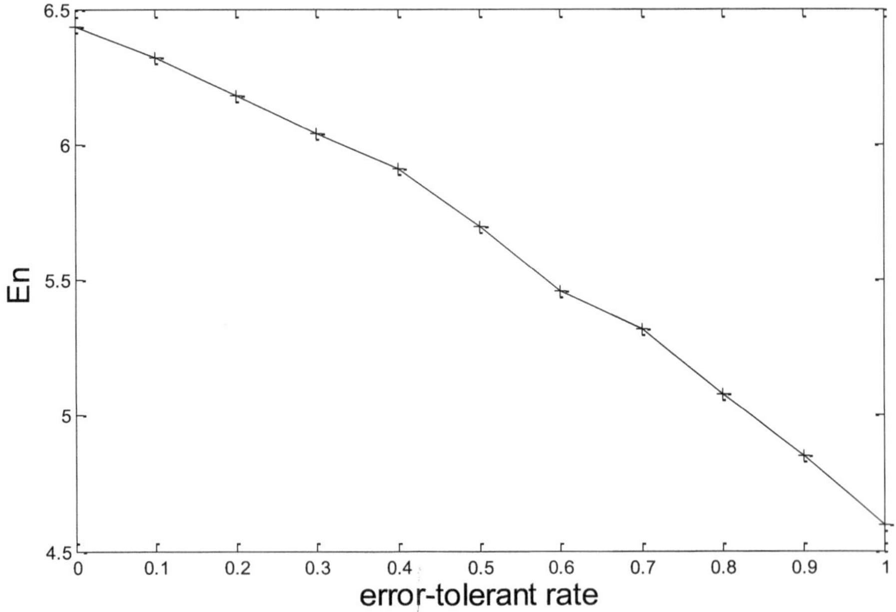

Fig. 2.15 Relationship between fault tolerance rate and E_n when fault rate is fixed

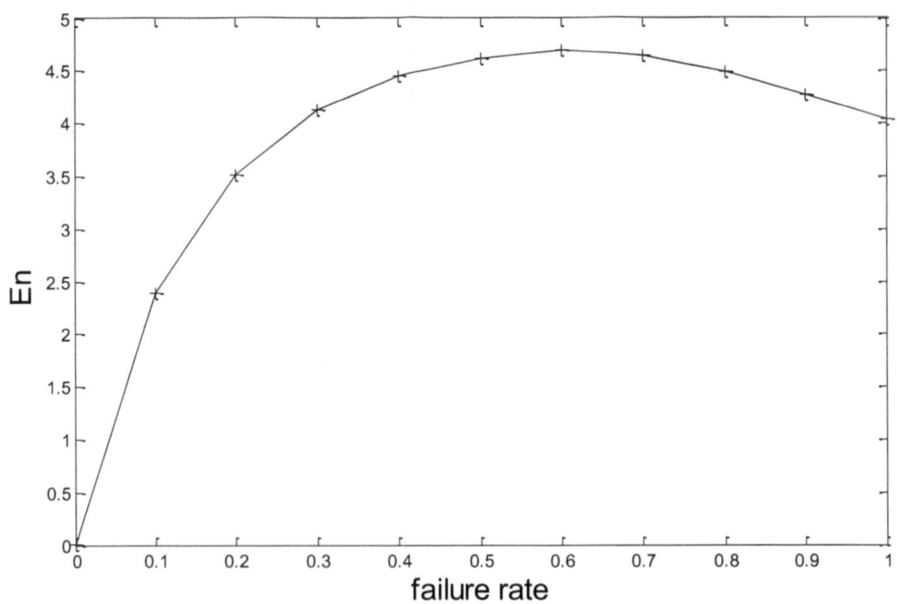

Fig. 2.16 Relationship between fault rate and E_n when fault tolerance rate is fixed

6. From Fig. 2.15, it can be seen that when the fault rate is fixed, there is an inverse proportional relationship between fault tolerance and E_n, that is, the higher the fault tolerance, the lower the E_n.
7. From Fig. 2.16, it can be seen that within a certain range, increasing the fault rate leads to a proportional increase in E_n, that is, the higher the fault rate, the higher the E_n. However, when the fault rate increases to a certain value, E_n decreases with the increase in fault rate.

2.5.2 Case Study: Analysis of Fault Propagation Model Considering Correlated Faults

The software of a mobile robot is selected as a case study to analyze the system. This software is responsible for completing the task of moving the robot from an initial location to a target location by acquiring information such as the image of the target location and the angle of the movement position. The control flow diagram and data flow diagram of the software are shown in Fig. 2.17.

The UML activity diagram of the mobile robot software is shown in Fig. 2.18.

The parameter settings of each component in the mobile robot software are listed in Table 2.2.

1. Selecting angle determination as the research object, and set its failure rate to $0.0846 + r*0.158 + j*0.178$, where the x and y coordinates represent the correlation coefficients r and j, respectively, with a value between 0 and 10, and the z coordinate represents the probability of normal operation of the system. From Fig. 2.19, we can see that the larger the correlation coefficient, the lower the probability of normal operation of the system, while the smaller the correlation coefficient, the higher the probability of normal operation of the system. Therefore, it can be concluded that the closer the correlation between components, the lower the reliability of the system.
2. After removing the impact of correlation, Fig. 2.20 shows the impact of fault rate and fault tolerance on the normal operation of the example system. In this figure, assuming that all components have the same fault tolerance and failure rate. It can be seen from the figure that the failure rate is inversely proportional to the probability of normal operation of the system, while the fault tolerance is proportional to the probability of normal operation of the system. Reducing the failure rate or increasing the fault tolerance can make the system more reliable.
3. Next, we observe the failure rate of each component. The failure rate is set from 0 to 1, and the relationship between the failure rate and the probability of normal operation of the system is analyzed. From Fig. 2.21, we can see that there is an inverse proportional relationship between the failure rate and the probability of normal operation of the system.

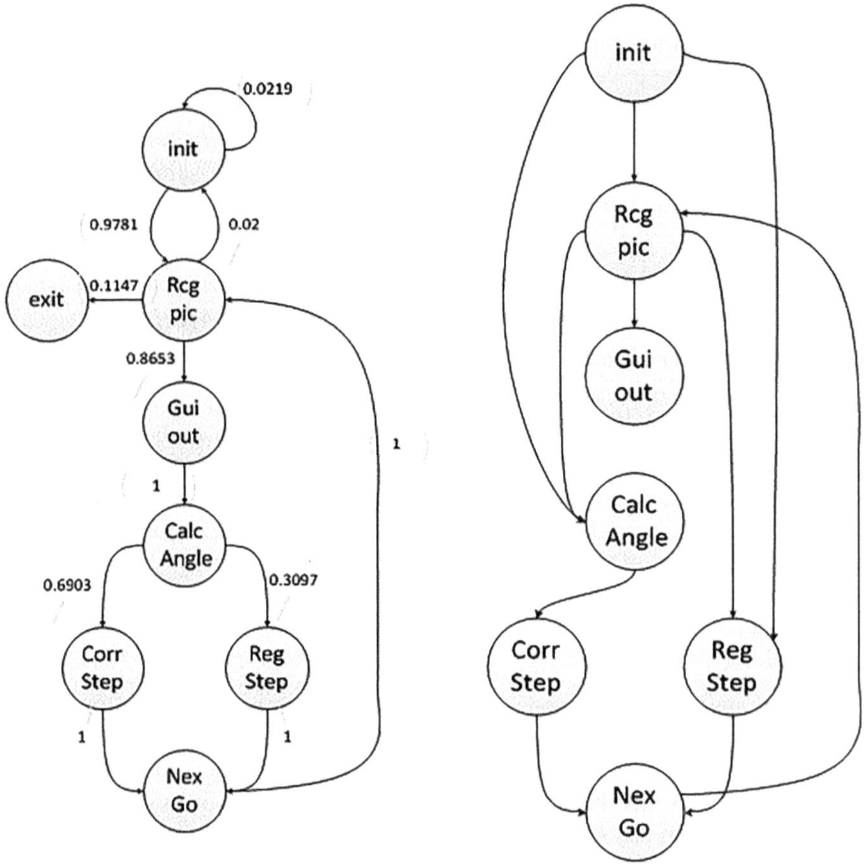

Fig. 2.17 The control flow and data flow diagrams of mobile robot software

4. Next, we observe the fault tolerance of each component. The fault tolerance is set from 0 to 1, and the relationship between the fault tolerance and the probability of normal operation of the system is analyzed. From Fig. 2.22, we can see that there is a proportional relationship between fault tolerance and the probability of normal operation of the system.

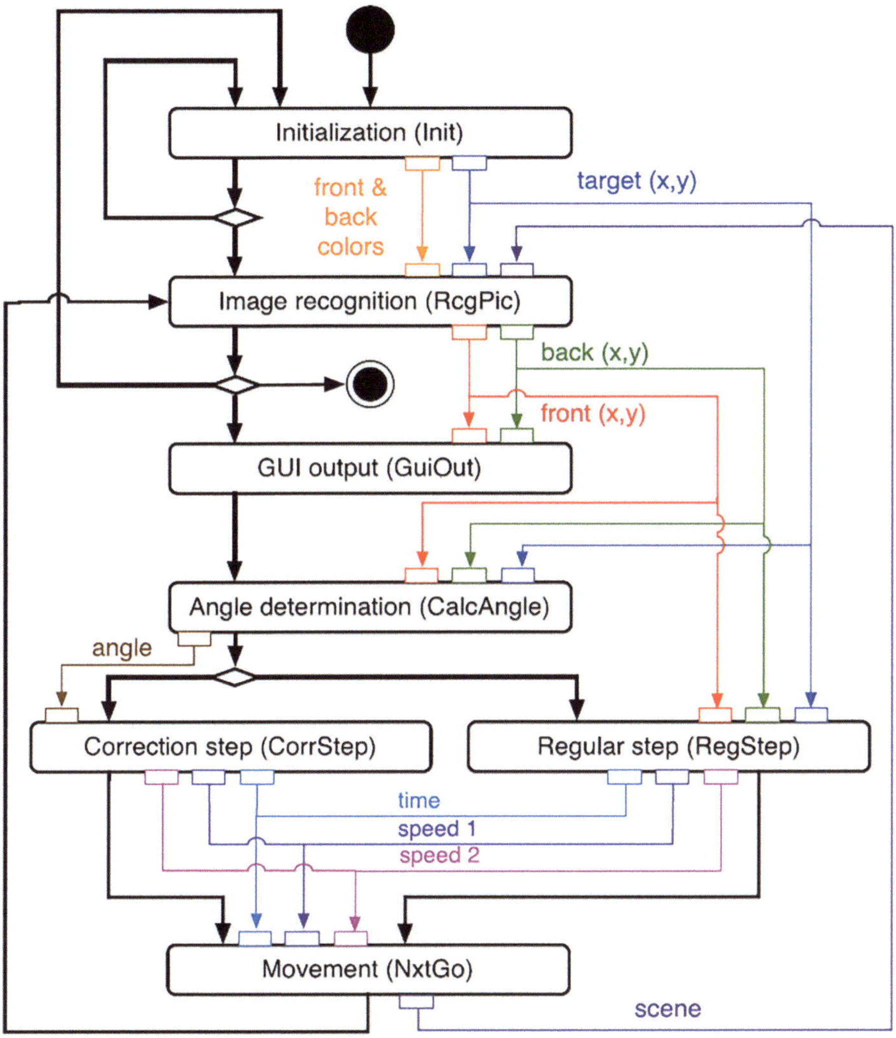

Fig. 2.18 UML activity diagram of mobile robot software

Table 2.2 Parameter table of each component in mobile robot software

Component	Parameters	
	failure rate	fault tolerance rate
Initialization Module	0	1
Image Recognition model	0.0208	0.073
GUI Output	0	0.9069
Angle Judgment model	0.0846	0.1739
Path Correction Module	0.1407	0.0506
Normal Path Module	0.0211	0.7857
Move	0.0371	0.0093
End	0	0

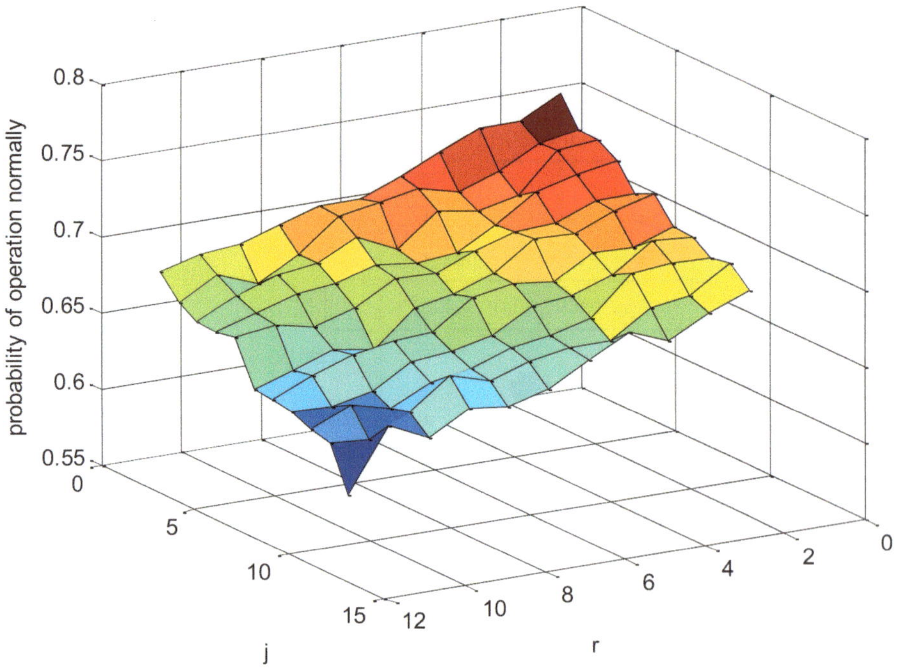

Fig. 2.19 The relationship between correlation coefficients *r* and *j* and the probability of normal operation

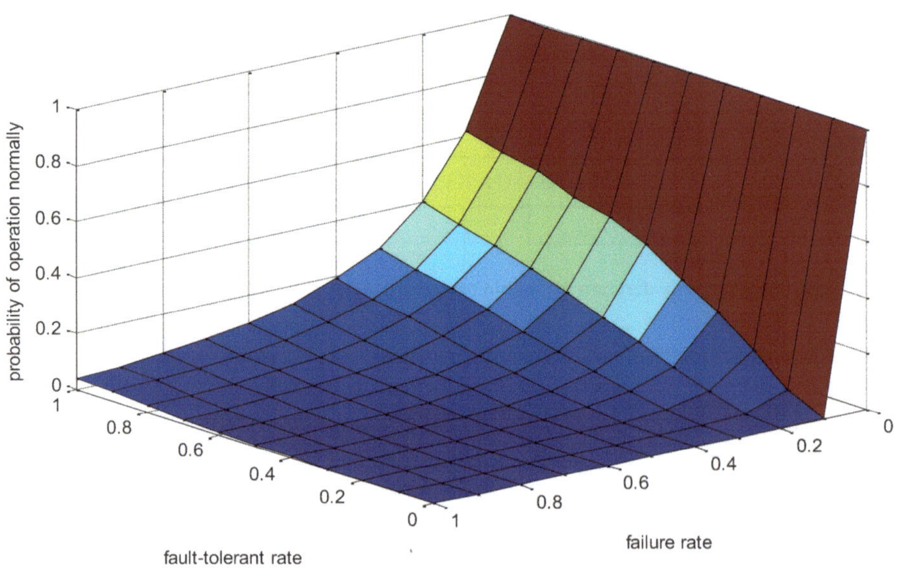

Fig. 2.20 The relationship between fault rate, fault tolerance, and probability of normal operation after removing the influence of correlation

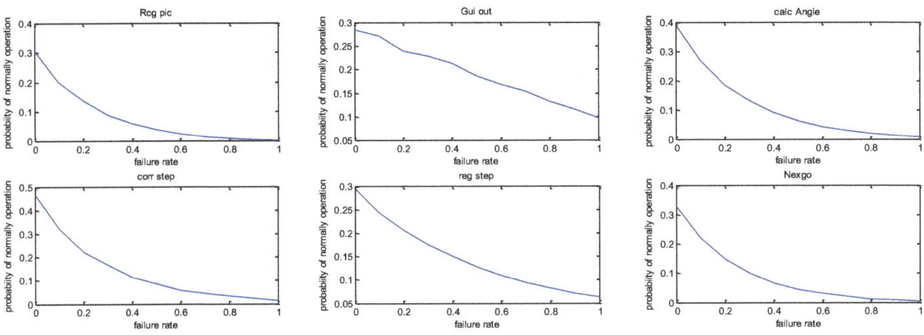

Fig. 2.22 The relationship between the fault tolerance of each component and the probability of normal operation of the system

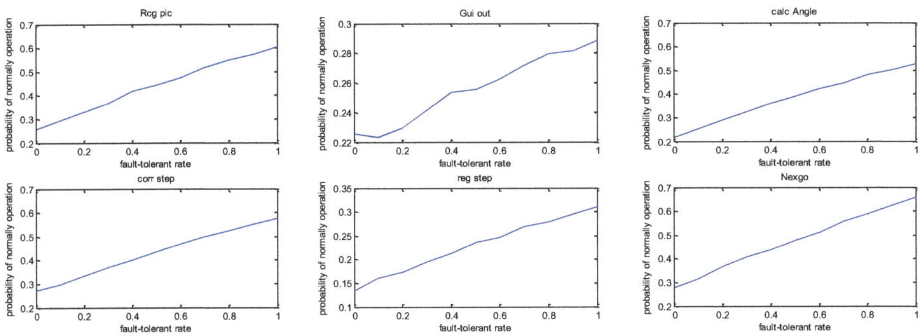

Fig. 2.21 The relationship between the failure rate of each component and the probability of normal operation of the example system

2.6 Chapter Summary

Fault research has long been a foundational topic in the field of software reliability. Different understandings and findings regarding the mechanisms of software faults have led to the development of various software reliability models, which in turn have facilitated the creation of techniques for analyzing, testing, and evaluating software reliability. This chapter provides a comprehensive summary of current research on software faults and further examines the fault mechanisms in complex software systems from the perspective of fault propagation. Additionally, it establishes corresponding models to serve as a conceptual foundation for subsequent technical discussions.

Chapter 3
Reliability Models for Complex Software Systems

A model represents a subjective interpretation by the modeler of the objective physical world. Reliability refers to a capability exhibited by an entity or system, and a reliability model is essentially a representation of this reliability as understood through human cognition. Researchers, based on their individual thinking and investigations, have developed diverse perspectives on the nature of reliability and the factors influencing it. By expressing these perspectives through quantitative or non-quantitative approaches, they have created a wide variety of reliability models. According to incomplete statistics, there are currently over 100 different software reliability models in existence.

Software reliability is a critical aspect of software quality. It is addressed in the majority of software quality models, and in some cases, reliability has evolved into a collective term encompassing various quality attributes. To develop a comprehensive understanding of the concept of software reliability, this chapter first provides an overview of quality models, followed by an introduction to several representative reliability models.

3.1 Software Quality Models

3.1.1 The Concept of Software Quality Models

Since the advent of the first general-purpose programmable computer, ENIAC, the quality of computer systems has been a central concern for developers and users alike. From the creation of the first assembly language program, "Initial Orders," to today's complex software systems comprising millions of lines of code, software quality has consistently remained a focal point for both developers and users.

To refine the concept of software quality and facilitate research activities, scholars have structured the concept into a comprehensive and systematic framework,

Y. Wang et al., *Reliability Technology for Complex Software Systems*,
https://doi.org/10.1007/978-981-95-3889-8_3

widely recognized as a software quality model. The primary objectives of such models are to aid in identifying software quality requirements, evaluating the quality of existing software systems, and predicting the quality of future software systems.

The evolution of software quality models fundamentally reflects the progression of people's understanding of software quality. It also represents the continuous expansion and refinement of the connotation and denotation of the concept of software quality.

According to ISO/IEC 25000:2005, a software quality model is defined as a framework comprising a set of defined characteristics and the relationships between them, designed to determine software requirements and assess software quality. This indicates that quality models serve as a critical foundation for identifying software requirements, establishing software metrics, and evaluating software quality. Typically, a software quality model is structured as a hierarchical collection of software quality attributes.

In the software domain, numerous quality models exist, differing in their terminology, definitions of quality attributes, and structural frameworks. Among these, the ISO/IEC series of quality standards are the most widely recognized and used in the industry, particularly the earlier ISO/IEC 9126 standard and its successor, the 25010 standard. Both models provide specific recommendations for quality frameworks, organizing software quality attributes into hierarchical classifications.

The IEEE 24765 standard further defines key terms related to software quality models, including the concepts of software quality factors, quality attributes, and metrics.

1. **Definition of Software Quality Factors**
   (a) Management-oriented attributes in software that contribute to quality.
   (b) High-level quality attributes.
2. **Definition of Software Attributes**
   (a) Software attributes refer to characteristics or a general term used to denote the values of quality factors, quality sub-factors, or metrics.
   (b) Characteristics or features that influence the quality of a product item.
   (c) Defines the degree of an attribute that reflects the quality a system or software must possess.
3. **Definition of Metric**
   (a) A metric is a quantitative measure of the degree to which a software item possesses a given quality attribute.
   (b) A function that takes software data as input and produces a single numerical output, which can be used to assess the degree to which a software item exhibits a specific quality attribute.

The related concept of "measure" is also defined here.

4. **Definition of Measurement**
   (a) Measurement is a process in which numbers or symbols representing attributes are assigned to objects in the real world according to clearly defined rules to indicate the characteristics of the objects.
   (b) In a more formal description, a measurement is defined as a mapping from the real world to the formal relation world. Therefore, measurement is the assignment of numbers or symbols to attributes of objects in the real world through this mapping to indicate the characteristics of the objects.

3.1.2 Development of Software Quality Models

Quality models have been a subject of research for several decades, with numerous scholars proposing a wide range of models. Rather than detailing each individual model, this section highlights key milestones in the development of software quality models from the perspective of structural evolution:

1. Hierarchical quality models.
2. Meta-model-based quality models.
3. Implicit quality models.

3.1.2.1 Hierarchical Quality Models

The first published software quality models date back to the late 1970s, when McCall, Boehm, Richards, Walter, and others described software quality characteristics and their decomposition processes. Their approaches were similar, employing a hierarchical method to break down software quality into factors such as maintainability and reliability. Over time, numerous variations of these models emerged, with the most popular being the FURPS model. This model categorizes quality into five key attributes: functionality, usability, reliability, performance, and supportability. Beyond hierarchical decomposition, the core idea of these layered models is to "decompose quality to a measurable level, enabling the evaluation of quality."

This layered quality model served as the foundation for the international standard ISO/IEC 9126, introduced in 1991. The standard established a systematic decomposition of quality characteristics and proposed several measurement criteria. While these characteristics did not encompass all aspects of software quality nor enable its complete measurement, the standard nonetheless became a widely recognized and influential benchmark for software quality, adopted by many countries and organizations as a practical quality guideline.

Two decades later, ISO introduced ISO/IEC 25010 as a successor to ISO/IEC 9126. The new standard revised the classification of some quality characteristics while preserving the fundamental concepts of layering and decomposition.

Some researchers have employed measurement standards to directly assess quality characteristics defined by ISO/IEC 9126 or similar frameworks. Franch and Carvallo modified the ISO quality model by assigning more specific measurement indicators and emphasized the importance of explicitly describing the "relationship between quality entities and characteristics." Van Zeist and Hendriks further extended the ISO model by incorporating additional indicators, such as average learning time. Similarly, Samoladas et al. expanded upon several quality characteristics from ISO/IEC 9126 and applied them to open-source software, aggregating measurement results into an ordered quantitative table based on quality characteristics.

These approaches collectively underscore the necessity of extending and adapting the ISO standard. However, many researchers acknowledge the significant challenges involved in using the standard to directly measure abstract quality characteristics.

Ortega, Perez, and Rojas developed a systematic quality model from a distinctly different perspective, using product effectiveness and efficiency as the foundational dimensions for constructing quality characteristics. Other characteristics, such as reliability and maintainability, can be derived by decomposing these two fundamental dimensions.

As hierarchical quality models were being developed, research tools were also being developed to integrate quality models and evaluation toolkits to improve the efficiency and accuracy of software quality assessment. In the Squale research project, researchers took a more comprehensive and integrated approach, developing an explicit quality model to describe ISO/IEC 9126 quality characteristics in a layered and decomposed process. This model includes formulas for synthesizing and standardizing measurement values for various indicators. Based on this quality model, they developed an automated tool for evaluating software product quality, which embedded the quality model and the measurement and evaluation process in the tool.

Hierarchical models represent the most widely adopted type of quality model today, yet they still face several challenges. Many researchers have highlighted that the principles underlying the decomposition of quality characteristics are often ambiguous. Furthermore, the resulting quality characteristics tend to lack specificity and cannot be directly measured, posing significant difficulties for the engineering application of such models.

Although subsequent standards, such as ISO/IEC 25010 and the measurement reference model in ISO/IEC 25020, introduced certain improvements, the lack of concrete operational measures has limited their practical effectiveness. Additionally, recent surveys reveal that fewer than 30% of commercial companies adopt these standard models, while over 70% develop their own quality characteristics and measurement standards based on the standard models.

The Pioneer of Software Quality Models

The first publicly disclosed software quality model was the renowned McCall model, introduced by McCall in the 1970s. This model was originally developed for the United States Air Force, with a primary focus on system developers and the processes involved in system development. McCall's quality model conceptualized software quality from three distinct perspectives: product revision, product transition, and product operation. Product revision refers to the software's capability to accommodate changes, product transition denotes its ability to adapt to new environments, and product operation pertains to the operational attributes of the software. These three overarching perspectives were subsequently decomposed and systematically redefined into 11 quality factors, 23 criteria, and associated indicators.

The McCall model is significant not only because it was the first formally proposed software quality model but also because its hierarchical approach to defining quality has profoundly influenced nearly all subsequent quality models.

In the McCall model, quality factors were established to characterize the external quality of software as perceived by users, as well as the internal quality prioritized by developers. Furthermore, the model introduced associated quality indicators and criteria to facilitate the systematic measurement and evaluation of software quality.

3.1.2.2 Quality Model Based on a Meta-Model

Since the 1990s, researchers have introduced more sophisticated methods for decomposing quality characteristics, leading to the development of more comprehensive software quality models based on meta-models. The concept of a meta-model in this context closely aligns with the meta-model concept in the Unified Modeling Language (UML), serving as a framework to describe the construction of effective quality models.

In the ESPRIT project REQUEST, researchers sought to establish explicit relationships between quality factors and measurements within hierarchical quality models. This effort led to the development of a constructive quality model known as COQUAMO. The researchers posited that quality factors are central to the quality model and should be evaluated differently at various stages of software development, necessitating the use of distinct measurement indicators.

Moreover, they categorized quality into general quality characteristics and specific software quality characteristics. For instance, reliability was identified as a general software quality characteristic, as it is critical for all software systems, whereas security was classified as a specific software quality characteristic, given that only certain types of software systems prioritize security.

The COQUAMO model placed particular emphasis on defining clear, quantitative relationships between measurable quality drivers and quality factors. This approach enabled the assessment of quality by quantifying the underlying causes that contribute to quality factors.

Another notable quality model is COQUALMO, which introduces the concept of a defect flow model. This model illustrates how defects are introduced during each stage of software development as a result of building and modifying the software product and how these defects are subsequently addressed through quality assurance techniques.

For instance, during the software requirements phase, requirement defects may arise from misunderstandings of user expectations. These defects are later identified and eliminated through technical reviews and testing processes in subsequent stages.

In addition to the models with explicit meta-models mentioned above, there are other quality models that also emphasize establishing statistical relationships, albeit without a clearly defined meta-model structure. These models rely on a set of measurements expected to influence specific quality factors.

For instance, the maintainability index (MI) is derived from a combined set of static code metrics and provides an indication of the system's maintainability level. Another example is the reliability growth model, which leverages statistical correlations between testing data and software reliability to assess and predict reliability trends.

With the advent of increasingly complex software systems, the intricacy of software quality has grown significantly. To address this, Kitchenham et al. developed the SQUID method, building on the COQUAMO model, to represent the structure of increasingly complex quality models through a more explicit meta-model.

While there is no definitive evidence supporting the conclusion of the REQUEST project—that "there are no general software product metrics that can truly and accurately predict the final quality of software"—this assertion remains valuable for current practices in modeling and analyzing software quality. Consequently, Kitchenham et al. recommend monitoring and controlling "internal" indicators that may influence "external" quality.

Their quality meta-model incorporates measurable properties, which can be either internal software attributes or quality sub-characteristics. Internal software attributes influence quality sub-characteristics, and both types of properties are directly measurable.

Bansiya and Davis developed the QMOOD (Quality Model for Object-Oriented Design) model for object-oriented software, building on the foundation of the Dromey model. They identified several indicators associated with component design to evaluate what they termed design properties. These properties directly influence various quality attributes, offering a structured approach to assess the quality of object-oriented designs.

Bakota et al. emphasized the probabilistic nature of their quality model and quality assessment approach. They introduced virtual quality attributes for internal software properties, similar to SQUID. The quality model relies on only nine low-level metrics, which are evaluated and aggregated into probability distributions. However, the model is heavily mathematical, making it difficult to understand and challenging to use these distributions for practical assessments.

The concept of quality is inherently complex and necessitates the incorporation of more structured elements in quality models, beyond merely abstract quality characteristics and indicators. Quality models should address both external quality, as perceived by users, and internal quality, as understood by developers.

3.1.2.3 Implicit Quality Model

The implicit quality model refers to quality models that do not fall under the hierarchical models or meta-models discussed earlier. Instead, these models are more akin to statistical models, quality analysis tools, and quality checklists.

Statistical models are utilized to characterize the properties of products, processes, or organizations, and to estimate and predict various types of quality factors. One of the most prominent examples is the reliability growth model, which extends the concept of hardware reliability models to software. The software reliability growth model involves observing and recording software failures, such as those identified during system testing, along with their occurrence times, to predict how failure behavior evolves over time.

Another example is the Maintainability Index (MI) model, a regression model based on static code metrics, which provides insights into the maintainability of a software system.

Many quality analysis tools are developed based on specific quality models. For instance, tools for error pattern recognition, such as FindBugs, Gendarme, or PC-Lint, categorize the problems they detect into various categories, either based on the quality factors they may affect or the severity levels of the issues. As a result, these tools inherently define much of the information we anticipate from quality models, such as quality factors, measurements, and their impacts.

However, the results provided by these tools are often ambiguous, which poses challenges for conducting a comprehensive quality evaluation.

Interface dashboards, such as QALab, Sonar, or XRadar, can utilize measurement data from these tools as input to provide an overview of a software system's quality data. However, these models often lack explicit connections between the metrics employed and the quality factors they aim to measure. Consequently, it becomes challenging to ascertain the precise impact of detected defects on software quality or to evaluate the rationale behind the selected metrics.

Finally, the checklists used in software development or technical reviews can also be considered a form of quality model. While checklists are typically not directly integrated into a formal quality model, they define various properties of software artifacts—most commonly source code—that directly influence certain software quality factors. However, during the creation of these checklists, only a few items are explicitly linked to software quality factors, with most items being based on experience or established practices. In this regard, a well-defined quality model can provide valuable guidance for improving checklists.

Throughout the evolution of software quality models, existing models have lacked well-defined decomposition criteria for breaking down the complex concept of "quality." Most models depend on the classification, hierarchical structuring, and decomposition of quality factors. However, this decomposition often lacks adherence to established guidelines and can be entirely arbitrary. Consequently, it becomes challenging for users of these models to continuously improve a specific quality attribute, such as usability.

Furthermore, the ambiguous decomposition methods employed in many quality models contribute to the overlap between different quality factors. In most cases, these overlaps are not explicitly addressed within the model, further exacerbating the issue. For instance, denial-of-service attacks are typically considered a security factor, but they are also influenced by availability, which is itself a subset of reliability. Similarly, code quality is a critical factor for maintainability, yet it is also regarded as an indicator of security.

Most quality model frameworks do not provide methods for users to implement quality assurance using the quality models, making it unclear how to effectively convey the quality model to project participants. In practice, a common method of communicating such information is to develop a guide. To ensure its effectiveness, the guide should be specific and detailed, with a well-organized document structure. Additionally, if the guide includes mandatory rules, it should provide clear and adequate justifications, enabling project personnel to understand the rationale behind the rules and follow them effectively.

Finally, software systems vary greatly, ranging from large-scale commercial information systems to small embedded controllers. These differences must be addressed in quality models through well-defined customization methods. Current quality models often fail to account for these factors, highlighting the need for developing customized quality models tailored to the unique characteristics of specific software system types.

Such tailored models would ensure that the quality factors relevant to these systems are adequately addressed. By incorporating these considerations, the effectiveness and applicability of software quality models can be significantly enhanced, enabling them to better meet the diverse needs of different software systems.

3.1.3 ISO/IEC 25010:2010 Introduction to Software Quality Models

ISO/IEC 25010 is a standard for software quality models published by ISO in 2011. Building on the foundation of the earlier ISO 9126 standard, ISO/IEC 25010 introduces significant structural adjustments and incorporates several additional components.

This standard has since become the most widely recognized and influential quality model, playing a pivotal role in the practical application of quality models across various industries. Given its importance, the following section provides a more detailed description of ISO/IEC 25010 and its key features.

ISO/IEC 25010 evolved from ISO/IEC 9126, with both models being hierarchical structural frameworks derived from the foundational McCall and Boehm hierarchical quality models.

The core principle of the ISO/IEC 25010 model is to apply classification techniques to break down the complex concept of software product quality into smaller, more manageable components. This decomposition is performed systematically, layer by layer, until it reaches a level where direct measurement can be applied.

The metamodel of the ISO/IEC 25010 quality model is represented by a UML class diagram, as shown in Fig. 3.1. The model uses a layered structure to decompose quality into various characteristics, which are further decomposed into sub-characteristics and sub-sub-characteristics. All of these characteristics are collectively referred to as quality factors, some of which are measurable, while others are not directly measurable and are replaced by other measurable quality properties.

It should be noted that the ISO/IEC 25010 model does not use the terms "quality factors" and "measurable" in the standard, but these terms are used here for better explanation of the model.

Using this layered structure, ISO/IEC 25010 defines two quality models: the product quality model and the use quality model. Additionally, ISO/IEC 25012 introduces a data quality model. These models function as a quality checklist, enabling software engineers to systematically and comprehensively address various aspects of software quality during the development process.

However, it is important to note that the standard does not provide explicit guidance on customizing quality models. In practice, it is essential to select and tailor appropriate quality characteristics based on the specific features of the software product—a process referred to as adapting the quality model. This adaptation should be documented and incorporated into the software quality plan.

Although the model described in ISO/IEC 25010 outlines the relationship between quality characteristics and measures, it does not provide detailed guidance on operational measurement. Such guidance is instead provided by the standard ISO/IEC 25040, which focuses on applying the quality model for measurement purposes.

The following sections introduce the product quality model and the use quality model as defined in ISO/IEC 25010.

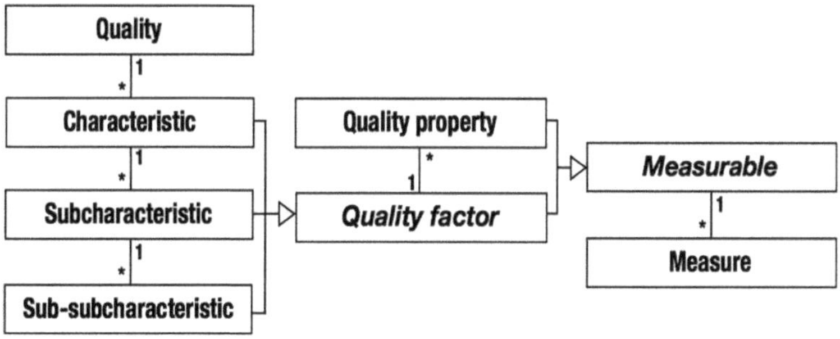

Fig. 3.1 Conceptual metamodel of ISO/IEC 25010

3.1.3.1 Product Quality Model

Each characteristic defined in ISO/IEC 25010 can be directly analyzed in software products. In addition, ISO/IEC 25010 claims that they should also be effective for system quality. The product quality model includes eight quality characteristics and 38 quality sub-characteristics, as shown in Fig. 3.2 of ISO/IEC 25010.

The eight characteristics defined by ISO/IEC 25010 comprehensively represent the quality attributes of software products. Functional suitability refers to the degree to which the product fulfills the functional requirements of users and customers. Performance efficiency pertains to the responsiveness of the product to user requests and its overall operational efficiency. Given that contemporary software products seldom operate in isolated environments, compatibility is defined as the quality attribute that ensures the product neither interferes with nor impedes the functioning of other products and facilitates seamless integration with them. Usability

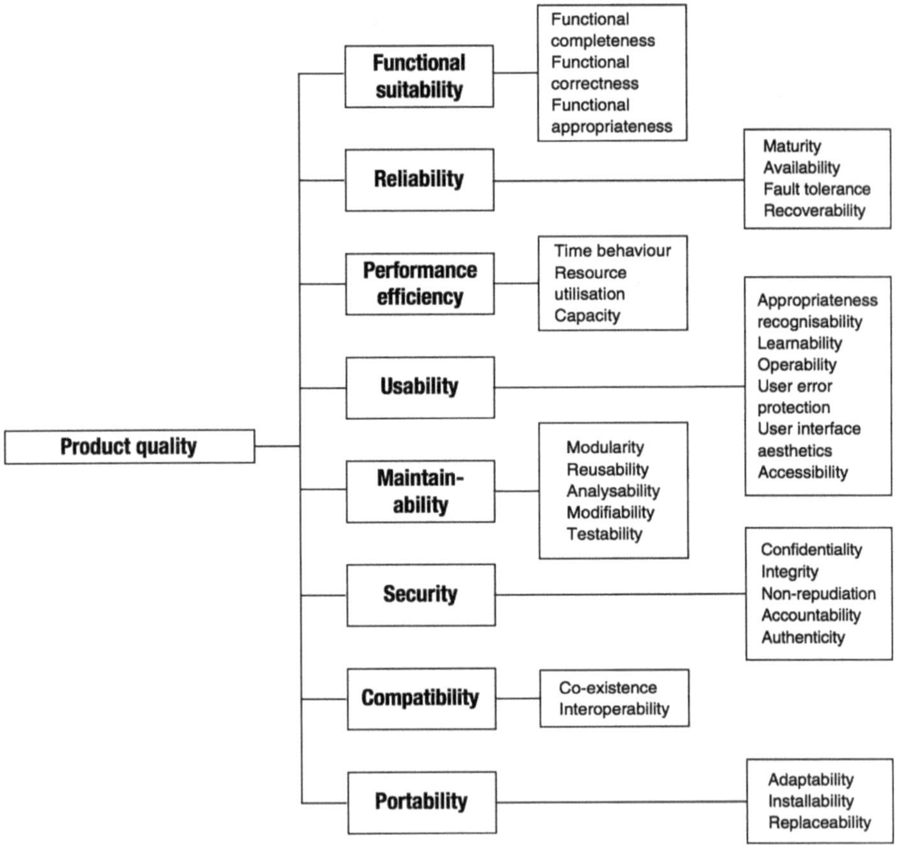

Fig. 3.2 Product quality model in ISO/IEC 25010

encompasses various aspects of the ease of using the system, including the speed at which users can learn to operate it and the appeal of its interface design.

When discussing software quality, reliability is often the first attribute that comes to mind. Reliability issues, along with performance efficiency, are typically the aspects most directly relevant to users. A product that experiences faults is perceived as unreliable by its users. As most modern systems are accessible via networks, security has become increasingly critical. Security involves ensuring data integrity and confidentiality, as well as verifying and protecting the identities of users. Maintainability, a characteristic targeted at developers, highlights the importance of designing and programming systems in a manner that facilitates ease of understanding, modification, and testing. Lastly, portability is another significant attribute for developers who need to adapt the product to various platforms, such as programming languages, operating systems, and hardware. Portability implies that necessary modifications can be implemented efficiently, and the product can be installed with minimal effort.

We will now provide a detailed introduction to these eight quality characteristics.

Functional Suitability

Functional suitability refers to the ability of a product or system to provide functions that meet both explicit and implicit needs under specified usage conditions. This characteristic is further decomposed into three sub-characteristics:

1. Functional completeness: The degree to which the set of functions provided by the software fully covers all specified tasks and user objectives.
2. Functional correctness: The extent to which the product or system delivers accurate results with the required level of precision and reliability.
3. Functional appropriateness: The degree to which the provided functions support the effective accomplishment of specific tasks and objectives. For instance, appropriateness is evident when only the essential steps required to complete a task are presented to the user, excluding any unnecessary or extraneous steps.

Reliability

Reliability refers to the extent to which a system, product, or component can perform specified functions under defined conditions within a designated time frame. Unlike physical components, software does not degrade over time; thus, software reliability primarily addresses failures resulting from errors in requirements, design, or implementation, as well as failures triggered by changes in context. Characteristics commonly associated with trustworthiness include availability and its intrinsic or external influencing factors, such as reliability (encompassing fault tolerance and recoverability), security (including confidentiality and integrity), maintainability, durability, and supportability.

Reliability encompasses the following sub-characteristics:

1. Maturity:The extent to which a system, product, or component satisfies reliability requirements under normal operating conditions. The concept of maturity can also be applied to other quality attributes, indicating the degree to which they meet required standards during normal operation.
2. Availability: The degree to which a system, product, or component is operational and accessible when needed. Availability is determined by a combination of maturity (which affects the frequency of failures), fault tolerance, and recoverability (which determines the duration of downtime following each failure).
3. Fault Tolerance: The capability of a system, product, or component to continue functioning as intended in the presence of hardware or software faults.
4. Recoverability: The degree to which a product or system can restore directly affected data and re-establish its expected operational state following an interruption or failure. For example, in the event of a failure, a computer system may experience downtime, the duration of which depends on its recoverability. Furthermore, recoverability encompasses another quality attribute, survivability, which refers to the extent to which a product or system can maintain its mission and deliver essential services even under attack.

Performance Efficiency

Performance efficiency refers to the extent to which resources are utilized effectively under specified conditions. These resources encompass other software products, the system's software and hardware configurations, and consumables (such as printer paper and storage media).

The three sub-characteristics of performance efficiency are as follows:

1. Time Behavior: This refers to the extent to which the product or system meets the requirements for response time, processing time, and throughput while performing its intended functions.
2. Resource Utilization: This denotes the degree to which the product or system efficiently utilizes the quantity and variety of resources required to perform its functions, in alignment with the specified requirements.
3. Capacity: This pertains to the extent to which the maximum allowable limits of the system's parameters satisfy the requirements. These parameters include, but are not limited to, the number of items that can be stored, the number of concurrent users supported, communication bandwidth, transaction throughput, and database capacity.

Usability

Usability refers to the extent to which a specific user can effectively and efficiently utilize a product or system to achieve defined goals within a specified usage context. Usability can be characterized or evaluated either as an independent product quality attribute or as a measurable subset of the broader concept of usage quality.

The six sub-characteristics of usability are as follows:

1. Appropriate Recognizability: The extent to which users can determine whether a product or system meets their needs. This characteristic depends on the ability to identify the suitability of the product or system's functionality based on its initial impression and any relevant documentation. Such documentation may include demonstrations, tutorials, user manuals, websites, or information displayed on the homepage.
2. Learnability: The extent to which specified users can achieve designated goals effectively, efficiently, safely, and satisfactorily by learning how to use the product or system.
3. Operability: The degree to which a product or system possesses attributes that facilitate ease of operation and control for users.
4. User Error Protection: The extent to which the system minimizes the likelihood of user errors and assists in preventing such errors during operation.
5. Aesthetics of the User Interface: The degree to which the user interface enhances user satisfaction and enjoyment in human-computer interaction. This includes properties of the product or system, such as the thoughtful use of colors and graphic design, that contribute to a pleasant user experience.
6. Accessibility. The degree to which the product or system can be used by a broad range of people with different characteristics and abilities to achieve specific goals in a specific usage environment.

Maintainability

Maintainability refers to the degree to which a product or system can be effectively and efficiently modified by its intended maintainers. These modifications encompass corrections, enhancements, or adaptations made to the software to accommodate environmental changes, as well as updates necessitated by evolving requirements or functional specifications. Additionally, modifications may be carried out by professional support personnel, business or operational staff, or even end-users.

Maintainability also extends to activities such as software installation, updates, and upgrades. It can be understood as the inherent capability of a product or system to support maintenance tasks, or as the quality of use experienced by maintainers in achieving the objectives of maintaining the product or system.

Maintainability encompasses the following five sub-characteristics:

1. Modularity: The degree to which a system or software program is composed of discrete components, such that changes to one component have minimal impact on others.
2. Reusability: The extent to which software assets can be utilized across different systems or leveraged in the creation of other assets.
3. Analyzability: A characteristic that measures the effectiveness and efficiency of the evaluation process. This includes the assessment of the potential impacts caused by anticipated changes to one or more components of the product or system, diagnosing the presence of defects or the underlying causes of faults, and identifying the components that need modification.
4. Modifiability: The degree to which a product or system can be modified effectively and efficiently without introducing defects or compromising its existing quality. Modifications include activities such as coding, design revisions, documentation updates, and validation of changes. Modifiability is influenced by the combined effects of modularity and analyzability, and it reflects a balance between changeability and stability.
5. Testability: The extent to which effective and efficient test criteria can be defined for a system, product, or component, and the degree to which tests can be executed to verify compliance with these criteria.

Security

Security refers to the extent to which a product or system safeguards information and data, ensuring that individuals or other products or systems have access to data only in accordance with their authorized type and level. Security encompasses not only the protection of data stored within the product or system but also the protection of data during transmission.

Security includes the following five sub-characteristics:

1. Confidentiality: The extent to which a product or system ensures that data can only be accessed by authorized individuals or entities.
2. Integrity: The extent to which a system, product, or component prevents unauthorized access or modification of computer programs or data. Integrity also encompasses the characteristic of immunity, which refers to the degree to which a product or system resists attacks.
3. Non-repudiation: The extent to which actions or events can be proven to have occurred, ensuring that they cannot be denied at a later time.
4. Traceability: The extent to which the behavior of an entity can be uniquely tracked and attributed back to that specific entity.
5. Authenticity: The extent to which the identity of a subject or resource can be verified as the claimed or required identity.

Compatibility

Compatibility refers to the extent to which a product, system, or component can exchange information and perform its required functions seamlessly with other products, systems, or components, while operating within the same hardware or software environment.

Compatibility includes the following two sub-characteristics:

1. Coexistence: The extent to which a product can effectively perform its required functions while sharing a common environment and resources with other products, without causing any adverse effects on other products.
2. Interoperability: The extent to which products or components within two or more systems can exchange information and utilize the exchanged information to perform their intended functions.

Portability

Portability refers to the degree of effectiveness and efficiency with which a system, product, or component can be transferred from one hardware, software, or other operational or usage environment to another. Portability can be understood either as the inherent capability of a product or system to support portability activities, or as a quality characteristic that reflects the ease of porting the product or system to achieve its intended purpose.

Portability includes the following three quality sub-characteristics:

1. Adaptability: The extent to which a product or system can effectively adapt to varying or evolving hardware, software, or other operational or usage environments. Adaptability encompasses the scalability of internal capacities, such as screen fields, tables, transaction volumes, and report formats. Adaptability modifications may be performed by professional support personnel, business or operational staff, or end-users.
2. Installability: The degree of effectiveness and efficiency with which a product or system can be successfully installed or uninstalled in a specific environment. If the product or system is intended for installation by end-users, installability may influence its resulting functional suitability and operability.
3. Replaceability: The extent to which a product can be substituted within the same environment by another specified software product serving the same purpose. Replaceability is particularly significant during software upgrades to new versions. It may incorporate attributes of installability and adaptability. Replaceability mitigates lock-in risks by enabling the use of alternative software products in place of the current one, such as through the adoption of standardized file formats.

3.1.3.2 Using a Quality Model

Product quality models directly describe the intrinsic characteristics of a product, whereas using quality models focus on the characteristics of the interaction between various stakeholders and the product. Among these stakeholders, the user is the most prominent. This is why "quality in use" is often associated exclusively with usability.

However, quality in use extends beyond usability to encompass the quality of maintaining or porting the product or its content. In this broader context, the terms "use" and "user" are interpreted in a wide-ranging sense.

The ISO/IEC 25010 standard defines quality in use through five key characteristics, as illustrated in Fig. 3.3.

The using quality model encompasses five quality characteristics and fourteen sub-characteristics: effectiveness, efficiency, satisfaction, avoidance of risk, and context coverage.

Among these, the two core characteristics are effectiveness and efficiency.

- Effectiveness refers to the degree to which a product helps users achieve their intended goals.
- Efficiency reflects the amount of resources required to accomplish these goals.

For example, in the context of a driver interacting with a car's navigation system, effectiveness concerns whether the system provides appropriate guidance to reach

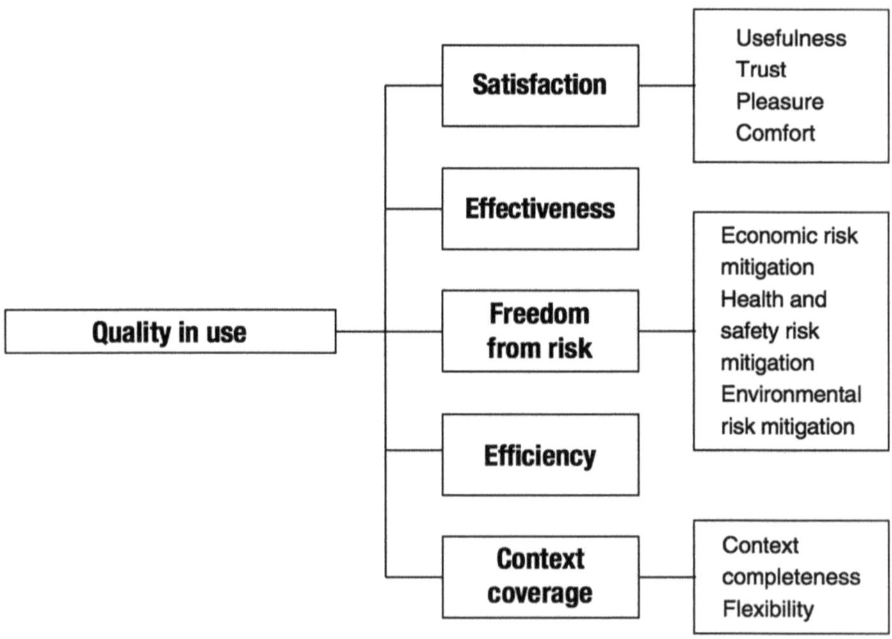

Fig. 3.3 Model of quality in use in ISO/IEC 25010

the desired destination. Efficiency, in this scenario, is represented by the number of button presses required to obtain this guidance.

For maintenance personnel, effectiveness relates to the successful completion of a maintenance task, while efficiency is characterized by the number of new faults introduced during the process and the effort expended to complete the task.

Quality in use extends beyond merely achieving goals with the appropriate use of resources. User satisfaction is also a critical quality characteristic, encompassing both credibility and hedonic qualities (such as the user's sense of pleasure and enjoyment).

In many cases, the avoidance of risk is equally important. The most significant aspect of this involves ensuring the safety of systems that may pose harm to individuals. Additionally, environmental and economic risks should be taken into account.

Finally, the ISO/IEC 25010 standard emphasizes that the model should incorporate appropriate context coverage. Context coverage includes not only the various usage scenarios of the product but also the product's capacity to respond flexibly to changes.

Below is an introduction to the five quality attributes in the quality model:

1. Effectiveness: Can the product help users achieve their goals? This is one of the core objectives of product design, focusing on the product's utility. Effectiveness ensures that users' needs are met, enabling them to accomplish the tasks they need to complete.
2. Efficiency: Can the product help users achieve their goals quickly and efficiently with appropriate use of resources? This attribute emphasizes the product's performance, aiming to save time and resources so users can achieve their goals as quickly as possible.
3. Satisfaction: Can the product make users feel satisfied and pleased? Satisfaction includes credibility and enjoyment. Credibility refers to users' trust in and perceived reliability of the product, while enjoyment refers to whether the product brings users pleasure or a sense of fulfillment.
4. Avoidance of Risk: Can the product prevent harm to people or the environment? This attribute covers safety and reliability. Safety refers to the product's ability to prevent harm to users and the environment. Reliability refers to the product's stability and durability, avoiding unexpected failures.
5. Coverage of Context: Can the product adapt to different usage scenarios and respond flexibly to contextual changes? This attribute includes adaptability and flexibility. Adaptability refers to the product's capability to suit various usage scenarios, while flexibility refers to its ability to handle changing situations while maintaining core functionality.

Satisfaction

Satisfaction refers to the extent to which user needs are fulfilled when using a product or system in a specified environment. For users who do not interact directly with the product or system, factors such as goal achievement and trust become the primary considerations. Satisfaction generally refers to the user's response during the interaction with the product or system, encompassing their overall attitude towards its use. By evaluating satisfaction, a more comprehensive assessment of quality is achieved, reflecting the user's overall experience and perception of the product or system.

Satisfaction includes the following four sub-characteristics:

1. Usefulness: The degree to which users are satisfied with achieving their practical goals, including both the results of use and the consequences of using the product or system.
2. Trustworthiness: The degree of confidence that users or other stakeholders have in the product or system performing as expected.
3. Pleasure: The extent to which users derive satisfaction from fulfilling personal needs and experiencing enjoyment. Personal needs may include the desire to acquire new knowledge and skills, as well as to evoke pleasant memories.
4. Comfort: The degree of user satisfaction with the physical comfort provided by the product or system..

Effectiveness

Effectiveness refers to the degree of accuracy and completeness with which users are able to achieve specific goals when using a product or system.

Avoidance of Risk

Avoidance of risk refers to the extent to which a product or system is capable of mitigating potential risks that could negatively impact economic status, human life, health, or the environment. Risk is a function of both the probability of a given threat occurring and the potential adverse consequences associated with that threat.

Avoidance of Risk includes the following three sub-characteristics:

1. Mitigation of Economic Risks: The extent to which a product or system can mitigate potential risks in the expected usage environment that could adversely affect financial status, business assets, reputation, or other resources.
2. Mitigation of Health and Safety Risks: The extent to which a product or system can mitigate potential risks that could harm human health and safety in the expected usage environment.

3. Mitigation of Environmental Risks: The extent to which a product or system can mitigate potential risks that could lead to property damage or adverse environmental effects in the expected usage environment.

Efficiency

Efficiency refers to the resource consumption related to the accuracy and completeness with which users achieve their goals.

Scenario Coverage

Scenario coverage refers to the degree to which a product or system can be used efficiently, with risk avoidance and user satisfaction, in both specified and non-specified usage scenarios.

Scenario coverage includes two sub-characteristics:

1. Scenario completeness: The degree to which a product or system can be effective, efficient, risk-avoiding, and satisfactory in all specified usage scenarios. Scenario completeness can be defined as the extent to which users can use a product to achieve specified goals in all expected usage environments. Usage scenarios include using small screens, low network bandwidth, non-professional users, and running in fault-tolerant mode.
2. Flexibility: The degree to which a product or system can be used efficiently, with risk avoidance and user satisfaction, beyond the initially specified range, and can obtain flexibility in terms of user groups, tasks, and cultures by adjusting the product. Flexibility enables the product to consider unforeseen circumstances. If the design of the product is not flexible, using it in an unforeseen environment may be unsafe. Flexibility can measure the degree to which a product can be used by other types of users by being modified to accommodate new types of users, new tasks, and new environments.

3.1.3.3 Explanation

In the context of quality models, scenario coverage includes two sub-characteristics that are evaluated when assessing a product or system. The first sub-characteristic, scenario completeness, focuses on the ability of the product or system to perform effectively, efficiently, and safely in all specified usage scenarios. The second sub-characteristic, flexibility, focuses on the ability of the product or system to be used beyond the initially specified range while still maintaining high levels of performance, risk avoidance, and user satisfaction. Flexibility also involves the ability of the product or system to be modified to accommodate new types of users, new tasks, and new environments that were not initially anticipated. By evaluating these sub-characteristics, a comprehensive assessment of the scenario coverage of the product or system can be obtained.

It is important to emphasize that measurement plays a critical role in quality control. Consequently, accurately measuring the most significant quality factors is essential for an effective quality assurance process and the successful fulfillment of engineering requirements.

ISO/IEC 25010 provides a comprehensive list of quality characteristics related to software products and their interactions through two quality models. It reminds the requirements engineers not to overlook any quality characteristic and helps quality engineers analyze the quality of a system. It has made some useful improvements over the older version, ISO/IEC 9126, such as including security as a separate characteristic and reducing the number of quality models from three to two. However, it is not clear when to use which model. It seems that most software companies only adopt the product quality model and only consider using the quality model when analyzing usability. Besides mentioning that "the product quality model focuses on the target computer system including the target software product, while the usage quality model focuses on the entire human-computer system including the target computer," this standard does not specify when to use which model. But it is important to understand the classification of these models. The standard describes a possible quality structure but it is not the only possible or reasonable structure.

3.2 Quality of Complex Software System

3.2.1 The Process of Forming the Complex Software System Quality

The interdependence of complex software systems is becoming increasingly stronger, including structural dependencies between components, data dependencies between operations, and interaction dependencies between people and systems, are all dependent on important quality. Due to these strong dependencies, any component or related service failure or quality performance below acceptable levels may result in unacceptable system quality. Typical quality issues that frequently occur in complex software systems include the following:

1. Poor maintainability and portability of software code.
2. Lack of sufficient testing for the software services or their combinations.
3. Performance and reliability issues during system operation that are significantly lower than the required levels.
4. Excessive consumption of hardware resources by the software system.
5. Poor usability of the human-computer interaction in the software system.
6. A mistake in one software component may cause a serious failure in the entire system.
7. Ineffective software development process and poor effect.

The quality challenges in complex software systems have given rise to new trends in software engineering, as illustrated below:

1. Development-Operations (abbreviated as DevOps) approach. DevOps is a culture, movement, or set of practices that emphasizes collaboration and communication between software developers and other information technology (IT) professionals, while automating the processes of software delivery and infrastructure changes. Its goal is to foster a culture and environment in which software building, testing, and deployment can be achieved more quickly, frequently, and reliably.
2. Service-Oriented Architecture (SOA). SOA is an architectural model that enables the distributed deployment, composition, and utilization of loosely coupled, coarse-grained application components over a network based on specific requirements. The service layer serves as the foundation of SOA, allowing direct invocation by applications and effectively managing artificial dependencies in interactions with software agents within the system. SOA is a coarse-grained, loosely coupled service architecture, where services communicate through simple, precisely defined interfaces, without involving underlying programming interfaces or communication models. SOA can be regarded as a natural evolution of the B/S model and XML/Web Service technologies.

   According to the definition of SOA, a service has four main attributes:

   (a) It logically represents a business activity that has a specified outcome.
   (b) It is self-contained and self-sufficient.
   (c) It provides services to its consumers as a black box.
   (d) It may include other underlying services.

3. Cloud Computing Platforms. Cloud computing is an internet-based computing model that delivers shared computing resources and data on-demand to computers and other devices.
4. Agile development process. Agile software development describes a set of principles for software development that enable software requirements and solutions to evolve through the collaborative effort of self-organizing and cross-functional teams. It advocates adaptive planning, evolutionary development, early delivery, and continuous improvement, and encourages quick and flexible responses to change. These principles support the definition and continuous evolution of many software development methods.
5. Global Software Engineering (GSE). GSE refers to software engineering performed in global, distributed settings across different geographic locations. Work can be done within a company (multi-site development) or collaboratively between two or more companies in different locations.
6. Model-Driven Software Engineering. Model-Driven Engineering (MDE) is a software development method that focuses on creating and utilizing domain models, which are conceptual models of all the subjects relevant to a particular problem. It emphasizes and aims to manage the abstract representation of knowledge and activities specific to a particular application domain, rather than

computational (i.e. algorithmic) concepts. MDE methods aim to increase productivity by maximizing compatibility between systems (through the reuse of standardized models), simplifying the design process (through the application of repeatable design patterns in the application domain), and promoting communication between individuals and teams working on systems (through the standardization of terminology and best practices used in the application domain). These models are developed through extensive communication between product managers, designers, developers, and users in the application domain. With the completion of the models, software and system development follows.

Additionally, future software quality engineering methods, technologies, and tools will undergo new developments and face emerging challenges in the aforementioned areas. Quality processes, measurements, and management practices will need to be applied across various components of complex software, including non-software components and the entire system. The evolution of systems during runtime, encompassing changes to deployment environments, networks, hardware, and services, will require more advanced runtime quality management tools. As a result, a broader spectrum of software quality factors and characteristics will be necessary, as runtime quality differs significantly from the quality observed during traditional development processes.

Currently, many complex software systems involve distributed remote development models, agile development methodologies, and architectures that integrate numerous third-party service systems. Within organizations utilizing these approaches, there will be a pressing need for consistent, unified, and precisely defined quality factors, characteristics, quality indicators, thresholds, and more accurate quality prediction and evaluation methods in the quality assurance process.

Furthermore, the increasing complexity of future software systems will lead to an explosion of quality-related data, including large volumes of quality test data for individual components, analysis and prediction based on this data, and subsequent evaluations. This will also involve extensive analysis of failure data emerging during software development, testing, and operation.

3.2.2 Complex Factors Influencing the Quality of Complex Software Systems

Complex software systems typically operate over extended periods and are prone to failures, overloads, and attacks during this time. Additionally, these systems must maintain robust performance under adaptive mechanisms, which are generally not managed by an external, standalone component, but are instead typically integrated and controlled by the system itself.

As the scale of complex software systems becomes increasingly large, new behavioral patterns may emerge, leading to the emergence of new quality attributes. For example, internet storms can only occur in large-scale internet systems.

Predicting and avoiding these phenomena requires the application of new theories and methods from statistical mechanics and probability theory.

In complex software systems, human factors are prominent, so quality factors should also consider the factors related to people and organizations (as shown in Fig. 3.4).

The two fundamental approaches to managing complexity in complex software systems are abstraction and integration. Abstraction is used to conceal relevant details, while integration is employed to break down and conquer the complexity. This is also the rationale behind Morin's advocacy for "dualistic logic" in understanding and addressing complexity. He argues that the new science of complexity will be founded on a dualistic logical framework of "imagination and verification, empiricism and rationalism." The insights derived from this understanding are crucial for overcoming reductionism and advancing our conceptualization of complexity. This suggests that the world is, in fact, a world fragmented by our thinking. To some extent, the world is interpreted (decomposed) into parts due to the nature of human subjective thought. Thus, the only way to reconstruct the world is by reassembling it from these fragmented logical points of thought.

From the perspective of complexity, logical points refer to the fixed and distinguishable states within thought. Only by fully recognizing these objectively existing "distinguishable states" can one progress toward a deeper understanding and move from fixed states to indeterminate ones. A common misconception about complexity is that it is synonymous with chaos. However, if the object of study is merely chaotic and no tangible "distinguishable states" can be identified, logical points for thought cannot be formed, and as a result, a theory cannot be developed. In scientific terms, indistinguishability is nearly equivalent to "unknowability." In his book

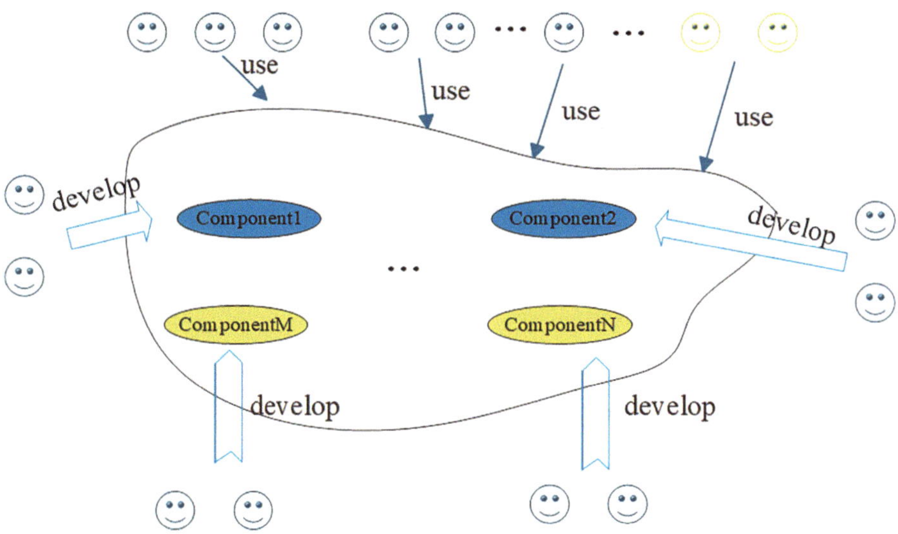

Fig. 3.4 Illustration of the development and use of complex software

Emergence, John Holland states: "The identifiable features and patterns are the key elements in studying emergent phenomena. Unless a phenomenon is identifiable and repeatable, I would not consider it an emergent phenomenon." The goal of complexity theory is to fully uncover the complexity of "distinguishable states," not to accept their chaos. Therefore, "reductionism" remains an unavoidable theme in the study of complexity science.

It is evident that, for increasingly complex software systems, the way people perceive and understand quality is also undergoing significant changes.

Firstly, there is a widespread awareness of quality.

When software was primarily used to meet the computational needs of researchers, the expectation was simply that the software could produce results quickly and accurately. Users were not particularly concerned with factors such as ease of use, comprehensibility, or ease of learning, as they were professionals. They might not even care if the software runs on another machine or is capable of addressing different problems. However, software users today are becoming more diverse. Many are not experts in computer software and may not even fully understand what software is, yet they still use it and have their own expectations and understanding of its quality. Software is now expected to be user-friendly, easy to learn, and aesthetically pleasing. It should meet all the users' needs, and if it falls short, an updated version should be released promptly to fulfill these needs, ideally at no cost. In short, modern software now carries the hopes of an ever-growing number of users, and these hopes are reflected in the quality expectations placed on software.

3.2.2.1 The Fragility of Software Quality

In the past, software operated on specifically designed hardware, used by a fixed group of users, and performed tasks that had already been predefined. It was akin to a person living in a familiar city, doing a job they excel at, and leading an orderly life—generally, such a situation was unlikely to result in any accidents. However, modern software often runs across a wide range of different hardware environments, is distributed in various locations, and is used in numerous unpredictable ways by individuals with diverse backgrounds and usage habits. It may even operate in extremely harsh physical environments. In such conditions, the software reveals its inherent fragility, where even the smallest oversight in design can lead to catastrophic failures. There are simply too many potential vulnerabilities where the software could be compromised.

3.2.2.2 The Contradiction in Software Quality

Software is becoming increasingly complex, while its lifespan continues to extend. This is because we desire the systems we have developed with significant "effort" (including time, money, and manpower) to serve us for as long as possible. Consequently, we continuously improve the software system's ability to adapt to its environment, learn, and correct errors autonomously. However, this leads to a paradox: software failure, or the persistent absence of software quality, has become the norm. This contradiction manifests in two ways: firstly, people's requirements for software quality are constantly evolving, including both functional and adaptability changes; secondly, the continuous updates to software often introduce new defects. As a result, software is used for extended periods in a "flawed" state.

3.2.2.3 The Helplessness of Software Quality

As previously mentioned, on one hand, software attracts significant attention and high expectations, while on the other, it demonstrates fragility that makes it susceptible to various influences. However, we still lack a clear and effective approach to constructing truly robust software that can meet everyone's needs. The current state of software quality is, in a sense, a result of this helplessness. The root of this helplessness lies in people's lack of a clear and precise understanding of what "perfect" software actually entails. In other words, it is not that we are incapable of creating high-quality software, but rather that we do not fully know what high-quality software should look like.

3.2.2.4 The Uncertainty of Software Quality

Complex systems often exhibit uncertainty, and complex software is no exception. To continuously meet users' needs, software systems typically possess adaptive characteristics, meaning they make corresponding adjustments in response to changes in the software environment. While the software code itself is deterministic, the software state exhibited during the execution of this code forms an extensive set that cannot be fully comprehended or controlled, leading to software behaviors that are often unpredictable. Additionally, because complex software has a large user base, each user has different expectations. As a result, the software may appear perfect to some users, while others may find it unsatisfactory. The same software, therefore, can present varying quality depending on the perspective of different users.

3.2.3 New Quality Characteristics of Complex Software Systems

People's demands for software quality represent an evolving process, which is clearly reflected in the ISO/IEC 25010 model. For instance, confidentiality and interoperability have been promoted from sub-characteristics at the third level of ISO/IEC 9126 to sub-characteristics at the second level. This change highlights the new characteristics arising from the increased networked nature of software and the dramatic expansion in the scale of software components.

This section focuses on introducing certain quality characteristics exhibited by complex software systems. Some of these characteristics have been part of the standards, albeit with slight differences in their interpretation, while others are newly emerging. These new quality characteristics are primarily reflected in the following aspects.

3.2.3.1 Availability

In reliability theory and engineering, the term availability is defined as follows. At the beginning of a task, when the task is requested at an unknown or random time, the system, subsystem, or device is in a specified state of operability and readiness for submission. Simply put, availability refers to the proportion of time that the system is in an operational state. This is typically described as the task capability rate. Mathematically, it can be represented as 100% minus unavailability. For example, a unit that is usable for 100 h out of a total of 168 h per week would have an availability of 100/168. However, typical availability values are often specified in decimal form (e.g., 0.9998). In high-availability applications, a metric known as "nine nines" is used, which corresponds to nine digits after the decimal point. Following this convention, "five nines" represents an availability of 0.99999 (or 99.999%).

3.2.3.2 Observability

Observability allows users to understand the internal state of the system by observing the performance of the interaction interface. In other words, it enables users to compare the current observed behavior with the task to be completed. If the user perceives that the system has not achieved the intended goal, they may adjust subsequent interactions accordingly. Observability encompasses five key principles: browsability, defaults, accessibility, persistence, and operational visibility.

1. Browsability allows users to understand the system's current internal state through the limited information provided by the interface. Due to the complexity of the problem, it is often not feasible to display all relevant information at once on the interface. In fact, the system typically limits the information displayed to a subset that is relevant to the user's current activity. For example, if the user is only interested in the overall structure of a document, they may not see the entire document but instead be presented with an outline. With such limitations, some information cannot be immediately observed and requires the user to explore further through additional browsing actions to uncover the desired details. Furthermore, browsing itself should not have side effects; that is, browsing commands should not alter the internal state.

2. The function of default values is to reduce the number of input operations required. As such, providing default values can be viewed as an error prevention mechanism. Default values can be classified into two types: static and dynamic. Static default values do not involve interaction sessions; they are either predefined within the system or assigned during system initialization. Dynamic default values, on the other hand, are set during the session, with the system configuring them based on the user's current input.
3. Accessibility refers to the possibility of transitioning from one state to another in a system. In other words, it is the ability to transit from one state to another through a series of actions. Accessibility also affects the recoverability mentioned below.
4. Persistence refers to the continuity of interaction response information and how users interact with these responses over time. While language in an interaction is not persistent, visible interaction responses can persist for a period, remaining available for future actions. For example, a sound emitted by a speaker to indicate the arrival of new mail delivers the message at that moment, but if the user is not paying attention, it may be overlooked. However, by using a persistent visible cue, such as a small dialog box, the message can remain for a longer period, ensuring it is noticed.
5. Operability visibility refers to the system's ability to inform the user about what is happening during the interaction process and to provide appropriate feedback at the right moment. In other words, the system should offer feedback for every user action, allowing the user to clearly determine whether the operation was successful or failed. Operability visibility requires that the system should not remain unresponsive after a user action, as this would leave the user uncertain about the next steps.

3.2.3.3 Responsiveness

Responsiveness reflects the frequency of communication between the system and the user. Response time is typically defined as the delay before the system responds to a change in state. In general, shorter delays or immediate responses are preferred, meaning that users can instantly observe the system's reaction. Even if a response is delayed, the system should notify the user that the request has been received and is being processed. Responsiveness is a key component of user experience.

3.2.3.4 Task Conformity

Task Conformity refers to whether the functions provided by the system to accomplish interactive tasks adhere to established standards. Users, having prior experience with interactions, often have certain expectations for these tasks. When the system's functions are standardized, users can more easily understand how the system supports these tasks and can more readily comprehend and use new features.

For example, standardized windows should include minimize, maximize, and close buttons, enabling users to easily perform window operations as part of their interaction tasks.

3.2.3.5 Recoverability

Recoverability refers to the ability of users to recognize errors and make corrections. These corrections can be made either forward or backward. Moving forward involves accepting the current state and progressing towards the target state, typically applied in situations where the effects of prior interactions are irreversible, such as when a file is permanently deleted. Moving backward involves undoing the effects of previous interactions and returning to a prior state.

Recovery can be initiated either by the system or by the user. System-initiated recovery involves concepts such as system fault tolerance, security, and reliability, while user-initiated recovery depends on actions determined by the user's intent.

Recoverability is closely related to accessibility. Without accessibility, users may find it difficult to transition from an erroneous or undesired state to the expected one.

When providing recovery capabilities, the recovery process should be aligned with the complexity of the task being restored. Generally, tasks that are easy to recover from are simple to implement, as they can be quickly restored even if errors occur. On the other hand, tasks that are difficult to recover from require more careful consideration, encouraging users to be more mindful during operations to avoid mistakes.

3.2.3.6 Adaptivity

Adaptivity refers to the automatic adjustment of processing methods, processing order, processing parameters, boundary conditions, or constraint conditions according to the characteristics of the processed data during the processing and analysis process, so that they can adapt to the statistical distribution characteristics and structural characteristics of the processed data to achieve the best processing results.

In computer science, the term "adaptation" typically refers to adaptive systems, which are interactive systems that adjust their behavior based on information obtained from their users and environment to better suit individual users.

Adaptive systems gather information through the analysis of user behavior, preferences, habits, and other aspects. Based on this data, they adjust their behavior, appearance, or other attributes to enhance the interaction between the system and the user. The goal of adaptive systems is to improve user experience and efficiency, making interactions more natural and seamless. These systems often employ techniques such as machine learning and artificial intelligence to better adapt to changes in user needs and the environment.

Complex software systems typically have long software engineering cycles. Before delivery, although requirement engineers, designers, and software developers may have completed the system, it is impossible to anticipate all user requirements. As a result, providing the optimal system configuration and behavior for each individual user is not feasible. Therefore, software must possess a certain level of adaptability to meet user needs in the following scenarios:

1. Active participation of software users: Users continuously raise new requirements and task demands during the use of the software.
2. Participation of potential users: The software encounters potential users that were not initially considered during its development.
3. New usage scenarios and environments: The software is deployed in a new scenario, or the system it operates within faces a new usage environment, including both user and natural environments.

3.2.3.7 Resiliency

The concept of software resiliency has not been around for a long time, and it does not originate specifically from complex software systems. Rather, this characteristic has become increasingly prominent within complex software systems in recent years.

For complex software systems, resiliency refers to the ability of the software system to respond to component failures and still provide the best possible service. For many software systems, failures themselves do not necessarily result in system crashes or complete failures, but rather cause a decrease in the performance of one or more system functions (degraded operations), which ultimately results in an inability to meet the needs of users. In this sense, resiliency and reliability share similar characteristics as both describe the impact of failures on user needs. However, software resiliency emphasizes the ability of the system to recover from certain types of failures and continue to maintain functionality from the customer's perspective after a system failure.

In complex software systems, there is another attribute related to resilience, namely survivability. Generally, survivability is the ability of a software system to maintain normal operation or continue to exist. However, this term may have a more specific meaning in certain contexts. For example, in IoT engineering, survivability is the quantitative ability of a system, subsystem, device, process, or program to continue to function during and after natural or human-induced disturbances.

Survivability is considered a subset of resilience. While resilience focuses on the ability of a software system to recover and continue providing services after a failure, survivability builds on this by emphasizing the causes of system failure. Survivability is more concerned with a system's ability to fulfill its mission in a timely manner in the presence of threats, such as attacks or large-scale natural disasters. It extends resilience by prioritizing the system's capacity to continue operating despite significant challenges.

This version refines the sentence structure for better clarity and flow, while maintaining a formal academic tone.

3.2.3.8 Dependability

Dependability is a collective term used to indicate availability and its influencing factors, including reliability, maintainability, and security. It is often used as a general description of non-quantitative terms.

In systems engineering, dependability refers to the measure of a system's availability, reliability, maintainability, and maintenance support performance, as well as, in certain cases, other characteristics such as durability, security, and confidentiality. In software engineering, dependability is the ability to provide trustworthy services that can be reliably relied upon over time. This also encompasses mechanisms designed to enhance and sustain the dependability of a system or software.

This version maintains the meaning while improving flow and clarity, ensuring the technical terms are presented in a more cohesive manner.

The International Electrotechnical Commission (IEC) develops and maintains international standards for the assessment of dependability and the entire lifecycle management of equipment, services, and systems through its Technical Committee TC 56. This provides a systematic approach and tools for the assessment of dependability and management throughout the complete lifecycle of equipment, services, and systems.

Dependability can be divided into the following three elements:

1. Attribute: A method for evaluating the dependability of a system.
2. Threat: Identifying potential factors or events that may affect the dependability of a system.
3. Means: Methods or approaches used to enhance the dependability of a system.

With the growing focus on fault tolerance and system reliability in the 1960s and 1970s, dependability emerged as one of the key measures for evaluating software reliability, as it incorporates additional factors such as security and integrity. In the early 1980s, Jean-Claude Laprie adopted the term "dependability" to encompass research on both fault tolerance and system reliability, without extending the inherent meaning of reliability.

Since then, the field of dependability has evolved into a globally active research area, with many well-known international conferences being held, especially the International Conference on Dependable Systems and Networks, the International Symposium on Reliable Distributed Systems, and the International Symposium on Software Reliability.

Traditionally, the dependability of a system includes aspects such as availability, reliability, and maintainability. However, since the 1980s, security and confidentiality have also been added to the measurement of dependability.

The current framework's flexibility encourages system architects to enable reconfiguration mechanisms of available and secure resources to support the most critical services, instead of over-configuring to build fault-tolerant systems.

With the popularity of networked information systems, accessibility has been introduced to place greater emphasis on the user experience.

To take into account the level of performance, measurability of executability is defined as "quantifying the degree to which an object system performs within a given period of time in the presence of failures."

3.2.4 *Quality Oriented Software Management Analysis*

First, multiple linear analysis is used to analyze process monitoring data to identify effective process factors that affect the final product quality, and important process factors that affect QCD software management measures are also discussed. Second, based on the process improvement model, i.e., the derivation process of the software management model, actual process monitoring data is analyzed, as shown in Fig. 3.5. Then, important process factors related to project management or affecting QCD measures are discussed, and their effects are demonstrated. Finally, process control data are analyzed from the perspective of software reliability measurement and evaluation in process control activities.

3.2.4.1 Process Monitoring Data

Predictive software quality management measures for QCD using process monitoring data are listed in Table 3.1. Five variables were detected in process monitoring for QCD software quality management measures, including peer review, development plan review, design completion review, test plan review, and test completion review, all of which are used as explanatory variables. The observed values of these five factors were standardized according to the size of the project development team. Three variables were used as the target variable, namely the number of faults detected in customer acceptance testing, the cost overrun rate, and the payment delay in days.

3.2.4.2 Analysis of Factors Influencing Quality

Based on canonical correlation analysis and the correlation analysis shown in Fig. 3.5, X_3 was selected as a key factor for estimating the software quality prediction model. Subsequently, single-factor regression analysis was conducted on the process monitoring data, as outlined in Table 3.1. Then, the single regression equation $\hat{Y}_q$ for predicting the number of software faults was derived using X_3, and the normalized single regression expression $\hat{Y}_q^N$ was obtained as follows:

$$\hat{Y}_q = 11.761 \cdot X_3 + 0.998 \tag{3.1}$$

$$\hat{Y}_q^N = 0{,}894 \cdot X_3 \tag{3.2}$$

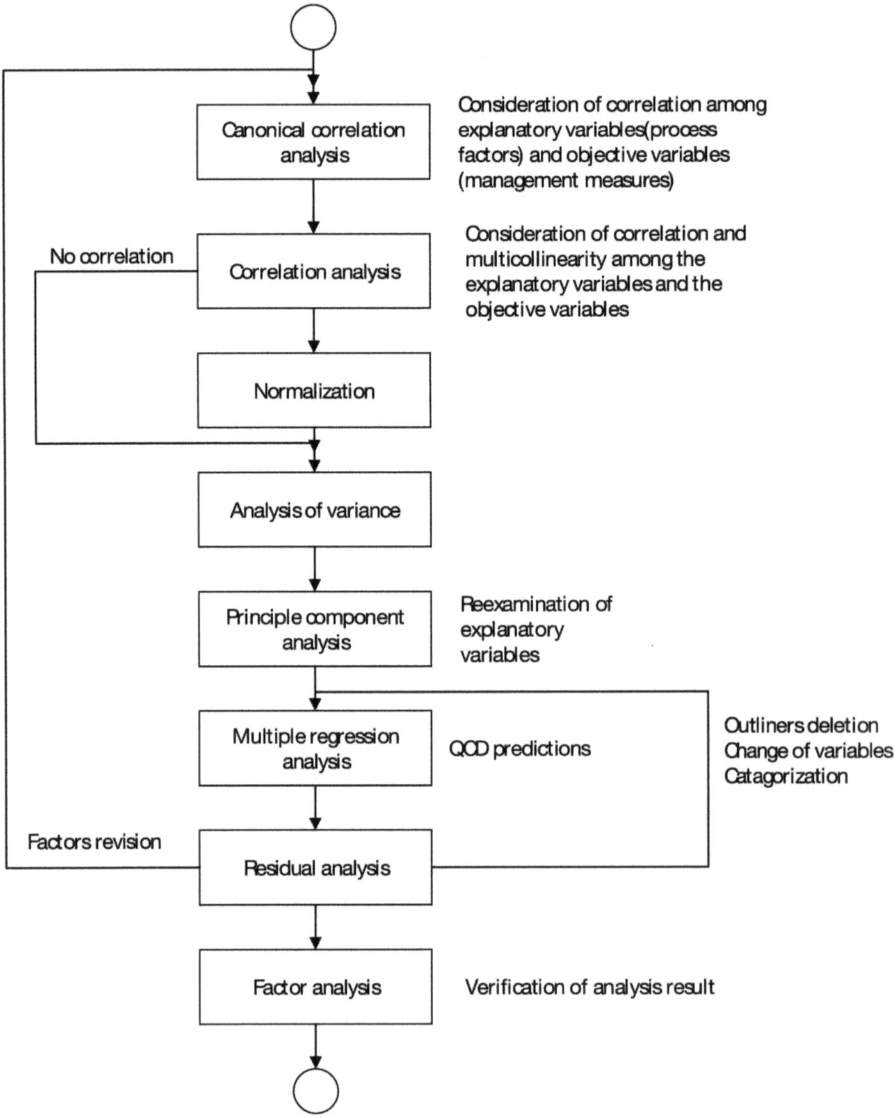

Fig. 3.5 Derivation process of software management model

The adjusted R-squared value, which takes into account the degrees of freedom, is 0.758, indicating that the derived linear quality prediction model is significant at the 1% level.

Similar to the discussion above regarding factors that affect the number of faults, canonical correlation analysis, correlation analysis, and principal component analysis can be used to select X_1 and X_5 as important factors for estimating the cost overrun rate and delivery delay in days. Then, multiple regression equations for

Table 3.1 Process monitoring data

Project number (Size of faults after development)	Inspection and review	Development plan X_2	Plan completion inspection X_3	Test plan review X_4	Test completion review X_5	Quality Y_q	Cost Y_c	Delivery Y_d
1.	0.591	1.181	0.295	0.394	0.394	4	1.456	28
2.	0.323	0.645	0	0.108	0.108	1	1.018	3
3.	0.69	0.345	0	0.345	0	0	1.018	4
4.	0.170	0.170	0	0.085	0	2	0.953	0
5.	0.150	0.451	0.301	0.075	0.075	5	1.003	0
6.	1/186	0.149	0	0.037	0.037	0	1	−8
7.	0.709	0	0	0	0	2	1.119	12

predicting the cost overrun rate $\hat{Y}_c$ were obtained using X_1 and X_5, and the normalized multiple regression expression is given by:

$$\hat{Y}_c = 0.253 \cdot X_1 + 1.020 \cdot X_5 + 0.890 \tag{3.3}$$

$$\hat{Y}_c^N = 0.370 \cdot X_1 + 0.835 \cdot X_5 \tag{3.4}$$

The adjusted R-squared value is 0.917, indicating that the derived cost overrun prediction model is significant at the 1% level.

Using the same approach as for predicting the cost overrun rate, multiple regression equations for predicting the delivery delay in days $\hat{Y}_d$ were obtained using X_1 and X_5, and the normalized multiple regression expression is given by $\hat{Y}_d^N$:

$$\hat{Y}_d = 24.669 \cdot X_1 + 55.786 \cdot X_5 - 9.254 \tag{3.5}$$

$$\hat{Y}_d^N = 0.540 \cdot X_1 + 0.683 \cdot X_5 \tag{3.6}$$

The adjusted R-squared value is 0.834, indicating that the derived delivery delay prediction model is significant at the 5% level.

3.2.4.3 Results of the Analysis of the Software Management Model

The software management model was derived through multiple linear regression analysis of actual process monitoring data. Based on this model, prediction models for final product quality, cost overruns, and delivery delays were developed, followed by high-precision quantitative evaluations. Furthermore, the software management model derivation process, as illustrated in Figs. 3.5, effectively facilitated software process improvement within the PDCA (Plan, Do, Check, Act) management cycle.

Furthermore, the design completion review plays a crucial role in influencing software quality. Therefore, in the early stages of software development projects, the outcomes of the design completion review within the process monitoring activities can be utilized to forecast the quality of the software product.

Secondly, the contract review and test completion review processes have a significant impact on the cost overrun rate and the duration of delivery delays. In other words, predicting cost overruns and delivery delays in the early stages of software development projects is challenging. However, analysis has shown that under the same process monitoring factors, these issues can become predictable.

3.2.4.4 Implementation of Process Management

Continuous Process Improvement

Through analysis of the software management model and factor analysis, it was found that the contract review process is closely linked to cost and delivery performance. To mitigate cost overruns and delivery delays, appropriate project management practices should be implemented to address critical issues identified during the contract review process.

The project management practices carried out in response to important issues identified during the contract review mainly include the following aspects:

1. Standardization of early decisions in the domain area.
2. Improvement of requirement specification technology.
3. Early decisions on development plans.
4. Improvement of project schedule management.
5. Improvement of testing technology.

Provide the risk ratio, i.e.

$$R = \sum_i \{\text{riskitem}(i) \times \text{weight}(i)\} \tag{3.7}$$

In formula (3.7), the risk assessment has a weight (w) for each risk item (i), and the risk ratio is between 0% and 100%. Based on the risk assessment checklist, project risks are identified through interviews. From the identified risks, the risk ratio is calculated using formula (3.7).

Implementation of Design Quality Evaluation

Similar to the cost and delivery measurement methods, there is also an important relationship between design quality review and software quality. In order to improve software quality, appropriate project management should be carried out in the design completion review, which is called design evaluation assessment?

Design evaluation assesses the following items based on the risk assessment checklist by the project manager, designers, and quality control department members. Based on the design evaluation below, it is determined whether the development can proceed to the next stage.

1. After requirement analysis, how many requirements are included in the requirement specification? Are the requirements (functional and non-functional) properly defined?
2. After basic design completion, are the requirements (functional and non-functional) adequately transferred from the user requirements to the design document without omitting any descriptions in the requirement specification?
3. Does the basic design document include the basic design?

After implementing plan review, design evaluation was conducted on several projects' design completion reviews. It was found that software quality had been improved, and the cost overrun rate and delivery delay had become more stable.

3.2.4.5 Software Reliability Assessment

Next, we will discuss the software reliability measurement and evaluation based on process monitoring data. The software reliability growth curve in process monitoring activities shows the relationship between the process control progress ratio and the cumulative number of detected failures (QCD issues) in process monitoring.

Here we discuss the NHPP-based software reliability growth model because its analysis and processing are relatively easy. Assuming that the observed test time data is continuous, the process control progress ratio is chosen as a substitute unit for the test time.

In order to describe the fault detection phenomenon at process control progress ratio t ($t > 0$), the counting process $\{N(t), t > 0\}$ is used to represent the cumulative number of detected failures up to progress ratio t. Thus, the fault detection phenomenon can be described as:

$$\Pr\{N(t)=n\} = \frac{\{H(t)\}^n}{n!}\exp\left[-H(t)\right](n=0,1,2,\ldots) \tag{3.8}$$

Where $H(t)$ is the expected value of $N(r)$ and is called the mean function of NHPP. Pr(A) in formula (3.8) represents the probability of event A. In this section, NHPP models are applied, including the exponential SRGM, the delayed S-type SRGM, and the logarithmic Poisson execution time model.

The software reliability evaluation method plays an important role in the quantitative evaluation of software reliability based on SRGM. The expected number of maintenance failures, $n(r)$, represents the potential number of failures in the software system at any test time t and is expressed as:

$$n(t) \equiv E\left[N(\infty)-N(t)\right] = E\left[N(\infty)\right]-H(t) \tag{3.9}$$

Where $E[A]$ represents the expected value of a random variable A, the instantaneous new MTBF (mean time between failures) is expressed as:

$$\mathrm{MTBF}_1(t) = \frac{1}{dH(t)/dt} \tag{3.10}$$

This is one of the measures using MTBF to substitute NHPP model.

In addition, the software reliability function represents the probability that no software failure occurs in the time interval $(t, t + x)$ $(t \geq 0, x \geq 0)$ assuming that testing or user operation continues until time t. If the counting process $\{N(t), t \geq 0\}$ follows the mean function $H(t)$ of NHPP, then the software reliability function is obtained as:

$$R(x|t) = \exp\left[-\left\{H(t+x) - H(t)\right\}\right] \tag{3.11}$$

Chapter 4
Reliability Assessment of Complex Software Based on Architecture

4.1 Reliability Assessment Based on Architecture

4.1.1 Assessment Process

Reliability assessment based on system architecture is typically regarded as a white-box evaluation method, primarily employed during the early stages of the software lifecycle. The foundational premise of this approach is that system failures are directly attributable to the failures of components within the architecture. By explicitly modeling and analyzing the relationship between these two types of failures, it becomes possible to assess the overall reliability of the system.

Generally, the architecture-based evaluation method involves the following key steps:

1. Characterize and describe the software architecture using a formal representation.
2. Define a set of fault-related parameters at the architectural level.
3. Derive or empirically determine the relationship between the defined parameters and system reliability.
4. Develop an evaluation model for assessing the reliability of the software system.

As early as 1975, Parnas and Shooman used the internal structure of software in the process of reliability assessment, and they believed that the internal structure of software has a significant impact on software reliability. Figure 4.1 shows the basic process of reliability assessment based on architecture.

Based on the above basic process, researchers have proposed various models to describe the software system structure based on their understanding. They have also described the parameters used in the evaluation, resulting in various kinds of evaluation methods. Depending on the model on which the architecture evaluation method is based, it can usually be classified into state-based evaluation and path-based evaluation. The specific classification is shown in Table 4.1.

Y. Wang et al., *Reliability Technology for Complex Software Systems*,
https://doi.org/10.1007/978-981-95-3889-8_4

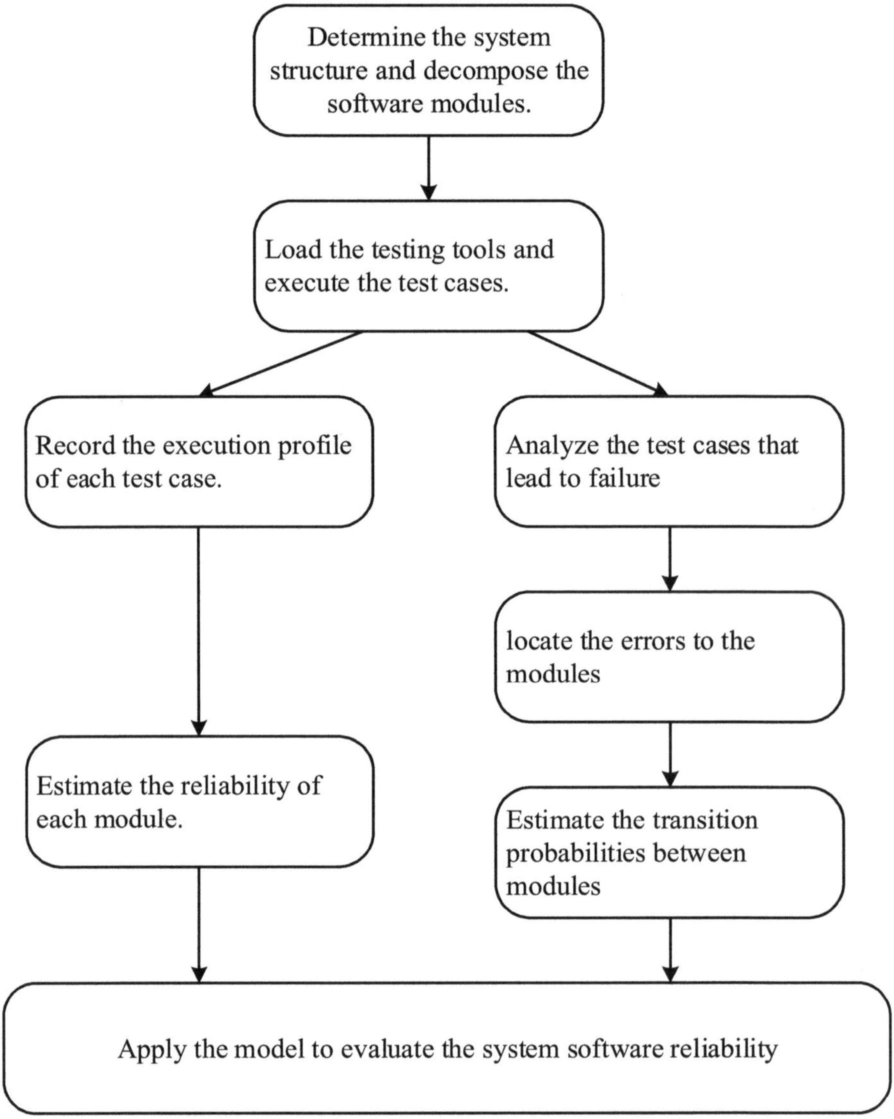

Fig. 4.1 The basic process of software reliability assessment

State-based models typically represent the system structure in terms of its components and the control transitions between them. The reliability of the entire system is then assessed by constructing an appropriate mathematical model. In contrast, path-based models generally involve calculating all possible execution paths of the system under evaluation. This calculation is typically performed either through simulation of the execution process or by deriving the execution paths using theoretical algorithms. Subsequently, a mathematical model is constructed to evaluate the system's reliability.

Table 4.1 Classification of reliability assessment models based on system architecture

		Parameter categories			Structural views		Interaction relationships between components		
Main classification	Model proposer	R	λ	$\lambda(t)$	Single view	Multiple views	Transition probabilities	Path frequency	None
State model	W. Wang	√	–	–	√	–	√	–	–
	Cheung	√	–	–	√	–	√	–	–
	Goseva-Popstojanova	√	–	–	√	–	√	–	–
	Reussner	√	–	–	√	–	√	–	–
	Roshande	√	–	–	√	–	√	–	–
	Gokhale	√	–	–	√	–	√	–	–
	Littlewood	–	√	–	√	–	√	–	–
	Laprie	–	√	–	√	–	√	–	–
Path model	Shooman	√	–	–	√	–	–	√	–
	Krishnamurthy and Mathur	√	–	–	√	–	–	√	–
	Yacoub	√	–	–	√	–	√	–	–
	Rodrigues	√	–	–	–	√	√		–
	Singh	√	–	–	–	√	–	√	–

4.1.2 *Main Methods for Reliability Assessment Based on Architecture*

4.1.2.1 State-Based Method

The state-based method represents the system structure through component transitions, and subsequently evaluates the overall system reliability by constructing a corresponding mathematical model. The two key factors in developing a state-based model are the reliability of individual components and the system's software architecture. Typically, the software architecture is represented by a control transition probability matrix, where the transition probabilities are largely determined by the operational profile. To more accurately reflect the system's usage, the control transition probability matrix is ideally derived from the operational profile. Below are brief descriptions of two representative methods.

Combination Model Method

The combinatorial model method employs an absorbing discrete-time Markov chain as the system architecture model and constructs a reliability evaluation model based on the transition probability matrix P.

The specific steps for constructing the Composite Model method are as follows: First, let p_{ij} represent the transition probability from component i to component j. Next, assume that the reliability of each node is denoted as R_i, and each node has two states: C (correct output) and F (failure). Assume that the system has a single input node and a single output node, where the input node is denoted as node 1 and the output node as node n. Considering the reliability of each node, the original transition probabilities can be adjusted to $R_i p_{ij}$, which represents the probability of control transferring from component i to component j, given that component i successfully executes. The system reliability is defined as the probability that no failure occurs from the input node to the output node. The method for calculating this is provided by the following formulas:

$$\boldsymbol{Q}_{n\times n} = \boldsymbol{D} \times \boldsymbol{P}$$

In the equation, $\boldsymbol{D}$ is $n \times n$ diagonal matrix.

$$d_{ii} = R_i$$

Calculate $\boldsymbol{S}_{n\times n} = (\mathbf{I} - \mathbf{Q})^{-1}$, and the system reliability is obtained by the formula $R = \boldsymbol{S}(1,n) \times R_n$.

Building upon the model described above, it can be further extended to accommodate different architectural styles, such as pipe-filter, backup fault tolerance, call-return, and batch processing. For each style, the state transition patterns should first be analyzed, followed by the calculation of the corresponding transition probabilities. Once the reliability of each architectural style is computed, a software system can be decomposed into a combination of these four architectural styles, thereby enabling the reliability assessment of heterogeneous software systems.

Hierarchical Model Method

This method also models the architecture based on Markov chain, but this model assumes that the software execution process is a continuous-time Markov chain (CTMC). In this method, the model uses m_{ij} and v_{ij} to represent the average execution time from component i to j and the failure probability of component i to component j respectively. The model assumes that the failure of each component follows a Poisson distribution with parameter λ_i. At the same time, the model assumes that the number of control transfers in the software is much greater than the number of failures. For the entire software system, its failures also follow a Poisson distribution. Thus, the method for calculating the software failure rate is:

$$\lambda_S = \sum_i a_i \lambda_i + \sum_{i,j} b_{ij} v_{ij}$$

where $a_i = \dfrac{\pi_i \sum_j p_{ij} m_{ij}}{\sum_i \pi_i \sum_j p_{ij} m_{ij}}$, which represents the percentage of time component i is executed in the total time; And $b_i = \dfrac{\pi_i p_{ij}}{\sum_i \pi_i \sum_j p_{ij} m_{ij}}$, which represents the transition frequency from component i to j. $\boldsymbol{\pi}_i$ is the steady-state Markov vector of the entire software system.

4.1.2.2 Path-Based Method

The primary difference between the path-based evaluation method and the state-based evaluation method lies in whether the execution paths of the software are taken into account when assessing its reliability.

In the architecture modeling technology that considers the software execution path, it is usually assumed that there are m possible paths for the software to execute, and for each path i, there is a frequency f_i for its execution and a probability of failure, q_i. Therefore, assuming that the software runs N times, the total number of failures can be calculated using the formula $n_f = \sum_{i=1}^{m} N f_i q_i$, where Nf_i represents the number of times path i is executed. The probability of software failure can be calculated using the formula $q_0 = \lim_{n \to \infty} \frac{n_f}{N} = \sum_{i=1}^{m} f_i q_i$.

Additionally, reliability evaluation can also be conducted based on the reliability of individual components. For a given test path, its reliability can be calculated, and then the overall software reliability can be obtained by running test cases and analyzing the data. Let R_m represent the reliability of component, assuming that the failure behaviors of components are independent of each other. Consider software P being executed once under test case tc. The execution sequence of all components during this process forms a sequence $M(P, tc)$. The path reliability of P can then be calculated based on tc and R_m. Let TS represent the set of test cases, and define $R_{tc} = \prod_{m \in M(P,tc)} R_m$, where the reliability of the software system is:

$$R = \frac{1}{|TS|} \sum_{tc \in TS} R_{tc}$$

When a cyclic path appears in the architecture, the assumption of independent failures implies that the reliability of a path is calculated by multiplying the reliability of each component by its execution frequency. However, the execution of a cyclic path leads to an infinite number of executions, which causes the path's reliability to continuously decrease, potentially approaching zero. This can reduce the accuracy of the reliability assessment. To address this issue, a common solution is that the failure behavior of components in a cyclic path no longer satisfies the independence assumption, and the reliability of the path should not be repeatedly multiplied. Instead, a component's Degree of Independence (DIO) is defined to represent the number of times the reliability is multiplied during repeated executions. When $DIO = 1$, the path is considered executed only once, while the maximum value of DIO is, indicating that the execution frequency of the path should be capped at n.

4.1.3 Challenges Encountered by Complex Software Systems

As software systems become increasingly complex, methods for reliability assessment based on system architecture are encountering new challenges.

4.1.3.1 Complexity of the Software Construction Process

The construction process of complex software systems is challenging to complete in a single phase, as it involves design, development, and deployment, and typically spans a long duration with numerous participants in various roles. In general, the occurrence and impact of software defects tend to be magnified throughout the development lifecycle. Consequently, the later a defect is detected, the higher the cost associated with its correction. The complexity of the construction process not only increases the likelihood of defect occurrence but also amplifies the propagation effects of these defects. Furthermore, it complicates the mechanisms behind defect generation, presenting significant challenges for both defect detection and correction. According to statistics, the requirements and design phases of complex software development are the most susceptible to defects, with the proportion of defects occurring in these stages ranging from 50% to 75% of the total.

4.1.3.2 The Complexity of Software Architecture Structures

The components of a complex software system consist of various heterogeneous autonomous systems. The internal structure of each autonomous system is intricate, and the interactions between these heterogeneous systems are highly complex. Each autonomous system operates in a distinct development environment and serves different stakeholder groups. This implies that the various parts of a complex software system possess diverse structures and failure modes, and cannot be simply treated as a black box with a single failure mode.

4.1.3.3 The Complexity of Software Failure Mechanisms

Firstly, the behavior of complex software systems is emergent, complex, and constantly changing, which makes it difficult to control and manage the system's overall behavior. Moreover, the failure of the various autonomous subsystems within the system has become commonplace, demanding improvements to traditional methods for measuring component failure behavior. Not to mention that traditional calculations of component failure behavior have limitations. Generally, component failure behavior is expressed as a constant, failure rate, or failure density function. These parameters are difficult to calculate and determine, even in simple software systems. For complex systems, due to the complexity of system behavior and interrelationships, these parameters become even more difficult to determine. Therefore, new measures are needed to replace traditional methods of representing component failure behavior.

4.1.3.4 Consideration of False Propagation

As described in Chap. 2, faults within complex software systems exhibit propagation and evolution characteristics, which affect the evaluation of techniques based on architecture. As software systems are interconnected, a fault in one component may propagate to other components directly or indirectly connected to it. However, due to the uncertainty, emergent behavior, and non-linear interactions of complex software systems, the propagation of faults between autonomous subsystems or components may not lead to the ultimate failure of the entire system. Therefore, when assessing the reliability of software systems at the architectural level, the impact of fault propagation on system reliability must be taken into account.

The use of architecture-based reliability analysis and evaluation has long been recognized as a viable approach. However, as the architecture of software systems becomes increasingly complex, relying solely on structural patterns and features to evaluate software reliability is no longer sufficient. This is due to the fact that the complex software systems has complicated interrelationships which leads to emergent behavior. Thus, studying the reliability of architecture from the perspective of the fault propagation at the structural level has become a current research hotspot.

4.2 Description of Software Architecture

Before introducing the reliability evaluation methods that consider the impact of fault propagation, a brief overview of the software architecture concepts involved in these methods will be provided.

4.2.1 Software Architecture

According to Bass, Clements, and Kazman, the architecture of a computer system or software program is a structure that encompasses one or more software components, the visible external attributes of those components, and the relationships between them. Software components come in various forms, ranging from simple components like classes or program modules in object-oriented design, to more complex middleware, which may include databases or systems that facilitate network configuration between clients and servers. To understand the interactions between components, it is essential to examine their attributes. The relationships between components can vary significantly, ranging from something as simple as one module calling a process in another, to something as complex as a database access protocol.

4.2.1.1 Software Architecture Types

Essentially, software architecture represents a structure. In various types of architectures, three key elements are present: components (a set of entities), connectors (a set of defined relationships), and attributes (the properties of components and connectors), which help distinguish the types of components and connectors. In an architecture, components are connected through connectors. Bass, Clements, and Kazman identified five typical basic architectural styles."

This version improves the flow and makes the text more concise while maintaining the original meaning. The term "architectural styles" is also more commonly used in academic discussions of software architecture.

1. Functional Structure

   Functional or processing entities constitute the components of the functional structure, while interfaces serve as the connectors, enabling functions such as "calling" other components or "transferring data" between them. The characteristics of the functional structure are used to describe the features of the components and the organization of the interfaces.
2. Implementation Structure

   In the implementation structure, any entity that packages functionality at various abstraction levels can be considered a component, such as packages, classes, objects, procedures, functions, and methods. The connectors in the implementation structure primarily involve the capabilities for controlling and transferring data, sharing data, and invoking other components. The characteristics of the implementation structure focus on quality attributes during the implementation process, such as maintainability and reusability.
3. Concurrency Structure

   Concurrency units' constitute the components of the concurrency structure, with parallel tasks or threads being examples of such units. The connector relationships in the concurrency structure include five types: cannot run with, send data to, must run with, synchronize with, and higher priority. The key characteristics of the concurrency structure are priority, execution time, and preemptive capability.

   This version improves the flow and makes the sentence structure more natural. It also avoids repetition and enhances clarity.
4. Physical Structure

   The deployment model in design and development is analogous to the physical structure, where physical hardware constitutes the components of the structure, and software resides on top of the hardware. The interfaces between hardware components form the connectors of the structure, with attributes including performance, bandwidth, capacity, and other properties.

5. Development Structure

   This structure defines the necessary information sources for components, software engineering processes, and other work products. The relationships between work products form the connectors of the structure, and the characteristics are used to identify the specific features of each element.

Each of the different structures mentioned above represents a distinct perspective of the software architecture, allowing software teams to model and construct the software based on various useful information.

4.2.1.2 Basic Elements that Compose the Software Architecture

Concept of Component

Components are generally regarded as modular, deployable, and replaceable units within a system. They encapsulate functionality and expose a set of interfaces. These interfaces ensure that components can be invoked and data can be transferred between them. A component may have multiple interfaces, which define how it communicates and collaborates with other components and entities outside the software boundary, such as other systems, devices, and users. Component interfaces can be linked to connectors, or they may interact directly with other components via global variables, parameter sets, messages, semaphores, and other mechanisms.

Software components have the following characteristics:

1. Components are modular units that are nearly independent, can be replaced by other components, perform specific functions, and are independently configurable. They are designed, developed, tested, and released independently.
2. Components exhibit strong internal cohesion, while their interactions with other components are loosely coupled, minimizing dependencies between software modules or parts.
3. The interface and implementation of components are separated. The interface defines the communication protocol between components, while the implementation refers to the internal design of the components.
4. Components should have clearly defined interface specifications to ensure they can be called by different systems in various environments.

Concept of Connector

Connectors, as structural elements of the software architecture, play a key role in facilitating interactions between components and governing the rules of these interactions. There are various ways in which components can interact with each other, including dependency relationships, synchronization, mutual invocation of methods or functions, message or semaphore passing, and the exchange and transformation of data. In the simplest cases, connectors between components may be direct connections, enabling direct interaction. However, in more complex scenarios, interactions between components must occur through connectors, which may take the form of pipelines, communication mechanisms, or communication protocols, among others.

Connectors and their reliability-related attributes include call probability and connector reliability. Call probability refers to the likelihood that a component exchanges information with another component through a specific connector. A component's interaction with a connector is actually carried out by invoking the interface connected to that connector. Therefore, by treating the call to the connector as a call to its associated interface, the call probability of the connector can be converted into the call probability of the connected component's interface, thereby standardizing the component's output as a call to the component's interface.

4.2.2 Introduction to Markov Chains

In 1896, the Russian mathematician Markov introduced an experimental model in which he defined the Markov process: when the 'future state' of the process depends only on the 'current state' and is independent of the 'past states,' the process is referred to as a Markov process. When the states of the Markov process are discrete, it is called a Markov chain. The theory of Markov processes has widespread applications across many fields, including physics, engineering, biology, and even social sciences. In commonly used Markov processes, both the state space and time parameters are discrete. Below is the mathematical definition of such a Markov process.

Definition 4.1: Let $\{X_1, X_2, \ldots, X_n\}$ be an ordered sequence of random variables in a Markov process, and let the state space $S = \{j_1, j_2, \ldots, j_n\}$ be a finite or countable set. For any integer n, if the conditional probability satisfies:

$$P\{X_n = j_n | X_1 = j_1 |, X_2 = j_2 |, \ldots X_{n-1} = j_{n-1}\} = P\{X_n = j_n | X_{n-1} = j_{n-1}\} \tag{4.1}$$

If the above condition always holds, the process is referred to as a Markov chain. Equation (4.1) is called the Markov property, or the memoryless property. This property implies that when the current state $X_{n-1} = j_{n-1}$ *of the process is known, the conditional probability distribution of the future state* $X_n = j_n$ *depends only on the current state* $X_{n-1} = j_{n-1}$ *and is independent of the past states* $X_1 = j_1, X_2 = j_2, \ldots, X_{n-2} = j_{n-2}$.

Definition 4.2: In a discrete-parameter Markov chain $\{X_1, X_2, \ldots, X_n\}$, the conditional probability is defined as follows:

$$P\{X_{m+1} = j | X_m = i\} = p_{ij}^{m,m+1} \tag{4.2}$$

Transition probability refers to the probability that a Markov chain $\{X_1, X_2, \ldots, X_n\}$ transitions from state i to state j under parameter m, which is also called a one-step transition probability. The transition probability is generally represented by $P_{ij}(m)$, where $i, j \in S$ and S is the state space of the Markov chain.

Transition Probability:

$$P\{X_{m+n} = j | X_m = i\} = p_{ij}^{(n)m,m+1} \tag{4.3}$$

The n-step transition probability of a Markov chain $\{X_1, X_2, \ldots, X_n\}$ refers to the probability that the Markov chain transitions from state i to state j after n steps under parameter m, represented by $P_{ij}(n|m)$, where $i, j \in S$ and S is the state space of the Markov chain.

Under normal conditions, the transition probability has the following properties: for any two states i and j in the state space S, it always holds that

1. $p_{ij}^{(n)m,m+1} \geq 0$
2. $\sum_{j \in S} p_{ij}^{(n)m,m+1} = 1, n = 1,2,3,\ldots$

Definition 4.3: For a homogeneous Markov chain $\{X_1, X_2, \ldots, X_n\}$ with state space $S = \{0, 1, 2,\ldots, n,\ldots\}$, a matrix is constructed by arranging the transition probabilities P_{ij} in ascending order for any two states i and j in S.

$$P = \begin{pmatrix} p_{11} & p_{12} & \cdots & p_{1j} & \cdots \\ p_{21} & p_{22} & \cdots & p_{2j} & \cdots \\ \vdots & \vdots & \vdots & \vdots & \cdots \\ p_{i1} & p_{i2} & \cdots & p_{ij} & \cdots \\ \vdots & \vdots & \vdots & \vdots & \cdots \end{pmatrix} \tag{4.4}$$

This matrix is called the one-step transition probability matrix, which can also be simply referred to as the transition probability matrix.

The transition probability matrix P has the following two properties:

1. $p_{ij} \geq 0$, which means all elements are non-negative.
2. $\sum_{j \in S} p_{ij} = 1$, which means the sum of each row is equal to 1.

4.2.3 *Absorbing Discrete-Time Markov Chains*

Now, we will introduce how to utilize absorbing discrete-time Markov chains (ADTMC) to represent the architecture of software, where the components are functional modules and the connectors represent control transitions between these components, forming the functional structure. The model assumes that the control transitions between components follow a Markov process, meaning that the execution of the next module depends solely on the current module and is independent of previous modules. This assumption may not hold true for all types of programs. Although assuming Markovian behavior at the instruction level presents certain challenges in modeling control flow, experiments in memory management and temporal scheduling conducted by researchers indicate that, from a macroscopic perspective (considering modules), this assumption holds for many programs. If the components are not modified, the transition probabilities between components in a given user environment will remain constant. In this context, we treat these probabilities as constants that characterize the features of the user environment and represent the user operation profile. The probability values are derived from the operation profile.

First, let's introduce the concept of discrete-time Markov chain (DTMC). A DTMC is represented by a one-step transition probability matrix $P = [p_{i,j}]$. P is a transition matrix, in which the sum of each row's elements is equal to 1, and each element has a value in the range of [0,1].

DTMCs can be divided into the following two types:

1. **Irreducible**. If each state can be reached from any other state, the DTMC is irreducible. It is usually used to represent the architecture of an infinite running program.
2. **Absorbing**. If there is at least one state where there is no transition to any other state, the DTMC is absorbing. Once a DTMC reaches an absorbing state, it will remain there forever. Absorbing DTMCs can be used to represent the architecture of terminating applications or instruction-operated applications.

Here, we consider a terminating application and represent its architecture using an absorbing discrete-time Markov chain (DTMC). A detailed introduction to absorbing DTMCs is provided below.

The one-step transition probability matrix for an absorbing DTMC is:

$$P = \begin{bmatrix} Q & M \\ 0 & E \end{bmatrix} \tag{4.5}$$

In the above equation, Q is a sub-transition matrix of size $(n - m) \times (n - m)$ where each row sums to <1. E is the $m \times m$ identity matrix. M is a $(n - m) \times m$ matrix representing the location of the m absorbing states in the Markov chain of n states.

The k-step transition matrix P^k for an absorbing DTMC is given by:

$$P^k = \begin{bmatrix} Q^k & M' \\ 0 & E \end{bmatrix} \tag{4.6}$$

The elements of matrix M' are uncorrelated, and the (i, j)th element of the matrix Q^k represents the probability of reaching transient state j from transient state i after k steps. As t approaches infinity, $tk = 0$ Q^k converges to a limit, and the inverse matrix $S_{n \times n} = (E - Q) - 1$ exists. The formula for calculating this matrix is given by:

$$S_{n\times n} = E + Q + Q^2 + Q^3 + \ldots = (E - Q)^{-1}$$

This inverse matrix represents the expected number of times a system starting in a transient state will visit each absorbing state before it terminates. It can be used to evaluate the performance of an absorbing DTMC and optimize its architecture.

$$S_{n\times n} = (E-Q)^{-1} = E + Q + Q^2 + \ldots = \sum_{i=0}^{\infty} Q^i \tag{4.7}$$

In this context, we use an absorbing DTMC to model software that consists of n components represented by the set $\{N_1, N_2, \ldots, N_n\}$. Without loss of generality, we assume the software has a single starting component N_1 and a single ending component N_n. We treat each component as a state in the Markov process, where the starting state is component N_1. We add two terminating states, C and F, representing correct output and failure, respectively, which are the rows and columns of the matrix M as mentioned earlier. The (i, j)th element of matrix Q in the above equation represents the probability of transitioning from component N_i to component N_j.

Therefore, the software architecture model is determined to be an absorbing discrete-time Markov chain (DTMC) with the following characteristics:

$$Q = \begin{array}{c} \\ N_1 \\ \vdots \\ N_i \\ \vdots \\ N_{n-1} \\ N_n \\ C \\ F \end{array} \begin{array}{c} \begin{array}{ccccccccc} N_1 & N_2 & \cdots & N_j & \cdots & N_n & C & F & \end{array} \\ \begin{bmatrix} 0 & P_{12} & \cdots & P_{1j} & \cdots & P_{1n} & 0 & 0 \\ \vdots & \vdots & \cdots & \vdots & \cdots & \vdots & 0 & 0 \\ 0 & P_{i2} & \cdots & P_{ij} & \cdots & P_{in} & 0 & 0 \\ \vdots & \vdots & \cdots & \vdots & \cdots & \vdots & 0 & 0 \\ 0 & P_{(n-1)2} & \cdots & P_{(n-1)j} & \cdots & P_{(n-1)n} & 0 & 0 \\ 0 & 0 & \cdots & 0 & \cdots & 0 & R_n & 1-R_n \\ 0 & 0 & \cdots & 0 & \cdots & 0 & 1 & 0 \\ 0 & 0 & \cdots & 0 & 0 & \cdots & 0 & 1 \end{bmatrix} \end{array} \tag{4.8}$$

In the above equation, Rn represents the reliability of component N_n. Equation (4.8) gives us the one-step transition matrix Q of size $n \times n$. This matrix represents the transition rates between the n components, i.e., the control transfer rates between the components.

$$Q = \begin{array}{c} \\ N_1 \\ \vdots \\ N_i \\ \vdots \\ N_{n-1} \\ N_n \end{array} \begin{array}{c} \begin{array}{ccccccc} N_1 & N_2 & \cdots & N_j & \cdots & N_n \end{array} \\ \begin{bmatrix} 0 & P_{12} & \cdots & P_{1j} & \cdots & P_{1n} \\ \vdots & \vdots & \cdots & \vdots & \cdots & \vdots \\ 0 & P_{i2} & \cdots & P_{ij} & \cdots & P_{in} \\ \vdots & \vdots & \cdots & \vdots & \cdots & \vdots \\ 0 & P_{(n-1)2} & \cdots & P_{(n-1)j} & \cdots & P_{(n-1)n} \\ 0 & 0 & \cdots & 0 & \cdots & 0 \end{bmatrix} \end{array} \tag{4.9}$$

4.3 The Description of Fault Propagation in Complex Software Systems

4.3.1 Basic Concepts of Fault Propagation

Fault propagation refers to the process by which failures in individual components lead to system failures. This process can be described as follows: if a defect exists in the code implementation of a component's internal activity—referred to as an internal error—the execution of this code will result in an internal fault within the component. Once this fault reaches the component's interface, it will cause the component to fail. However, if the fault is concealed by other internal operations (fault tolerance), the component will not fail. Similarly, in a component-based system, if a component's fault propagates to the system's interface, it will cause the system to fail; however, if the fault is masked by the instructions of other components (fault tolerance), the system will not fail.

From the perspective of systems science, the propagation of software failures is a complex process that occurs between system entities. These entities include physical components, processes running on one or more CPUs, data objects in databases, functions within programs, or statements in a program. In complex software systems, the failure of one entity, due to internal or external factors, may lead to the failure of other entities, a process known as fault propagation.

This book will provide a detailed discussion of the architecture-based fault propagation model. To begin, we will first briefly explore the architecture-based fault propagation model.

Building upon the architectural model, a set of characteristic metrics is proposed to describe the fault propagation behavior. In the field of software engineering, most classical fault propagation methods are based on fault injection or error injection techniques, combined with further mathematical analysis and estimation. Researchers commonly employ probability models to describe fault propagation, which are well-suited for quantitative analysis. The definition and construction of the relevant parameters in these models reflect the researchers' understanding of the factors that influence fault propagation behavior within the context of software architecture.

In general, studying the fault propagation probability as one of the attributes of the architecture can reflect the probability of an error in component A propagating to component B during runtime, which can be expressed as:

$$\mathrm{EP}(A,B) = \mathrm{Prob}\left([B](x) \neq [B](x') \mid x \neq x'\right) \tag{4.10}$$

The fault propagation probability describes the likelihood of error propagation from component A to component B, under the condition that component A calls component B with a probability of 1. To make the model more reflective of real-world scenarios, we can introduce the unconditional fault propagation probability, which is defined as:

$$E(A,B) = \text{EP}(A,B) \times T(A,B)$$

The unconditional fault propagation probability from component *A* to component *B* is determined by the conditional fault propagation probability, EP(*A*, *B*), and the probability transition matrix, *T*(*A*, *B*). The elements of the probability transition matrix represent the probability of the connector being activated, that is, the probability of component *A* calling component *B*.

Base on the previous analysis, we can introduce the concept of error penetration rate, which represents the probability of an error in the input signal penetrating to the output signal. For each input-output signal pair, there is an associated error penetration rate, which is the conditional probability of an error in the output signal given that an error exists in the input signal. The expression for the error penetration rate is given by:

$$0 \le P_{i,k}^{M} = \Pr\{\text{err in o / p}\, k | \text{err in o / p}\, i\} \le 1 \tag{4.11}$$

Based on the fault penetration rate, the relative penetration rate and unweighted relative penetration rate are proposed. The fault penetration rate applies to a specific input-output pair, whereas the relative penetration rate refers to the probability of a module's fault being transmitted from its input to its output. That is:

$$0 \le P^{M} = \left(\frac{1}{m} \cdot \frac{1}{n}\right) \sum_{i} \sum_{k} P_{i,k}^{M} \le 1 \tag{4.12}$$

The relative penetration rate may not always be able to reflect the overall error penetration rate of a module, but it can serve as an abstract measure to characterize the relative error penetration capability between modules. Therefore, it is necessary to introduce the unweighted relative penetration rate, which is defined as:

$$0 \le P^{M} = \sum_{i} \sum_{k} P_{i,k}^{M} \le m \cdot n \tag{4.13}$$

As another important factor influencing fault propagation, fault tolerance was not reflected in the previous models. However, during the fault propagation process, faults may be masked by fault-tolerant modules within the software, preventing further propagation. Therefore, in addition to the concepts introduced earlier, two parameters—fault intensity and fault tolerance capacity—are necessary to represent the fault tolerance capability of the components.

As observed, the above discussion employs a probabilistic approach to calculate the likelihood of fault propagation, but it is challenging to determine the specific path of fault propagation. Some studies have utilized the ant colony algorithm to identify the most robust fault propagation path by calculating the fault propagation capability. Based on small-world networks, the concept of fault diffusion strength is introduced to represent the ability of faults to propagate along a particular edge. The greater the fault propagation strength, the easier it is for faults to spread along that edge, thereby increasing the scope of their impact.

Indeed, the accuracy and precision of the architecture-based fault propagation model are influenced by the architectural information of the system. Muhammad Shafique et al. modeled both fault masking and fault propagation attributes, while Abdelmoez addressed the issue of fault masking but did not delve deeply into it. In contrast, Muhammad Shafique et al. argued that the fault propagation parameter is a product of the error masking parameter, which depends on structural parameters, control flow graphs, data flow graphs, and the execution probabilities of basic modules. From the perspective of fault masking, the fault propagation probability estimate is derived through experiments, algorithms, and mathematical analysis.

By considering both fault masking and fault propagation attributes, researchers can develop more accurate and comprehensive architectural-based fault propagation models, which can improve the reliability and security of complex software systems.

In addition to using probabilistic models to study fault propagation, some researchers have also applied statistical methods to represent fault propagation models as $y = f(X + Z)$, where f represents the system function or propagation function, X is the input vector of the system, Z is the error vector of the X vector, and y is the output of the system. This method uses mathematical statistics to expand $f(X)$ using Taylor series and derive the mean and variance of y when estimating or calculating the error Z's dispersion distribution.

Devesh Bhatt applied interval mathematics theory and proposed an interval-based data flow model for quantitatively describing fault propagation. This model uses a set of intervals to represent input and output variables, and applies interval arithmetic operations to analyze and propagate the uncertainty and errors in the input variables to the output variables.

These statistical and interval-based methods can provide different perspectives and approaches for studying fault propagation in complex software systems, and can complement the probabilistic-based models mentioned earlier.

4.3.2 Basic Definition of the Fault Propagation Model

This section starts from the perspective of signals and evaluates the reliability of software systems by characterizing the propagation of faults in the architecture. By modeling the propagation of faults in the architectural model, researchers can identify potential failure scenarios and evaluate the impact of failures on the system's overall behavior.

The use of signal-based models provides a systematic and structured approach to studying fault propagation, which can help in identifying the critical components and input-output pairs that need to be further analyzed and tested to ensure the reliability of the system. These models can also be used to develop fault-tolerant strategies and optimize the system's architecture to minimize the impact of failures.

Overall, the signal-based modeling approach is a useful tool for analyzing and mitigating software failures in complex systems, and can provide valuable insights into the fault propagation behavior of the system.

Definition 4.4 Signal: Information related to a component's input and output, including control signals, call signals, data signals, etc.

Definition 4.5 Fault Signal and Normal Signal: A signal is considered a fault signal if its value lies outside the normal value range (as specified by requirements); otherwise, it is classified as a normal signal.

Definition 4.6 Associated Component of Signal *i*: If signal *i* can influence the behavior of a component or affect the signal value produced by component *A*, then *A* is termed the associated component of signal *i*.

Definition 4.7 Reachable Signal: During a single execution of the software, if signal *i* can influence signal *j* through the signals produced by the associated components of *i*, then signal *j* is considered reachable from signal *i*, and signal *j* is termed the reachable signal of *i*.

Definition 4.8 Propagation Path r_{ij} Between Signal *i* and Signal *j*: If signal *i* can reach signal *j*, then the propagation path r_{ij} can be represented by a directed path Pa(C,S,E), where C represents the associated component of signal *i* between signal *i* and signal *j*. S represents the signal transfers between components. E is a set of directed edge, which represents the direction of the signal transferring between signals. In a complex embedded system, there might be more than one path between signal *i* and signal *j*, therefore we specify that comparing to other paths between signal *i* and signal *j*, propagation path r_{ij} is considered a different path if there exists a different signal or component in r_{ij}.

4.3.3 *Fault Propagation Between Two Signals*

Based on the above definitions, a method for representing fault propagation between two signals can be given. Since there are two types of signals: fault and normal, a set $S_i = \{1,0\}$ is defined to represent the state of signal *i*, where S_i is defined by Eq. (4.14):

$$S_i = \begin{Bmatrix} 0(\text{signal } i \text{ is a normal signal}) \\ 1(\text{signal } i \text{ is a fault signal}) \end{Bmatrix} \tag{4.14}$$

For component *A*, if signal *i* is an input signal of *A*, signal *j* is the output signal of *A*, then for the input and output of a component, there will be four kinds of following situations:

1. Scenario 1: $S_i = 0$, $S_j = 0$ indicates that no fault occurred in component *A*, or any internal faults in *A* did not affect its output signal. Both the fault rate and fault tolerance rate are 1.

2. Scenario 2: $S_i = 0$, $S_j = 1$ indicates that signal i caused a fault in component A, resulting in a faulty output signal j. The fault rate of component A increases, while its fault tolerance rate decreases.
3. Scenario 3: $S_i = 1$, $S_j = 0$ indicates that the fault tolerance mechanism within component A corrected the faulty input signal, resulting in a correct output signal. The fault tolerance rate of A increases, while its fault rate remains unchanged.
4. Scenario 4: $S_i = 1$, $S_j = 1$ indicates that component A failed to correct the faulty input signal, producing a faulty output signal. Both the fault rate and fault tolerance rate decrease.

Therefore, by considering these four scenarios and the fault and fault tolerance rates of each component, researchers can evaluate the reliability and fault tolerance capability of the system.

Assume that the fault of a signal is caused solely by component failure. To quantify the four situations mentioned above, we propose the fault rate and fault tolerance rate, as defined below:

Definition 4.9 Fault Rate *f*: The fault rate of component A is the probability of fault output when giving normal input to A. the formular of calculating fault rate of component A is:

$$f_A = \frac{m}{n} \tag{4.15}$$

In the equation, n is the n normal input signals of component A while m is the m fault output signals of component A.
If component A is hardware, then the fault rate of A can be calculated by Eq. (4.16):

$$f_A = 1 - R_A(t) \tag{4.16}$$

$R_A(t)$ is the reliability function of hardware component A.

Definition 4.10 Fault Tolerance Rate *g*: The probability that component A produces a normal output signal when given a fault input signal is called the fault tolerance rate of component A, which is calculated as:

$$g_A = \frac{m}{n} \tag{4.17}$$

where N is the number of fault modes of component A, and m is the number of fault modes that can be handled by the fault tolerance mechanism in component A.
This Eq. (4.17) shows the relationship between the fault tolerance rate and the fault modes of component A. By increasing the number of fault modes that can be handled by the fault tolerance mechanism, the fault tolerance rate of component A can be improved.

Definition 4.11 Component State Transition Matrix M_A: The state transition matrix of a component is represented by a 2 × 2 matrix, defined as:

$$M_A = \begin{bmatrix} S_{00} & S_{01} \\ S_{10} & S_{11} \end{bmatrix} \tag{4.18}$$

The input signal of component A is i, and the output signal is j.

$$S_{00} = \text{Prob}\left(S_j = 0 | S_i = 0\right); S_{01} = \text{Prob}\left(S_j = 1 | S_i = 0\right)$$

$$S_{10} = \text{Prob}\left(S_j = 0 | S_i = 1\right); S_{11} = \text{Prob}\left(S_j = 1 | S_i = 1\right)$$

Therefore, the formula for calculating the state transition matrix is:

$$M_A = \begin{bmatrix} S_{00} & S_{01} \\ S_{10} & S_{11} \end{bmatrix} = \begin{bmatrix} 1 - f_A & f_A \\ g_A & 1 - g_A \end{bmatrix} \tag{4.19}$$

Definition 4.12 Signal State Transition Matrix M_{ij}: If signal i and signal j are reachable, then the signal state transition matrix is defined as:

$$M_{ij} = \begin{bmatrix} \text{Prob}\left(S_j = 0 | S_j = 0\right) & \text{Prob}\left(S_j = 1 | S_i = 0\right) \\ \text{Prob}\left(S_j = 0 | S_i = 1\right) & \text{Prob}\left(S_j = 1 | S_i = 1\right) \end{bmatrix} \tag{4.20}$$

where $\text{Prob}(S_j = 0 | S_i = 0)$ represents the probability that signal j is in a normal state given a normal input signal i, $\text{Prob}(S_j = 1 | S_i = 0)$ represents the probability that signal j is in a fault state given a normal input signal i, $\text{Prob}(S_j = 0 | S_i = 1)$ represents the probability that signal j is in a normal state given a fault input signal i, and $\text{Prob}(S_j = 1 | S_i = 1)$ represents the probability that signal j is in a fault state given a fault input signal i.

This matrix describes how the state of signal j changes based on the state of signal i, which is important in analyzing fault propagation and evaluating the reliability of the system.

Assuming that there are n components $(1, 2,\ldots, n)$ on a certain execution path t_{ij}, the state transition matrix of the path from signal i to signal j can be obtained as:

$$M_{t_{ij}} = M_1 M_2 M_3 \ldots M_n \tag{4.21}$$

For example, when $n = 2$, we have:

$$\begin{aligned} M_{t_{ij}} = M_A \cdot M_B &= \begin{bmatrix} S_{00} & S_{01} \\ S_{10} & S_{11} \end{bmatrix} \begin{bmatrix} S'_{00} & S'_{01} \\ S'_{10} & S'_{11} \end{bmatrix} \\ &= \begin{bmatrix} S_{00}S'_{00} + S_{01}S'_{10} & S_{00}S'_{01} + S_{01}S'_{11} \\ S_{10}S'_{00} + S_{11}S'_{10} & S_{10}S'_{01} + S_{11}S'_{11} \end{bmatrix} \end{aligned} \tag{4.22}$$

Where:

$$S_{00}S'_{00} + S_{01}S'_{10} = \text{Prob}\left(S_j = 0 | S_i = 0\right)$$

$$S_{00}S'_{01} + S_{01}S'_{11} = \text{Prob}\left(S_j = 1 | S_i = 0\right)$$

$$S_{10}S'_{00} + S_{11}S'_{10} = \text{Prob}\left(S_j = 0 | S_i = 1\right)$$

$$S_{10}S'_{01} + S_{11}S'_{11} = \text{Prob}\left(S_j = 1 | S_i = 1\right)$$

4.4 System Reliability Evaluation Model Based on Fault Propagation

This section introduces a reliability evaluation model based on fault propagation (REMBFP), which uses reliability as a measure of software reliability. From the perspective of software users, software reliability can be defined as follows: In a specific user environment, given a typical input set, it is the probability that the software produces the correct output. Based on this definition, the input-output state matrix for components can be defined, taking into account the fault propagation model. Additionally, the potential range and impact of fault propagation on the control transfer path after a component failure can be considered based on the system architecture, and a fault propagation intensity matrix can be defined. Finally, by combining the component input-output state matrix with the system's fault propagation intensity matrix, the system's input-output state matrix can be computed, and the system reliability can be obtained.

4.4.1 Component Input-Output State Matrix

In Sect. 4.3.1, the concept of fault propagation was defined, and the process of component failure leading to system failure was described from the perspective of fault propagation. The process is as follows: If there is a defect in the implementation of a component's internal activity code—an internal error—the error will result in an internal fault of the component when this part of the code is executed. Once this fault reaches the interface of the component, it will cause the failure of the component. However, if the fault is masked by other internal activity instructions, the component will not fail. Similarly, if a component failure occurs within a component system and the fault propagates to the system interface, the component failure will cause system failure, while if the fault is masked by instructions from other components, it will not cause system failure.

Firstly, distinguish between normal input and faulty input for a component. Based on this distinction, the process can be described using the component's input-output states as follows:

1. Normal input and faulty output describe the scenario where an internal defect in the component, when executed, causes a fault that reaches the component's interface.
2. Normal input and normal output describe the scenario where an internal defect in the component, when executed, causes a fault that is masked by other internal activities within the component, preventing the fault from reaching the component's interface.
3. Faulty input and normal output describe the scenario where a fault generated by the component reaches its interface, but the fault is masked by instructions from other components.
4. Faulty input and faulty output describe the scenario where a fault generated by the component reaches its interface and is not masked by instructions from other components. In this case, when the fault propagates to the system interface, it will cause the system to fail.

The traditional architecture-based software reliability analysis model (ABSRM) describes the reliability attributes of components in a relatively simple manner, using a single value to represent the component's reliability. This value is typically a constant reliability or failure rate, or a time-based failure density function. Similarly, in reliability evaluation models that account for fault propagation, a single probability model for fault propagation is often proposed. This model is combined with the original component reliability measure to derive a unified component reliability metric. For example, the concept of corrected component reliability refers to the probability that a component will not fail and will not have faults propagated to it by other components. Additionally, some studies have redefined component reliability as the probability of a component not failing or failing but being corrected by an internal fault-tolerant mechanism.

Building upon the fault propagation model introduced in Chap. 2, the input-output state matrix of a component can be defined to represent the changes in its input-output states. This attribute not only extends the mathematical description of component reliability but also provides a mathematical representation of the fault propagation process discussed earlier, reflecting the quantitative relationship between component failures and system failures. The input-output state matrix of a component is defined as follows:

$$A_i = \begin{bmatrix} \text{Prob}_i(C|C)\,\text{Prob}_i(C|I) \\ \text{Prob}_i(I|C)\ \text{Prob}_i(I|I) \end{bmatrix} = \begin{bmatrix} S_{00} & S_{01} \\ S_{10} & S_{11} \end{bmatrix} \tag{4.23}$$

where:

- $\text{Prob}_i(C|C)$ represents the probability of a normal output of component i given a normal input.

- $Prob_i(C|C)$ represents the probability of a fault output of component i given a normal input.
- $Prob_i(I|C)$ represents the probability of a normal output of component i given a fault input.
- $Prob_i(I|I)$ represents the probability of a fault output of component i given a fault input.

To calculate the elements of matrix A_i, $Prob_i(CI)$ is equal to the failure rate f_i of component i, as defined in Definition 4.9. $Prob_i(IC)$ is equal to the fault tolerance rate g_i of component i, as defined in Definition 4.10. The formula for calculating g_i is:

$$g_i = \frac{\rho_i}{n_i} \tag{4.24}$$

where ρ_i is the probability of a fault occurring in component i, and n_i is the number of fault-tolerant mechanisms implemented in component i.

In this formula, n_i represents the number of possible fault modes in component i, and ρ_i represents the number of fault modes that can be handled by the fault-tolerant mechanism designed for component i. For each component, the possible fault modes can be analyzed during the software design phase, and the number of fault modes that can be addressed by the fault-tolerant mechanism can be obtained from the design documentation. By calculating the component's fault tolerance rate, both of these values can be determined during the software design phase.

The input-output state matrix of component i can be represented as follows:

$$A_i = \begin{bmatrix} 1-f_i & f_i \\ \rho_i & 1-\rho_i \end{bmatrix} \tag{4.25}$$

Based on the input-output state matrix of component i, the reliability R_i of the component can be calculated using the following formula:

$$R_i = P_{ci} * (1-f_i) + P_{ii} * \rho_i \tag{4.26}$$

where P_{ci} represents the probability of a correct input to the component, and P_{ii} represents the probability of an incorrect input to the component. f_i is the failure rate of the component, and ρ_i is the probability of a fault occurring in the component and being handled by the fault-tolerant mechanism designed for the component.

From this equation, we can obtain the input-output state matrix A of the entire system:

$$A = A_1{}^* A_2{}^* \ldots{}^* A_n$$

where A_1, A_2, …, A_n are the input-output state matrices of the components in the system. This formula represents the joint probability of the input-output state of the system, and considers the fault propagation process among components. The calculation of the system's reliability is based on the input-output state matrix A, and the system's reliability can be obtained by calculating the probability of the system providing the correct output given a typical input set.

$$A = \begin{bmatrix} 1-f & f \\ \rho & 1-\rho \end{bmatrix} \tag{4.27}$$

Given that the probability of the correct input to the system is P_c and the probability of the incorrect input is P_i, the reliability R of the system can be calculated using the following formula:

$$R = P_c * (1-f) + P_i * \rho \tag{4.28}$$

where f represents the system failure rate, and ρ represents the probability of a fault occurring in the system and being handled by the fault-tolerant mechanisms of the system. This formula reflects the quantitative relationship between system faults and system failures, and can be used to evaluate the reliability of the system.

The calculation of P_c and P_i is as follows: Let S be the total number of inputs in the typical input set of the system, with S_c being the number of normal inputs and S_i being the number of abnormal inputs. P_c and P_i can be calculated using the following formulas:

$$\begin{cases} P_c = \dfrac{S_c}{S} \\ P_i = \dfrac{S_i}{S} \end{cases} \tag{4.29}$$

where P_c represents the probability of a correct input to the system, and P_i represents the probability of an incorrect input to the system. These probabilities can be calculated based on the proportion of normal and abnormal inputs in the typical input set of the system.

Here is an important property of the input-output state matrix:

Property 4.1: If there are n components N_1, N_2, …, N_n on an execution path, the path can be treated as a whole, and the input-output state matrix of the path can be expressed as the product of the input-output state matrices of the components along the path, i.e., $A_1 \times A_2 \times \ldots \times A_n$, as shown in Fig. 4.2.

This property implies that the input-output state matrix of a system can be calculated by multiplying the input-output state matrices of its components along the execution path. This simplifies the calculation process of the system's reliability and facilitates the analysis of the fault propagation process among components.

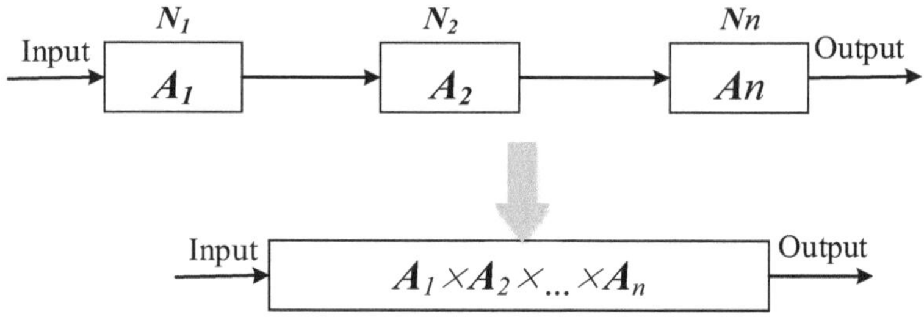

Fig. 4.2 Input-Output state matrix of execution path

4.4.2 *Fault Propagation Strength Matrix*

4.4.2.1 Introduction of Fault Propagation Strength Matrix

The absorbable DTMC to represent the architecture of software can be described not only using the one-step transition probability matrix, as shown in formula (4.8), but also using a control transition graph $G(N,E,P)$. Here N is the set of all nodes, with each node representing a component in the software; E is the set of all connecting edges in the directed graph, with edge (N_i, N_j) representing a possible control transition from component N_i to component N_j; P is the set of all connecting edge weights in the directed graph, representing the probability of control transitions between components, where P_{ij} represents the probability of control transition from component N_i to component N_j.

When a component in the software fails, the fault gradually propagates to other related components. During this propagation process, the fault preferentially spreads along edges with higher control transition probabilities, where these probabilities are represented as elements in the one-step transition probability matrix of the DTMC. However, in addition to considering the control transition probabilities between components, it is also important to take into account the degree of the components. This is because, in practical applications, the risk of low-frequency large-scale failures is comparable to the cumulative risk of high-frequency small-scale failures.

During the evaluation process, the input-output matrix of each component must be combined with the mathematical description of the software architecture. Using matrix theory, the input-output state matrix of the entire software system can be calculated, from which the system reliability can be derived. In general, the input-output state matrix of a control transition path that is difficult to propagate but has a large impact on the system's reliability will have a greater effect on overall system reliability than that of a path that is easy to propagate but has a smaller impact. To account for this impact, fault propagation strength is introduced as the weight of the connecting edges in the control transition graph, which modifies the system architecture model determined in Chap. 2. The larger the propagation strength, the more easily the fault spreads through the edge, increasing the scope of its impact. A simple control transition graph is constructed here to illustrate fault propagation strength, as shown in Fig. 4.3.

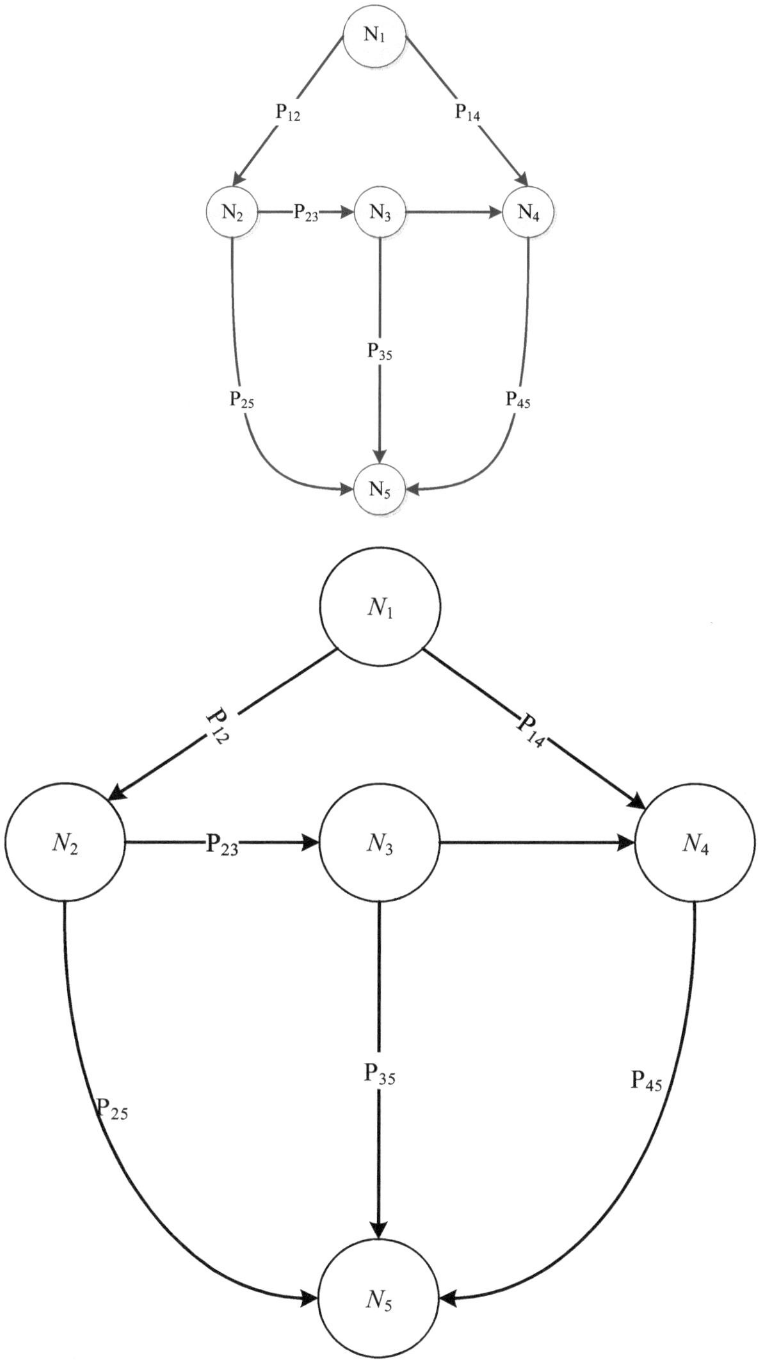

Fig. 4.3 A simple control transition graph

The weight on each edge (edge: denoted by E_i between two components) in Fig. 4.3 represents only the probability of control transition from one component to another, without considering the range of fault signal propagation if an edge E_i fails. However, in practical applications, the risk of low-frequency large-scale faults is comparable to the sum of the risks of high-frequency small-scale faults, which is also closely related to the reliability of the system. Therefore, the degree of each node should also be considered. For example, if we compare $E(N_1, N_2)$ with $E(N_1, N_4)$, assuming that $P_{12} < P_{14}$, and that the degree of node N_2 is greater than that of node N_3, the impact on the system when the fault propagates along $E(N_1, N_2)$ may be greater than that when the fault propagates along $E(N_1, N_4)$. By considering both the control transition probability and the component degree, the definition and calculation formula of fault propagation strength are as follows.

Definition 4.13 Fault Propagation Strength: The weight of the connecting edges in the control transition graph, which represents the probability of fault propagation through this edge and the range of affected components. The larger the fault propagation strength, the easier the fault spreads through this edge and the larger the affected range. The calculation formula is as follows:

$$k_{i,j} = w_p P_{i,j} + w_d d_j / \sum_{l \in F} d_l \tag{4.30}$$

In the formula, w_p and w_d represent the weight coefficients corresponding to the control transition probability and the degree of the node, respectively. F is the set of all components that the fault may propagate to from component i. $P_{i,j}$ represents the probability of control transition from component i to component j after i is executed. d_j represents the degree of component j, which is calculated as the number of possible future control transitions from that component.

Based on the component fault propagation matrix, the Q matrix in formula (4.9) can be transformed into matrix K, which is the fault propagation strength matrix of the software. The calculation formula of matrix K is:

$$K = (1 - Q)^{-1}$$

Where I is the identity matrix. Each element K_{ij} in matrix K represents the probability of component j being affected when component i fails, and calculates the possibility and range of fault propagation between all components. Matrix K can be used to evaluate the reliability and analyze faults of the software, providing strong support for software development and maintenance.

$$K = \begin{matrix} \\ \\ N_1 \\ \vdots \\ N_i \\ \vdots \\ N_{n-1} \\ N_n \end{matrix} \begin{bmatrix} N_1 & N_2 & \cdots & N_j & \cdots & N_n \\ 0 & K_{12} & \cdots & K_{1j} & \cdots & K_{1n} \\ \vdots & \vdots & \cdots & \vdots & \cdots & \vdots \\ 0 & K_{i2} & \cdots & K_{ij} & \cdots & K_{in} \\ \vdots & \vdots & \cdots & \vdots & \cdots & \vdots \\ 0 & K_{(n-1)2} & \cdots & K_{(n-1)j} & \cdots & K_{(n-1)n} \\ 0 & 0 & \cdots & 0 & \cdots & 0 \end{bmatrix} \quad (4.31)$$

The values of all non-zero elements in the matrix are calculated according to formula (4.30).

4.4.2.2 The Concept of Weight in Fault Propagation Strength

In statistical theory and practice, weight refers to a numerical value that indicates the importance of each evaluation index (or evaluation item) and represents the varying roles that each evaluation index plays in the overall assessment. There are various types of weights, each with different mathematical characteristics and meanings. Generally, the following types of weights are identified:

1. According to the form of the weights, they can be divided into absolute weights and relative weights. Relative weights, also known as proportional weights, more intuitively reflect the role of weights in the evaluation.
2. According to the method of weight formation, they can be divided into artificial weights and natural weights. Natural weights are derived from the transformation of statistical data and the synthesis of statistical indicators, also referred to as objective weights. Artificial weights are determined based on the research purpose, the connotation of the evaluation index, and subjective analysis and judgment, reflecting the importance of each indicator. These are also known as subjective weights.
3. According to the numerical characteristics of weight formation, weights can be classified as qualitative weighting and quantitative weighting. The combination of qualitative and quantitative weighting methods in statistical comprehensive evaluation generally yields better results.

There are several methods for determining weights, with the most commonly used methods in practical engineering being the statistical average method, the coefficient of variation method, and the analytic hierarchy process (AHP). The statistical average method calculates the arithmetic average of the relative importance coefficients assigned by selected experts to each evaluation index, with the resulting average serving as the weight for each index. This method falls under artificial weighting methods. The coefficient of variation method directly uses the information contained in each index to calculate the weight of that index. This is considered an objective weighting method. The analytic hierarchy process (AHP), also known as the construction weighting method, organizes the complex evaluation object into a structured hierarchical system. In this system, pairwise comparisons and judgments are made between the evaluation items to calculate the relative importance coefficients, which correspond to the weights of the items.

The concept of component fault propagation strength fully accounts for the range of impact of fault propagation. This concept combines the control transition probabilities between components with the degree of the components, and their respective weights, wp and wd, reflect the relative importance of control transition probability and component degree in influencing system reliability within the current architecture. For a specific architecture, the values of these two weights are fixed.

4.4.2.3 Calculation of Weights in Fault Propagation Strength

The coefficient of variation method is an objective weighting method that directly uses the information contained in each index to calculate its weight. The basic approach of this method is as follows: in the evaluation index system, the greater the difference in the values of the index—in other words, the more difficult the index is to achieve—the more it reflects the differences between the evaluated entities. For example, when evaluating the economic development status of different countries, per capita gross national product (GNP) is chosen as one of the standard evaluation indicators because it not only reflects the economic development level of each country but also indicates the degree of modernization. If the per capita GNP of each country shows little variation, then this index loses its value as a measure of modernization and economic development.

Since the units of measurement for each index in the evaluation system differ, it is not appropriate to directly compare their differences. To eliminate the influence of differing units, the coefficient of variation is used to measure the extent of variation in the values of each index. The formula for the coefficient of variation for each index is:

$$V_i = \frac{\sigma_i}{\overline{x}_i} \quad (i = 1,2,\ldots,m) \tag{4.32}$$

In the formula, V_i represents the coefficient of variation or standard deviation coefficient of the i-th index; σ_i represents the standard deviation of the i-th index; x_i represents the average value of the i-th index; m represents the total number of indicators. The formula for calculating σ_i is:

$$\sigma_i = \sqrt{\frac{\sum_{j=1}^{n}\left(x_{i_j} - \overline{x}_i\right)}{n}} \tag{4.33}$$

where n represents the sample size and x_{ij} represents the value of the i-th index in the j-th sample. The standard deviation calculates the dispersion of the index values around the mean, which reflects the degree of variability of the index values. By using the standard deviation and the mean, the coefficient of variation can be calculated to measure the degree of variability of the index values relative to their mean.

In the formula, x_{ij} represents the different values of the i-th index; n represents the total number of values taken by the i-th index, and this variable may change accordingly depending on the specific index.

Based on the coefficient of variation of each index, the weight of each index can be determined as:

$$W_i = \frac{V_i}{\sum_{i=1}^{m} V_i} \tag{4.34}$$

The formula for calculating the coefficient of variation V_p and the weight W_p is as follows:

$$\begin{cases} V_p = \dfrac{\sigma_p}{\overline{x}_p} \\ W_p = \dfrac{V_p}{V_p + V_d} \end{cases} \tag{4.35}$$

In the formula, V_p and V_d represent the coefficient of variation of $P_{i,j}$ and d_j in formula (4.30) respectively. σ_p represents the standard deviation of $P_{i,j}$ in formula (4.30), while x_p represents the average value of $P_{i,j}$. W_p represents the weight of $P_{i,j}$ in formula (4.30). To calculate σ_p and x_p, firstly the standard deviation and average value of the control transition probability for each component are calculated, and then the overall values are averaged across all components.

$$\begin{cases} V_d = \dfrac{\sigma_d}{\overline{x}_d} \\ W_d = \dfrac{V_d}{V_p + V_d} \end{cases} \tag{4.36}$$

In the formula, V_p and V_d represent the coefficient of variation of $P_{i,j}$ and d_j on the set Fdl in formula (4.30) respectively. σ_d represents the standard deviation of d_j on the set Fdl in formula (4.30), while x_d represents the average value of d_j on the set Fdl. W_d represents the weight of d_j on the set *Fdl* in formula (4.30). To calculate σ_d and x_d, first the out-degree of each component is calculated, and then σ_d and x_d are computed based on these out-degree values, which are used as the standard deviation and average value of d_j on the set Fdl in formula (4.30).

4.4.3 System Reliability Evaluation Method Based on Fault Propagation

Building on the previous two sections, a system reliability evaluation method based on fault propagation is proposed. This method uses the system's reliability as a metric for assessing its overall reliability, incorporating the fault propagation model discussed earlier. As outlined in Sect. 4.1, the key to this method lies in calculating the system's input-output state matrix.

The basic concept of the system reliability evaluation method based on fault propagation can be illustrated using a simple control transition graph, as shown in Fig. 4.4.

From Fig. 4.4, this software can be represented as $G(N,E,P)$, where $N = \{N_1,N_2,N_3,N_4\}$, $E = \{(N_1,N_3),(N_3,N_4),(N_1,N_2),(N_2,N_4)\}$, and $P = \{P_{13},P_{12},P_{34},P_{24}\}$. Assuming that the input-output state matrix for each component is A_1, A_2, A_3, and A_4, respectively, there are two paths in the software, namely $L_1 = (N_1,N_3,N_4)$ and $L_2 = (N_1,N_2,N_4)$, and these two paths are mutually exclusive. Similarly, the diffusion intensity on each edge can also be obtained.

According to Property 4.1 (see Sect. 4.4.1), if we treat paths $L1$ and $L2$ as a single entity, the input-output state matrix for these two paths are $A_1 \times A_3 \times A_4$ and $A_1 \times A_2 \times A_4$, respectively. By considering the fault propagation matrix on each path, the overall input-output state matrix for the software can be expressed as $A = A_1 \times K_{13} \times A_3 \times K_{34} \times A_4 + A_1 \times K_{12} \times A_2 \times K_{24} \times A_4$. The system reliability value can be obtained from matrix A as $R = P_c \times A_{11} + P_i \times A_{21}$. The above calculation process can be transformed into matrix operation.

The fault propagation intensity matrix K' can be derived as:

$$K' = \begin{bmatrix} 0 & K_{12} & K_{13} & 0 \\ 0 & 0 & 0 & K_{24} \\ 0 & 0 & 0 & K_{34} \\ 0 & 0 & 0 & 0 \end{bmatrix} \tag{4.37}$$

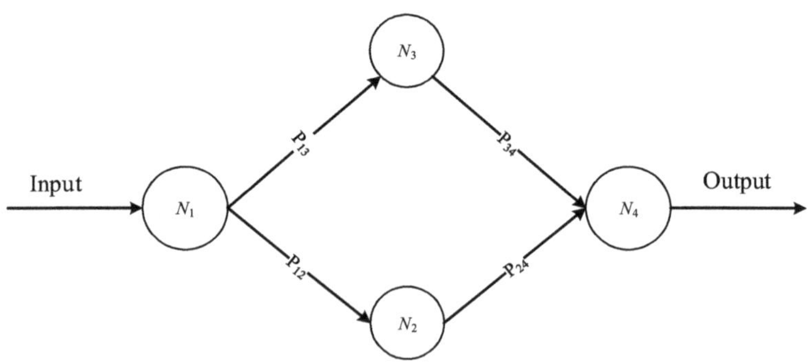

Fig. 4.4 Simple control transition graph

Next, we define a diagonal matrix D', where the elements on the diagonal are the input-output state matrices A_1, A_2, A_3, and A_4 for each component:

$$D' = \begin{bmatrix} A_1 & 0 & 0 & 0 \\ 0 & A_2 & 0 & 0 \\ 0 & 0 & A_3 & 0 \\ 0 & 0 & 0 & A_4 \end{bmatrix} \tag{4.38}$$

This matrix D' is used to represent the input-output states for all components in the system.

Let $Q' = D' \times K'$,then we have:

$$Q' = \begin{bmatrix} 0 & A_1 \cdot K_{12} & A_1 \cdot K_{13} & 0 \\ 0 & 0 & 0 & A_2 \cdot K_{24} \\ 0 & 0 & 0 & A_3 \cdot K_{34} \\ 0 & 0 & 0 & 0 \end{bmatrix} \tag{4.39}$$

Let $S' = Q' + Q'^2$, then we have:

$$S' = \begin{bmatrix} 0 & A_1 \cdot K_{12} & A_1 \cdot K_{13} & A_1 \cdot K_{12} \cdot A_2 \cdot K_{24} + A_1 \cdot K_{13} \cdot A_3 \cdot K_{34} \\ 0 & 0 & 0 & A_2 \cdot K_{24} \\ 0 & 0 & 0 & A_3 \cdot K_{34} \\ 0 & 0 & 0 & 0 \end{bmatrix} \tag{4.40}$$

$S'(i,j)$ takes into account the input-output state matrices of each path from component i to component j, either in two steps or two steps before (excluding the input-output state matrix of component j). which is the summation over each path. Therefore, $S'(i,j) \times A_j$ is the input-output state matrix of component i to component j. Thus, $S'(i,j) \times A_4$ is the input-output matrix for the entire system. This result is consistent with the analysis based on paths as described above.

Generalizing to more complex software structures, according to matrix calculation theory, a more general algorithm can be proposed to compute the overall system reliability.

An $n \times n$ fault propagation intensity matrix is presented, which characterizes the fault propagation intensity between nnn execution components.

$$K = \begin{array}{c} \\ \\ N_1 \\ \vdots \\ N_i \\ \vdots \\ N_{n-1} \\ N_n \end{array} \begin{bmatrix} N_1 & N_2 & \cdots & N_j & \cdots & N_n \\ 0 & K_{12} & \cdots & K_{1j} & \cdots & K_{1n} \\ \vdots & \vdots & \cdots & \vdots & \cdots & \vdots \\ 0 & K_{i2} & \cdots & K_{ij} & \cdots & K_{in} \\ \vdots & \vdots & \cdots & \vdots & \cdots & \vdots \\ 0 & K_{(n-1)2} & \cdots & K_{(n-1)j} & \cdots & K_{(n-1)n} \\ 0 & 0 & \cdots & 0 & \cdots & 0 \end{bmatrix} \tag{4.41}$$

In order to obtain the input-output state matrices for paths and the entire system through matrix operations, the above matrix needs to be transformed by multiplying each component's input-output state matrix with the corresponding row in the matrix, resulting in a $2n \times 2n$ matrix. To do this, we define a matrix $D_{n \times n}$, where $D_{n \times n}(i, i)$ is the input-output state matrix A_i for component i. Multiplying this matrix with the fault propagation intensity matrix for the system yields the following $2n \times 2n$ matrix, K_A:

$$K_A = \begin{array}{c} \\ \\ N_1 \\ \vdots \\ N_i \\ \vdots \\ N_{n-1} \\ N_n \end{array} \begin{bmatrix} N_1 & N_2 & \cdots & N_j & \cdots & N_n \\ 0 & A_1K_{12} & \cdots & A_1K_{1j} & \cdots & A_1K_{1n} \\ \vdots & \vdots & \cdots & \vdots & \cdots & \vdots \\ 0 & A_iK_{i2} & \cdots & A_iK_{ij} & \cdots & A_iK_{in} \\ \vdots & \vdots & \cdots & \vdots & \cdots & \vdots \\ 0 & A_{(n-1)}K_{(n-1)2} & \cdots & A_{(n-1)}K_{(n-1)j} & \cdots & A_{(n-1)}K_{(n-1)n} \\ 0 & 0 & \cdots & 0 & \cdots & 0 \end{bmatrix} \tag{4.42}$$

Treating the input-output state matrix A_i in matrix K_A as a whole, the element $K_A{}^k(i,j)$ in the k-step transition matrix $K_A{}^k$ represents the sum of the input-output state matrices for paths from component i to component j (excluding component j) after k steps. If we let the Neumann series $S_A = E + K_A + K_A{}^2 + K_A{}^3 + \ldots = \sum_{k=0}^{\infty} K_A{}^k$, then $S_A(i,j)$ represents the sum of the input-output state matrices for all paths from component i to component j (excluding component j) reached by several control transfers starting from component i.

Since the values of the elements in matrix K_A are all within the range of [0,1], and the sum of each row element is no >1, this ensures the spectral radius of matrix K_A, $\rho(K_A)$, is <1. Therefore, it can be concluded that the Neumann series $S_A = \sum_{k=0}^{\infty} K_A{}^k$ converges and satisfies:

$$S_A = \sum_{k=0}^{\infty} K_A{}^k = (E - K_A)^{-1} \tag{4.43}$$

Based on the above, it can be concluded that the input-output state matrix of the system is expressed as:

$$A = S_A(1,n) \times A_n \tag{4.44}$$

Here, $S_A(1, n)$ represents the sum of the input-output state matrices for all paths from component 1 to component n (excluding component n) reached by several control transfers starting from component 1, and A_n is the input-output state matrix for component n.

In this equation, A represents the input-output state matrix for the entire system; $S_A = (E - K_A)^{-1}$; A_n is the input-output state matrix for component n.

By utilizing the input-output state matrix of the system, along with Eqs. (4.28) and (4.29), the system reliability can be determined as:

$$R = P_c * A_{11} + P_i * A_{21} \tag{4.45}$$

Here, A_{11} is the input-output state matrix for component 1, and A_{21} is the input-output state matrix for component 2.

The algorithm mentioned above can be summarized as following six steps:

1. Let $K(n, n) = 0$
2. Defined $D_{n \times n}$, where $D_{n \times n}(i, i) = A$
3. Let $K_A = D \times K$
4. $S_A = (E - K_A)^{-1}$
5. $A = S_A(1, n) \times A_n$
6. $R = P_c * A_{11} + P_i * A_{21}$

This chapter provides a case study of implementing the above algorithm using the Python programming language.

4.5 A Case Study

To clearly illustrate the methods presented in this chapter, an example is analyzed in this section. Due to space limitations, a relatively simple software system is selected to demonstrate the feasibility and implementation process of the methods. Additionally, two other methods are included for comparison to highlight the approach discussed in this chapter. The quality of the results does not necessarily reflect the superiority or inferiority of the methods, but rather aims to help readers better understand the differences between these techniques.

4.5.1 Introduction to the Example Software

The software selected for this example is a digital media file transmission application. It utilizes satellite communication as the transmission method, enabling the reception, processing, and feedback of media program reception status (via automatic dialing through modems). The system supports unmanned, automated recording, with the recorded results either being automatically restored to magnetic tapes required by analog broadcasting systems, or seamlessly integrated with mainstream non-linear digital broadcasting systems at the receiver's end. This also significantly reduces the workload on the receiving side.

Figure 4.5 presents the external interface of the media recording software.

The main functional modules of this software include the authorization parsing module, data file merging module, command execution module, file cleaning module, recording module, log recording module, viewing module, and information feedback module. Consequently, the software is divided into eight components based on these functional modules, with component numbers ranging from 1 to 8.

Based on expert evaluations, the reliability of each component can be determined, with the specific data provided in Table 4.2.

Similarly, based on the software's operational profile and expert experience, the transition probabilities between components can be determined. These transition probabilities serve as the data foundation for subsequent model calculations, with the specific data listed in Table 4.3.

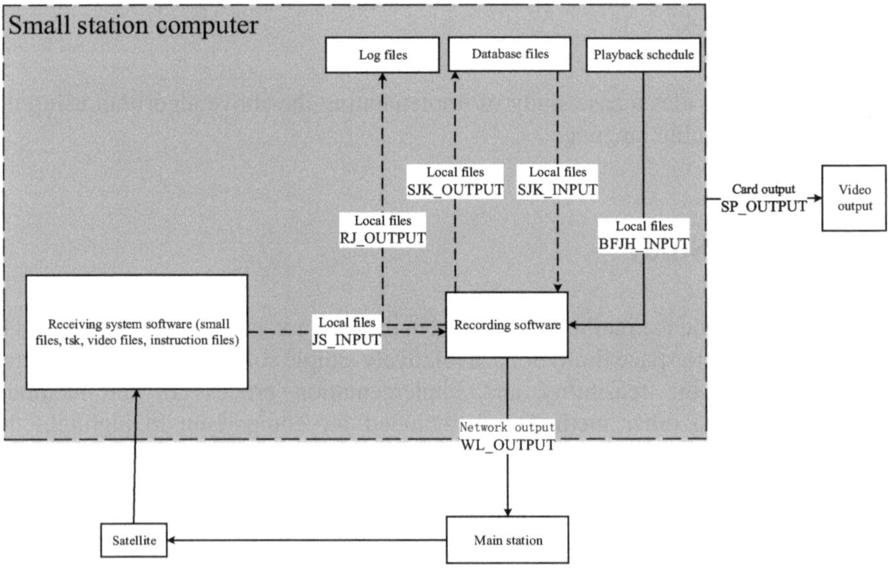

Fig. 4.5 The external interface of the media recording software

Table 4.2 Reliability of each component

Component	Reliability
1	0.998
2	0.996
3	0.998
4	0.857
5	0.997
6	0.925
7	0.996
8	0.709

Table 4.3 Transition probabilities between components

Components	1	2	3	4	5	6	7	8
1	0	0.3	0.05	0.05	0.3	0.2	0.1	0
2	0	0	0.05	0.05	0.56	0.2	0.14	0
3	0	0.225	0	0.1	0.225	0.225	0.225	0
4	0	0.35	0.05	0	0.2	0.23	0.17	0
5	0	0	0.1	0.05	0	0.3	0.25	0.3
6	0	0.25	0.05	0.05	0.25	0	0.1	0.3
7	0	0.2	0.1	0.05	0.25	0.2	0	0.2
8	0	0	0	0	0	0	0	0

4.5.2 Exponential Distribution Model Based on Failure Data

Typically, during the stable usage phase of software, if defects are not addressed once discovered, the software's entire lifecycle follows an exponential distribution. In 1983, Misra applied the exponential distribution model to estimate the defect occurrence rate for the ground system software of the U.S. Space Shuttle. In this section, this model, along with failure data obtained during the testing process, is used to assess the reliability of the tested software. The results from this model serve as the benchmark for comparing the results of the Cheung model and the evaluation model proposed in this chapter, thereby determining which model's results are closer to the benchmark.

The basic assumptions of this model are as follows:

1. The workload throughout the entire testing phase is uniformly distributed.
2. Defects that have been discovered are not corrected.
3. The i-th software defect follows an exponential distribution lifetime, with the reliability given by $R(t_i) = e^{-\lambda t_i}$, where λ is the parameter, and t_i represents the time between the $(i-1)$th and the i-th failure.
4. The software operates in the manner it is expected to.
5. All errors have the same severity level, and the probability of error occurrence is identical.
6. Failures are independent of each other.

7. Data requirements: complete failure data, specifically the failure interval time or cumulative failure time.

Under the above assumptions, using the methods of reliability engineering, the time x_i when the i-th failure occurs, starting from the $(i-1)$th failure, is a random variable that follows an exponential distribution with λ as the parameter. Its density function is

$$f(x_i) = \lambda e^{-\lambda x_i} \tag{4.46}$$

Assuming there are n failures in the software, the maximum likelihood function is

$$L(x_1,\ldots,x_n) = \prod_{i=1}^{n} f(x_i) \tag{4.47}$$

Taking the logarithm on both sides of Eq. (4.47) and using the maximum likelihood estimation method, the estimated value of the model parameter can be obtained as:

$$\begin{cases} \dfrac{d\ln L}{d\lambda} = \dfrac{n}{\lambda} - \sum\limits_{i=1}^{n} x_i = 0 \\ \hat{\lambda} = n / \sum\limits_{i=1}^{n} x_i \end{cases} \tag{4.48}$$

Where $x_i = t_i - t_{i-1}$, which is the failure interval time.

From this, the estimated values of the reliability parameters can be obtained as follows.

Fault rate is:

$$Z(x) = \hat{\lambda} \tag{4.49}$$

Reliability function is:

$$R(t) = \exp\left(-\hat{\lambda} t\right) \tag{4.50}$$

In this example, the software was tested using reliability testing methods, and the resulting data is presented in Table 4.4.

Based on the model calculation method and the failure data of the tested software, the failure rate of the tested software can be calculated as $\hat{\lambda} = 0.00395$. Therefore, the reliability function of the tested software is $R(t) = \exp(-0.00395t)$. Knowing that the expected execution time of the tested software is $t = 60$ s, this value can be substituted into the reliability function as an estimate of the overall reliability of the software within the specified operating conditions, and the reliability of the software can be calculated as 0.7890.

Table 4.4 Failure data of the tested software

Failure number	Failure interval time	Failure number	Failure interval time	Failure number	Failure interval time
1	165	11	172.2	21	417.3
2	122.2	12	90.5	22	298.5
3	149.8	13	619.8	23	294.2
4	147.7	14	266.7	24	317.9
5	258.8	15	342.1	25	276.7
6	114.9	16	106.8	26	604.2
7	13.3	17	46.5	27	133.3
8	332.5	18	66.2	28	214.2
9	237.9	19	158.2	29	267.6
10	390.2	20	723.9	30	

4.5.3 Cheung Model

The Cheung model is a classical architecture-based reliability assessment model, proposed by Cheung in 1980, and is a software reliability model based on Markov chains. This model assumes that erroneous outputs from one component are not tolerated by downstream components. The specific assumptions are as follows:

1. The reliability between components is independent of each other.
2. The control transfer process between components is a Markov process.
3. There is a completely reliable connection logic between components.

In the Cheung model, a directed graph GGG is used to represent the control flow structure of the software, where each node N_i in the graph corresponds to a component. A directed edge (N_i,N_j) represents a control transfer from component N_i to component N_j. For each directed edge (N_i,N_j) in the graph, the probability P_{ij} used in the Cheung model denotes the probability of the control transfer (N_i,N_j) occurring when the current executing component is N_i. The transfer probability P_{ij} represents the branching characteristics at the output of component N_i. Let R_i denote the reliability of component N_i. Based on the directed graph, the model constructs the component transition probability matrix MMM for the software, where $M(i,j) = R_iP_{ij}$. The elements in the transition probability matrix represent the probability of producing a correct output from the currently executing component and transferring control to the next component.

The calculation process of this model is shown below:

1. Let $M(n,n) = 0$
2. Calculation $S_{n\times n} = (I - M)^{-1}$
3. $R = S(1,n) \times R_n$

Where $R = S(1,n) \times R_n$ represents the reliability of the software, which indicates the probability of the software generating the correct output through a series of control transitions starting from the initial component N_i. Based on the component

Table 4.5 The component transition probability matrix

Component	1	2	3	4	5	6	7	8
1	0	0.2994	0.0499	0.0499	0.2994	0.1996	0.0998	0
2	0	0	0.0498	0.0498	0.5578	0.1992	0.1395	0
3	0	0.2246	0	0.0998	0.2246	0.2246	0.2246	0
4	0	0.3	0.0429	0	0.1714	0.1971	0.1457	0
5	0	0	0.0997	0.0499	0	0.2991	0.2493	0.2991
6	0	0.2312	0.0463	0.0463	0.2313	0	0.0925	0.2775
7	0	0.1992	0.0996	0.0498	0.249	0.1992	0	0.1992
8	0	0	0	0	0	0	0	0

reliabilities provided in Table 4.2 and the transition probabilities between components in Table 4.3, the component transition probability matrix M of the software can be derived, as shown in Table 4.5.

The above calculation process can be implemented using Python code. After running the program, the reliability of the tested software is obtained as 0.6129.

4.5.4 System Reliability Evaluation Model Based on Fault Propagation

After analyzing the relationships between components in the tested software system, the in-degree and out-degree of each component can be determined. By combining the component transition probabilities from Table 4.3 with Eq. (4.30), the fault propagation intensity matrix of the tested software can be derived. The fault propagation intensity matrix of the tested software is presented in Table 4.6.

First, based on historical test data and expert experience, the error rate for each component is determined. According to expert assessments, the specific values for the error rate and fault tolerance rate of each component in the tested software are provided in Table 4.7.

According to Table 4.7 and the formula, the input-output state matrix of each component in the tested software can be calculated.

Firstly, let $k_{n \times n} = 0$, then use a Python program to implement steps 2–4 in the algorithm, and finally, according to the formula $A = S_{1,n} \times A_n$, the input-output state matrix A of the system can be obtained, which is:

$$A = \begin{bmatrix} 0.7322 & 0.2678 \\ 0.7322 & 0.2678 \end{bmatrix}$$

In this example, the ratio of the error rate f_i to the fault tolerance rate g_i of each component is very close, so for the elements in the input-output state matrix A_i of each component, the ratio between the first row elements $A_i(1,1)$ and $A_i(1,2)$ is very close to the ratio between the second row elements $A_i(2,1)$ and $A_i(2,2)$. Moreover,

Table 4.6 The fault propagation intensity matrix of the tested software

Component	1	2	3	4	5	6	7	8
1	0	0.22	0.11	0.11	0.23	0.19	0.14	0
2	0	0	0.12	0.12	0.38	0.21	0.18	0
3	0	0.20	0	0.15	0.21	0.22	0.22	0
4	0	0.26	0.12	0	0.20	0.22	0.20	0
5	0	0	0.16	0.14	0	0.27	0.25	0.18
6	0	0.21	0.12	0.12	0.22	0	0.15	0.18
7	0	0.19	0.14	0.12	0.22	0.20	0	0.13
8	0	0	0	0	0	0	0	0

Table 4.7 Error rate and fault tolerance rate of each component

Component	Error rate	Fault tolerance rate
1	0.002	0.998
2	0.006	0.997
3	0.003	0.994
4	0.250	0.976
5	0.081	0.571
6	0.043	0.778
7	0.003	0.978
8	0.260	0.608

since the input-output state matrix *A*' of a path is the product of the input-output state matrix of all components on that path, the ratio between $\frac{A'(1,1)}{A'(1,2)}$ and $\frac{A'(2,1)}{A'(2,2)}$ is also very close. This will result in the ratio of $\frac{A(1,1)}{A(1,2)}$ and $\frac{A(2,1)}{A(2,2)}$ in the input-output state matrix *A* of the entire system (the sum of the input-output state matrices of all possible paths) being very close or even the same. Therefore, although the first and second rows of the input-output state matrix *A* of the system calculated in this example are the same, it is still reasonable.

In the actual project, 181 test cases were designed, including 56 normal test cases and 125 abnormal test cases. Therefore, $P_c = 0.3$, $P_i = 0.7$.

Substituting the values into the input-output state matrix, the reliability of the tested software can be calculated as $R = P_c * A_{100} + P_i * A_{10} = 0.7322$.

The following two conclusions can be drawn from the above experiment:

1. Architecture-based reliability evaluation is only an auxiliary method for reliability evaluation and can only be applied in the early stages of software development. Therefore, the evaluation results may not be very accurate.
2. Similarly, according to the results calculated by the model, it can be seen that the evaluation model considering fault tolerance and fault propagation obtained in this experiment is closer to the results of the exponential distribution model calculation than the evaluation method assuming independence between component reliabilities.

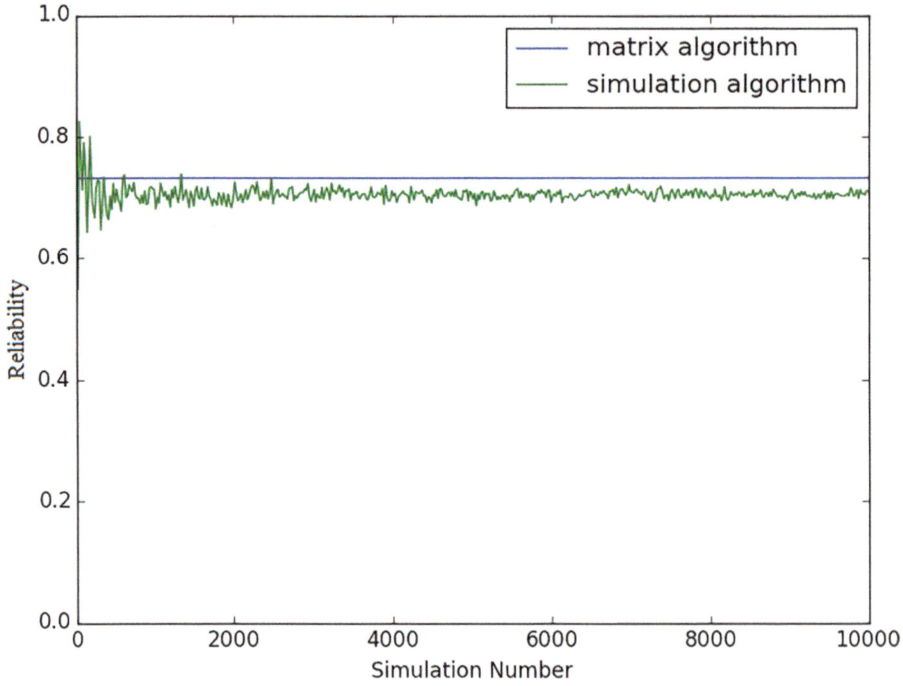

Fig. 4.6 Simulation experiment results

4.5.5 Comparison Through Simulation Experiment

In this section, a system simulation algorithm is used to conduct simulation experiments on the tested software. The simulation results are compared with the model calculation results from Sect. 4.5.4 to verify the system reliability evaluation method based on the fault propagation model proposed in this chapter. The fault propagation intensity matrix of the tested software and the input-output state model of each component can be computed based on Tables 4.6 and 4.7. Using these data and the simulation algorithm introduced in this chapter, 500 simulation experiments were performed on the tested software, with the number of simulations increasing by 20 in equal increments each time, leading to a maximum of 10,000 simulations. The simulation results are shown in Fig. 4.6.

As shown in the figure, when the number of simulations is small, the results are unstable and exhibit significant fluctuations. As the number of simulations in each experiment increases, the results progressively align more closely with the theoretical calculations of the algorithm proposed in this chapter. This confirms the validity of the system reliability evaluation model based on the fault propagation model introduced in this chapter.

Analysis of the errors in simulation experiment result:

The error in the simulation experiment results arises because the number of simulations in each experiment is finite, and it is impossible to conduct an infinite number of experiments to cover all possible execution paths. As a result, when there are execution paths with very low probabilities, it is highly likely that the simulation will not cover all of these paths, leading to discrepancies between the simulation results and the theoretical calculations.

4.6 Chapter Summary

Conducting software reliability analysis and evaluation based on architecture aligns with systems thinking. The reliability of a software system is closely related to failures in software behavior, which, in turn, is determined by its architecture. Identifying the factors that influence and constrain software reliability within the software architecture, along with their underlying patterns, and subsequently analyzing or evaluating reliability based on these findings, represents a robust technical approach. The method introduced in this chapter is developed by examining the propagation patterns of faults within the architecture, serving as a valuable exploration and research direction.

Chapter 5
Data-Based Assessment of Complex Software Reliability

With the advancement of information technology, software products have gradually become an indispensable part of both production and daily life. As the size and complexity of software products continue to grow, the management of software quality and reliability becomes increasingly challenging. This is particularly true in industries such as aerospace, transportation, and healthcare, where software quality issues arising from increased complexity may pose significant risks to human life, property, and even national security.

Software reliability prediction is a qualitative and quantitative prediction technique for software reliability indicators. Currently, there are two commonly used methods: one is to construct an analytical model through mathematical fitting based on software testing data, which mainly collects data and establishes models through reliability growth experiments. The other method is based on software measurement. First, software measurement data is obtained during the software development/testing process, and then appropriate calculation methods are selected to calculate the relevant indicators of software reliability.

The first method mentioned above is based on the software reliability growth model. In reliability growth experiments, software failures are observed, and the experimenters analyze these failures to identify their causes, which are typically defects within the software. By utilizing the defect data collected, the development, usage, and debugging of the software are guided, leading to a continuous improvement in system reliability.

In the early days, software reliability assessment mainly focused on predicting software defects. This is due to the fuzziness of the concept of software quality, and researchers focused only on software defects related to software quality. Later, as software quality models such as the McCall software quality model and the Boehm software quality model were proposed, the concepts of software quality and reliability gradually became clearer. The second method mentioned above is based on such quality models.

The ISO/IEC 9126 software quality model, which currently serves as the international standard for software quality evaluation, along with the latest ISO/IEC

Y. Wang et al., *Reliability Technology for Complex Software Systems*,
https://doi.org/10.1007/978-981-95-3889-8_5

25010 software quality model, has been widely applied in research related to software quality evaluation. In China, the ISO/IEC 14598 and ISO/IEC 9126 standards were introduced in 2002 and 2006, respectively. As a result, national standards such as GB/T 18905 "Software Engineering—Product Evaluation" and GB/T 16260 "Software Engineering—Product Quality" were developed. Following 2010, China further established a series of standards equivalent to the adoption of SQuaRE, including the GB/T 25000 series standards.

Many software quality models provide specific methods for evaluating and measuring software reliability. Software quality standards typically present a hierarchical model of software quality. These models decompose software quality into characteristics, sub-characteristics, and measurement elements. By calculating or counting the measurement elements, the score for a particular sub-characteristic of the software is determined. Then, by applying weighted values at each level, the overall software quality evaluation result is obtained.

For example, GB/T 25000-10:2016 "Systems and Software Engineering—Systems and Software Quality Requirements and Evaluation (SQuaRE)—Part 10: System Quality" defines software reliability as "the ability of a system, product, or component to perform its specified functions under specified conditions within a specified period of time." Reliability is further decomposed into maturity, availability, fault tolerance, recoverability, and compliance. Maturity is defined as "the degree to which a system, product, or component meets reliability requirements during normal operation," and is further broken down into sub-items such as "the difficulty of detecting defects," "the difficulty of troubleshooting failures," and "the proportion of test cases executed in virtual environments," among others. Specific measurement elements are provided under these sub-items. By calculating the scores for these sub-items and applying a weighted aggregation at each level, the overall software reliability score is determined. Other software quality models also employ similar methods for evaluating the reliability of software products.

For complex systems, the significance of software quality and reliability, coupled with the challenges of conducting reliability evaluations based on quality standards, has become a primary contradiction. The main difficulties in evaluating the quality of complex software based on quality standards are as follows:

1. The complexity of software systems: Complex software systems are difficult to evaluate due to their inherent complexity and the interactions between different components.
2. Lack of comprehensive and reliable data: The reliability evaluation of software systems relies on comprehensive and reliable data, and it is often difficult to collect and analyze such data effectively.
3. Changing software requirements: Software requirements are often subject to changes during the development process, which makes it difficult to establish reliable evaluation criteria.
4. The subjectivity of software quality: Software quality is a subjective concept, and different stakeholders may have different opinions on what constitutes high-quality software.

5. The need for specialized knowledge: The evaluation of software quality requires specialized knowledge and expertise, which may not be readily available to all stakeholders involved in the evaluation process."

   (a) How to define and break down the reliability of complex software is still under debate and there is no consensus on whether existing standards should be followed. Although there are both general quality standards, such as the GB/T 25000 series, and quality standards specific to particular areas, such as GB/T 30961-2014 "Embedded Software Quality Measurement," there is still a lack of quality standards for complex software systems. There is currently no unified standard for the quality evaluation dimensions and evaluation timing of complex software systems, which makes quality evaluation work difficult for complex software systems.
   (b) The feedback for improving development activities based on reliability evaluation results is weak. The definition of quality measurement elements (sometimes also referred to as "measurement elements") is mostly in the form of the ratio of two measurement items. For example, the maturity, one of the sub-characteristics of reliability, is measured using the fault repair rate, mean time between failures (MTBF), periodic failure rate, and test coverage rate. This evaluation method requires software reliability testing activities, and the results of the evaluation cannot provide improvement suggestions to software developers, as developers cannot directly improve reliability indicators such as the fault repair rate.
   (c) The measurement elements defined by the standards do not provide sufficient coverage of software reliability. In software development, a significant amount of software measurement data is generated. From a temporal perspective, activities during the software demonstration, requirements engineering, software design, coding, testing, usage, and maintenance phases all influence the quality and reliability of the software product. In fact, some of the products resulting from these activities can be considered part of the software product itself, and these activities generate substantial amounts of measurement data. However, relying on software quality standards for reliability evaluation only requires the collection of quality measurement elements as defined by the standards. This approach neglects many factors that could impact software reliability, thereby compromising the accuracy of the reliability evaluations and diminishing the effectiveness of the feedback provided to the development process.

Given the issues outlined above, this chapter will introduce methods for performing reliability evaluations based on data. The overall approach to evaluating software reliability based on data is to employ data analysis techniques to uncover the quantitative relationships between software metrics and software reliability, while identifying the most significant factors that impact software reliability.

A data-driven software reliability evaluation is a hierarchical evaluation system that assesses software reliability by examining software metrics, software quality characteristics, and ultimately software quality. Establishing a robust software reliability evaluation system plays a significant role in ensuring software reliability. The results of software quality and reliability assessments are critical for identifying weak areas in software quality and for resource allocation throughout the software lifecycle. Moreover, software product users rely on these evaluation results as the foundation for several important decisions, such as improving product quality, making large-scale purchases, or monitoring contract performance. Consequently, a well-established software reliability evaluation system not only helps reduce software development and testing costs but also enhances the quality and reliability of software products, thereby increasing customer trust and market competitiveness. Additionally, it can raise awareness among both industry professionals and users regarding the importance of software quality.

Due to the inherent characteristics of data-driven reliability evaluation models, these models can not only assess the reliability of software products but also identify specific metrics that influence software reliability through the analysis of evaluation results, thus facilitating improvements in software reliability. However, since there are significant differences in reliability evaluation across different organizations and types of software products, these differences are reflected both in the characteristics of the software's measurement data and in the specific needs for software reliability evaluation. Therefore, it is essential to collect historical data (for training data models) and develop evaluation models tailored to the specific context.

5.1 Data-Driven Evaluation Model Framework

The previous section discussed the limitations of traditional software reliability assessments and those based on software quality standards. The following will provide a comprehensive description of the data-driven software reliability assessment model.

As shown in Fig. 5.1, data-driven reliability assessment can be broadly divided into four parts. The process involves four main stages: data collection and preprocessing, feature selection and extraction, predictive model establishment, and reliability assessment and optimization. Data collection and preprocessing serve as the foundation of the entire process, and it primarily involves steps such as identifying data sources, collecting data, data cleaning, data transformation, and dataset partitioning. Feature selection and extraction aim to identify and extract features from the dataset that are relevant to software reliability. This typically includes steps like feature selection, feature transformation, and feature construction. The establishment of a predictive model focuses on forecasting and analyzing software reliability, with commonly used models including regression models, classification models, and clustering models. Finally, reliability assessment and optimization involve evaluating and refining the predictive model's performance, which typically encompasses model evaluation, model selection, and model optimization steps. By following these steps, data-driven reliability assessment can more accurately predict and

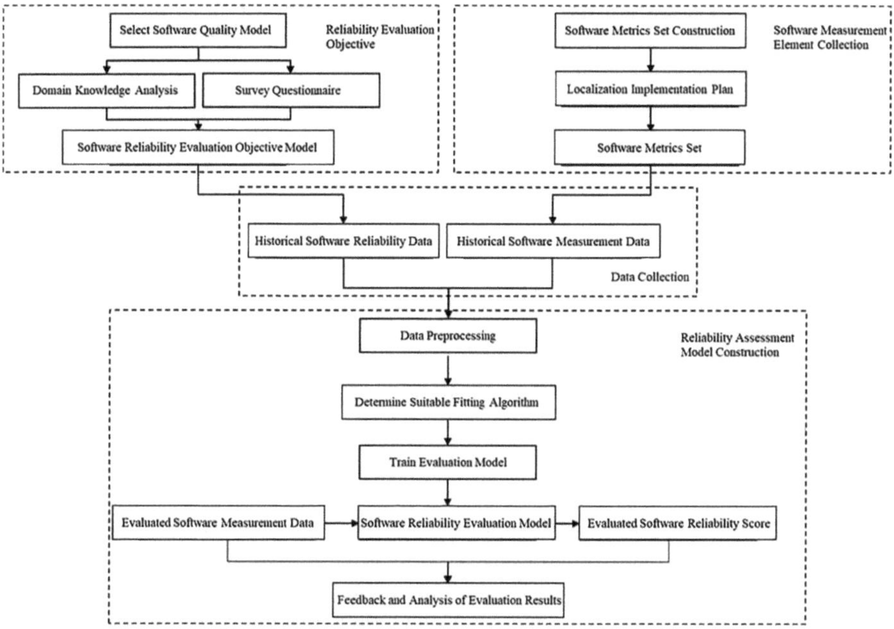

Fig. 5.1 Overall Framework of Data-Driven Software Reliability Assessment

evaluate software reliability, thereby improving the efficiency and quality of software development and testing.

1. Construction of Reliability Evaluation Objectives

   In data-driven software reliability assessment, the first step is to construct the reliability evaluation objectives. As mentioned earlier in this chapter, different software development organizations and software product types have varying objectives for reliability evaluation. Regarding quality standards, GB/T 25000.10-2016 "Systems and Software Engineering—Systems and Software Quality Requirements and Evaluation (SQuaRE)—Part 10: System" decomposes software reliability into five sub-characteristics: maturity, availability, fault tolerance, recoverability, and compliance. This represents a universal model for the decomposition of software reliability. However, in GB/T 30961-2014 "Embedded Software Quality Metrics," the sub-characteristic "availability" is excluded, likely due to the inherent characteristics of embedded software products. Other standards also define sub-characteristics differently.

   Moreover, different software development organizations and product types have distinct understandings and definitions of software reliability, as well as differing needs for reliability assessment. These differences primarily manifest in two ways: first, in the evaluation objectives—what factors are considered to constitute software reliability; and second, in the weights assigned to the sub-characteristics—certain sub-characteristics are more important for different types of software. Therefore, in data-driven software reliability evaluation, the first task is to construct the objective model for reliability evaluation. The purpose of this model is to define the evaluation objectives within a specific scope, that is, "what is being evaluated?"

The objective model typically includes the elements to be evaluated, which are usually the sub-characteristics of software reliability. These sub-characteristics can be derived from a quality standard, engineering expertise, or practical requirements. Additionally, the objective model should also define the weights corresponding to these sub-characteristics, which serve as the basis for weighted aggregation, ultimately leading to the calculation of the software reliability assessment score. This part of the model is generally obtained through methods such as analysis, interviews, questionnaires, and other approaches (as shown in Fig. 5.2).

2. The second part of data-driven software reliability assessment involves the construction of a software metric element set. Software metrics refer to quantifiable indicators of software products, development processes, and their inherent properties, which can be utilized in software quality management to evaluate and enhance software quality. Software metrics can be understood as attributes of the software product itself, analogous to a person's height, weight, or age. These metrics are typically obtained through the measurement of the software product or process. Some metrics, such as defect density, are considered to directly reflect software quality or reliability, whereas the relationship between most software metrics and software reliability remains unclear and requires further data collection and analysis to be determined. The preliminary construction of a software metric element set serves as the foundation for software project metrics. In practical implementation, not all projects or organizations have the capability to collect such extensive metric data; therefore, localized measures must be adopted to develop tailored implementation plans that support the actual collection of software metrics.

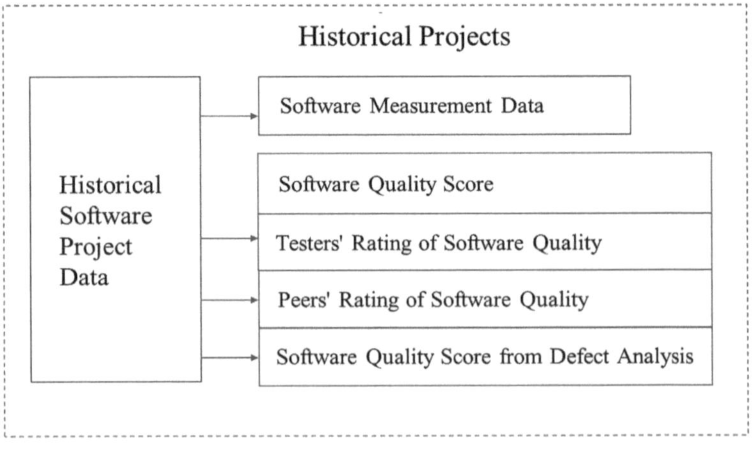

Fig. 5.2 Data Source of Historical Project

3. The third part involves the collection of actual software historical data based on the reliability assessment target model and the customized metric collection plan developed in the first two parts. The sources of software metrics typically include the software's source code, process documentation, project control data, and so on. The sources for reliability assessment targets are historical reliability data from previous projects, which may relate to defects discovered during software testing or the results of reliability testing conducted on past projects. In projects that lack reliability testing, such data are often collected through ratings provided by relevant stakeholders. Experience shows that in practice, software reliability assessments based on historical data typically spend the majority of their time collecting this data. This phase often faces issues such as insufficient data volume, low data quality, and disorganized data formats. Current software engineering activities provide weak support for metric element sets, with software development organizations focusing primarily on the final product and neglecting the importance of process products and data. As a result, many specific issues arise during actual data collection, which require detailed analysis and resolution.
4. After completing the data collection for historical projects (a process that typically accounts for about 70% of the time in quality assessment projects), the collected data must be processed and analyzed. The goal is to establish a quantitative relationship between historical project software metric data and reliability data through training on historical data. Project reliability data is typically available at the project level, and in some cases, it may need to be obtained at the system level. In contrast, metric data differs in that code metrics are usually defined at the function, file, or project level. For object-oriented software, these metrics are typically defined at the method, class, or project level, meaning that for a software project, its code metric elements may be presented as a very large matrix. Process metrics usually depend on the process control systems of individual organizations. Since the data is collected from historical projects, it is necessary to consult the organization's process database to obtain the corresponding process metrics. Document metrics, or process product metrics, are typically the most challenging for reliability assessors due to their limited understanding of specific projects. Document measurement often requires the help and continuous communication with project stakeholders. Furthermore, due to variations in organizational rules and styles for writing documents, the results of document measurements can exhibit domain-specific characteristics.

The description of software metrics above is intended to highlight one issue: the actual software metric data obtained often has a very complex data structure, which contrasts sharply with the reliability historical data at the other end of the evaluation model. Code metrics may be represented as a matrix with thousands of elements, each differing significantly in scale. Process metrics may consist of numbers that are not particularly realistic. Document metrics may take the form of defect reports shaped by the organization's own characteristics. On the other hand, the reliability data output by the model might only consist of a reliability parameter, such as MTBF, failure rate, or a project's reliability rating (e.g., excellent, good, fair, or poor). At the input side, it is necessary not only to eliminate noise—i.e., metrics unrelated to software reliability—but also to standardize the "uneven" metric data into a common dimension. This process often requires data compression, and a key research focus is how to preserve as much of the original data's information as possible during this compression.

Through data compression and preprocessing, we obtain the training data for the reliability evaluation model, which is divided into an input set and an output set. The input set consists of software metrics (compressed values), while the output set is made up of software reliability data. At this stage, a suitable algorithm must be selected for modeling. Common approaches include regression models, neural networks, and others. Each method has its own advantages and limitations, and in practice, they are typically selected and optimized through experimentation. Once the model is built, data analysis can be used to identify the software metric elements most closely related to the output (often through correlation analysis). These selected metrics form a subset of the previously constructed software metric elements. This subset can guide the measurement of new software projects and provide feedback following the evaluation. After the evaluation, it is possible to analyze which metrics influence the software's reliability and offer recommendations for improving software development activities. Naturally, this model is not finalized after a single training phase. By continually accumulating historical data, the training set can be expanded. The larger and higher quality the training data, the more accurate the resulting reliability evaluation model.

In the following sections, we will introduce specific methods and case studies related to the framework outlined above. These methods are well-established techniques in statistics or recognized engineering practices, and the aim is to demonstrate their application in data-driven complex software reliability evaluation.

5.2 Software Metrics that Impact Reliability Assessment

Measurement is a method of quantifying the attributes of entities based on established rules, typically encompassing the following elements:

1. Identifying the entity to be measured;
2. Selecting the specific attribute to be measured;
3. Assigning a corresponding measurement value to the entity;
4. Establishing the mapping relationship between the entity and its measurement value.

Software measurement refers to the quantification of indicators related to software product projects, development processes, and the properties of the products themselves. Through various metrics, it is possible to quantify different characteristics of the software lifecycle, such as software size, structural complexity, and data complexity. Therefore, software measurement is a critical factor in the evaluation, management, and enhancement of software product quality and reliability. Software quality models typically decompose software quality into measurable metrics, providing a framework for using these metrics to assess, manage, and improve software quality and reliability.

Software measurement was first introduced in 1958, while the concept of software science emerged in the 1970s. Halstead made a significant contribution to the development of this concept, emphasizing that, beyond theoretical research methods, software measurement should also prioritize engineering practice. Boehm argued that, in addition to qualitative research, the study of software properties must be quantifiable. It was from this perspective that software measurement as a discipline came into being.

The primary task of software measurement is to define, quantify, measure, and scientifically analyze the processes and products of software engineering. The main goal of these efforts is to extract software engineering from ambiguous qualitative processes, improving software design, development, and management through data-driven, visual, and traceable methods. This aligns with the integration of theoretical methods and engineering practices in the field of software measurement. Theoretically, software measurement relies on established rules and theories, and its representation typically takes the form of data (such as indicators, parameters, and ratings). Practically, obtaining software measurements necessitates the use of testing tools, auditing systems, and the actual involvement of engineers. Software measurement is an engineering practice guided by scientific principles. Today, software measurement has evolved into a critical research area within software engineering.

Furthermore, both in engineering practice and academic research, numerous researchers have proposed various software metrics, which may influence software quality and reliability. After obtaining the numerical results of software measurements, some researchers opt to use decision-making methods, such as the Analytic Hierarchy Process (AHP) and Fuzzy Comprehensive Evaluation Method, to calculate the weights for assessing software quality. Other researchers apply statistical methods, such as Support Vector Machines, Bayesian Networks, Markov Decision Processes, or Neural Networks, to develop software reliability evaluation models.

This chapter will explore software reliability assessment from a data-driven perspective, introducing various metric elements that may influence software reliability. These metric elements are primarily derived from relevant research and engineering experience in software reliability assessments. We will present key software metric elements and their acquisition methods, focusing on code metrics and process metrics. For code metrics, we will introduce commonly used measurement tools to facilitate automated data collection. For process metrics, we will outline a general analytical approach.

5.2.1 Establishment of Software Metric Set

Overall, the development of software measurement is relatively mature and systematic, yet the connection between these metrics and the data-driven software reliability assessment introduced in this chapter remains unclear. Specifically, is it valid to assess software reliability and evaluate software quality based on software code metrics data? To address this, two questions are posed, and relevant research is conducted to verify the feasibility of the proposed methodological framework. It should be noted that the literature review is not confined to software reliability assessment but also includes research on methods and models for software quality evaluation based on metrics. Given the inseparable relationship between software reliability and software quality, this chapter asserts that reliability assessment falls under the broader context of software quality evaluation and prediction.

The first question is: In relevant research, which software metrics have been used by researchers as inputs for software reliability and quality assessment models?

Typically, data-driven quality/reliability assessment models involve multiple inputs, and these input elements are often selected from a specific set by researchers. Based on existing research, these sets can be classified into two categories: software metrics, including code metrics, project metrics, and testing metrics; and software quality characteristics or their sub-characteristics. The distribution of these inputs is illustrated in Fig. 5.3.

Software code metrics refer to metrics that can be directly derived from the code. Studies that use code metrics as inputs for evaluation models offer a wide range of code metrics. In addition to widely used software metric sets such as McCabe's structural complexity metrics, Halstead's text complexity metrics, Henry-Kafura's

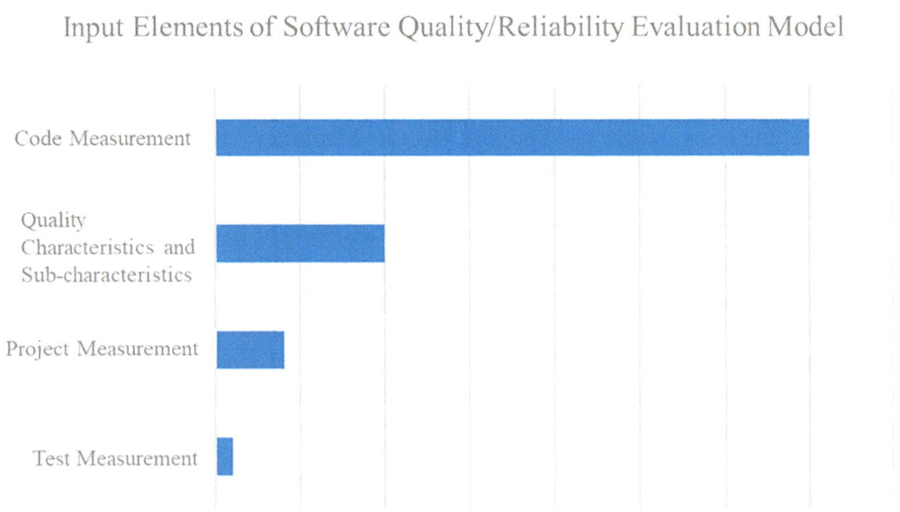

Fig. 5.3 Commonly used inputs for evaluation models in existing research

information flow metrics, C-K object-oriented metrics, and MOOD object-oriented metrics, many researchers have also proposed their own combinations of software metrics.

In these studies, Cyclomatic Complexity is the most frequently used software metric in neural network-based evaluation models, appearing in 40% of code metric combinations. The most commonly used metric set is the C-K object-oriented metric set, which includes Weighted Methods per Class (WMC), Depth of Inheritance Tree (DIT), Number of Children (NOC), Coupling Between Objects (CBO), Response for a Class (RFC), and Lack of Cohesion of Methods (LCOM). Additionally, it is noteworthy that Halstead's text complexity metric set has been widely applied. While the Halstead metric set includes many metrics, the four primary ones—number of unique operators $\eta1$, number of unique operands $\eta2$, total number of operators N1, and total number of operands N2—are the most commonly used. The rationale behind this selection is that, within a software complexity metric set, metrics derived from the primary metrics do not convey more software attribute information than the primary metrics themselves, as illustrated in Fig. 5.4.

The second question is: Why can the metrics selected in the first question be used as input elements for software quality and reliability evaluation models?

As discussed above, the input elements are divided into two categories: software metrics and software quality characteristics, or their sub-characteristics. Since software quality characteristics and their sub-characteristics are derived directly from software quality models, this question focuses on the rationale for using software metrics to evaluate software quality and reliability.

Some researchers have provided the rationale for using software metrics based on their definition. The 'Standard for Software Quality Metrics Methodology, IEEE Std 1061-1992, 1993' defines a metric as a function that takes software data as input and produces a single value, which can be used to explain the degree to which a

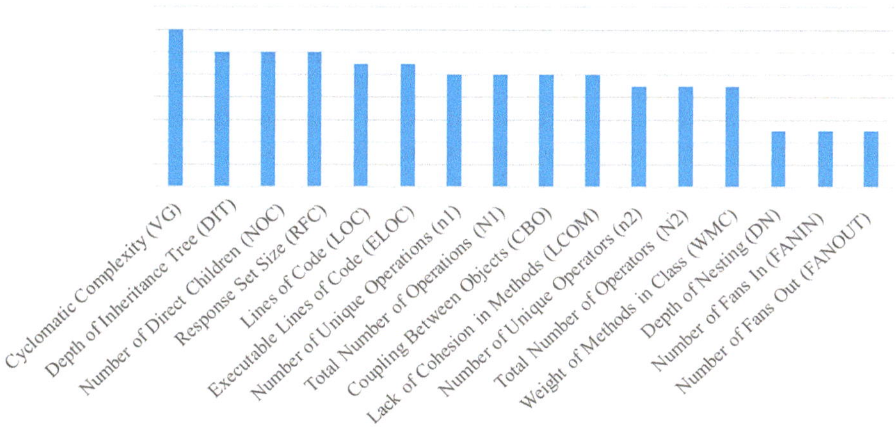

Fig. 5.4 Commonly used code metrics in quality assessment

given attribute influences software quality. It has been argued that software quality metrics refer to the quantitative methods for measuring attributes that impact software quality, while software metrics involve the quantification of internal attributes that affect software quality. Quality factors such as reliability and functionality cannot be directly measured and must be indirectly assessed through the measurement of internal software attributes. From a practical perspective, the use of software metrics facilitates adjustments to the development process. Typically, quantifiable metrics are employed to predict the quality of a software system, and this prediction is then used to guide adjustments in the development process to enhance the overall quality of the software. However, some critics question this approach, arguing that while software metrics, such as the number of operators and operands, may be useful for predicting software quality, they do not always accurately reflect the true quality of the software.

As can be seen, there is a solid theoretical and practical foundation for using software metrics to evaluate software quality and reliability. The data-driven software reliability assessment introduced in this chapter is built upon these foundations.

5.2.2 Code Metrics and Related Tools

Software code metrics refer to the measurement of software source code through static analysis. The source code is the core of the software, and most quality issues can be traced back to it; therefore, software code metrics are closely linked to software quality. During the era of structured programming, the most commonly used software code metrics were those that described the size and complexity of functions. The most common size metric was lines of code (LOC), while many researchers have proposed collections of software code metrics to describe code complexity, such as McCabe's cyclomatic complexity metric and Halstead's text complexity metric. These size and complexity metrics, developed for structured programming languages, are collectively referred to as traditional metrics. With the advent of object-oriented programming, many researchers have introduced collections of object-oriented metrics.

Overall, traditional software code metrics can be categorized into the following types: first, size metrics, with the most commonly used being lines of code, which can be further subdivided into total lines of code, executable lines of code, nonempty lines of code, and comment lines; second, data flow metrics, such as the number of input variables, output variables, fan-in, and fan-out of a module; third, structural complexity metrics, with McCabe's complexity metrics primarily including cyclomatic complexity and basic complexity, along with direct structural measurements such as branch count, condition count, multiple condition count, and decision count; fourth, text complexity metrics, referring to Halstead's text complexity metrics, which include unique operator count ($\eta 1$), unique operand count ($\eta 2$), total operator count (N1), and total operand count (N2), among others.

With the development of object-oriented technology, object-oriented languages have become widely adopted. Object-oriented systems place emphasis on the relationships between entities. The mechanisms in object-oriented systems also differ from those in traditional software systems, utilizing concepts such as abstraction, encapsulation, and polymorphism. Both the entity relationships and the new mechanisms contribute to the fact that object-oriented systems will lead to new developments in software metrics.

The C-K method is a metric for classes. Characteristics of a class, such as methods, inheritance, coupling, and cohesion, are unique to object-oriented systems and are the primary focus of metrics for such systems. The C-K method defines six metrics for object-oriented systems: Weighted Methods per Class (WMC), Depth of Inheritance Tree (DIT), Number of Children (NOC), Coupling between Objects (CBO), Response for Class (RFC), and Lack of Cohesion in Methods (LCOM). The MOOD method introduces six metrics, focusing on four aspects: encapsulation, inheritance, coupling, and polymorphism. For encapsulation, two indicators are proposed: Method Hiding Factor (MHF) and Attribute Hiding Factor (AHF); for inheritance, two indicators are proposed: Method Inheritance Factor (AIF) and Attribute Inheritance Factor (MIF); for coupling, the Coupling Factor (CF) is proposed as an indicator; and for polymorphism, the Polymorphism Factor (PF) is proposed as an indicator.

Due to the large size and high complexity of software source code, the acquisition of code metrics in practical projects heavily depends on metric tools. To obtain, analyze, and apply code metrics, reliance on existing automated tools is essential. Without testing tools, the analysis and application of software metrics cannot be effectively carried out.

This section primarily introduces several tools, including LDRA Testbed, QA-C, Parasoft C/C++test, and Polyspace

1. LDRA Testbed includes modules for static analysis, dynamic analysis, graphical display, and software testing at both the unit and integration levels. Static analysis primarily encompasses functions such as programming standards, software metric analysis, quality standards verification, and static data flow analysis. By consulting the official documentation, the software metric elements primarily supported by LDRA Testbed have been identified, as shown in Table 5.1.
2. QA-C divides the metrics into three groups: function metrics are generated for all functions using full definitions, and file metrics are generated for each analyzed file. Project metrics are computed once for the entire project and are generated through cross-module analysis. Similarly, QAC++ also divides the metrics into three groups: function metrics for all functions using full definitions, class metrics for each defined class, and file metrics for each analyzed file. All metric standards, except for LIN and TPP, exclude code that is disabled during the evaluation of preprocessor directives. QA's metrics are categorized into four parts: function-based metrics, file-based metrics, class-based metrics, and project-level metrics..

Table 5.1 The code metrics supported by the LDRA testbed

LDRA testbed
Control flow node metrics
LCSAJ density metrics
Fan-in/fan-out metrics
Loop depth metrics
McCabe cyclomatic complexity metrics
Halstead software science metrics
McCabe's essential complexity metrics
Comment line matrices
Code reachability metrics

Function-based metrics are applied to the source code of individual functions and include 34 metric elements such as Akiyama's Criterion (AKI), number of backward jumps (BAK), number of functions called from a function (CAL), and knot density (KDN). File-based metrics consist of 32 metric elements, including embedded programmer months (BME), comment-to-code ratio (CDN), and number of functions in a file (FNC). Class-based metrics are based on the source code of classes and include 8 metric elements, such as coupling between objects (CBO) and number of immediate children (NOC). Some of these indicators also require running cross-module analysis programs. Project-level metrics consist of 4 metric elements, including number of functions across the project (NEA). Some of QA-C's code metrics are listed in Table 5.2.

3. Parasoft C/C++ is an integrated solution that offers features such as static analysis, comprehensive code review, unit and component testing, and runtime error detection. The tool includes 49 built-in metrics, such as McCabe Cyclomatic Complexity, MOOD metrics, and others.
4. PolySpace is a tool used for analyzing software runtime processes. It consists of two modules: PolySpace Bug Finder and PolySpace Code Prover. The measurable metrics in PolySpace are categorized into project metrics, file metrics, and function metrics, with a total of 27 metric elements.

5.2.3 Software Process Metrics

This section will primarily discuss issues related to process metrics, including their impact on quality, the types of process metrics, and how to acquire them.

Process metrics are a subset of software metrics, primarily referring to the measurements taken throughout the entire software development and maintenance process for all products. The role of process metrics is to understand and manage the software process, to document and comprehend the software process during evaluation, and to manage the software process through control, prediction, and optimization. These metrics also assist management in making informed decisions. Software process metrics provide a more intuitive understanding of software

Table 5.2 Some code metrics supported by helix QAC

Abbreviation	Name	Explanation	Level
STAKI	Akiyama α criterion	This metric is the sum of Cyclomatic complexity (STCYC) and function call count (STSUB)	Function-based metrics
STAV1	Function average statement count (variant 1)	The average number of operations and operators per statement in the function body. STAV1 = (STFN1 + STFN2)/STST1	Function-based metrics
STAV2	Function average statement count (variant 2)	The average number of operations and operators per statement in the function body. STAV2 = (STFN1 + STFN2)/STST2	Function-based metrics
STAV3	Function average statement count (variant 3)	The average number of operations and operators per statement in the function body. STAV3 = (STFN1 + STFN2)/STST3	Function-based metrics
STBAK	Number of Backjumps in function	Jumps are not recommended in programs, especially backjumps. If possible, structured code should be redesigned	Function-based metrics
STCAL	Number of different function calls	This metric calculates the number of function calls within a function	Function-based metrics
STCYC	Cyclomatic complexity	Cyclomatic complexity is calculated as the number of decision points plus 1	Function-based metrics
STELF	Number of Else-if statements without an Else clause	Counts the number of if-else-if structures without an else clause. Calculated by counting all if statements without an else	Function-based metrics
STFN1	Number of operator occurrences in function	This metric is the Halstead count (N1) of operators in the function	Function-based metrics
STFN2	Number of operand occurrences in function	This metric is the Halstead count (N2) of operands in the function	Function-based metrics
STG TO	Number of go to statements	Some uses of go to are merely for error handling, but they should be avoided as much as possible. The plum hall guidelines state that go to should not be used.	Function-based metrics
HLB	Halstead error count	The error count provided by Halstead (B)	Function-based metrics
STKDN	Intersection density	This is the number of intersections per executable line of code, calculated as follows: STKDN = STKNT/STXLN, when STXLN is 0, this value is zero	Function-based metrics
STKNT	Number of intersections	This is the number of intersections in the function. Intersections represent the number of times control structures are traversed, caused by break, continue, goto, or return statements in control structures	Function-based metrics

attributes and performance, process efficiency, and various product information, thereby helping to better manage and optimize the software process and ultimately improve the quality of software products.

Software process metrics can be further categorized into measurements of the software process itself and measurements of the process products.

Process quality and product quality are generally regarded as two key aspects of software product quality management. Process quality is composed of work quality, design quality, development quality, and other factors across different stages of the software product lifecycle. The primary goal of software process metrics is to improve the quality of the software process. These process qualities are typically further subdivided into specific process metric objectives, such as management requirements, project estimation, development of monitoring plans, objective evaluation of work products, baseline establishment, and the institutionalization of management processes. These objectives are then broken down into specific practices, which correspond to the actual process metric elements. The activities involved in process metrics include data collection, data verification, numerical conversion, metric analysis, and decision-making based on metrics.

The improvement of software process metrics and process quality often depends on the combination of various methods. For example, some organizations first analyze the requirements of CMMI3 in the field of metric analysis, and then utilize specific metric analysis techniques (such as GQ(I)M) to construct metric models based on these requirements. Such approaches typically provide a mapping of metric objectives to indicators, along with specific measurement items, such as the defect discovery phase, defect injection phase, modules containing defects, actual defect count in software modules, defect severity, defect types, defect discovery time, and defect closure time.

Another aspect of software process metrics is the measurement of process products, which are often in the form of documents. In the engineering practices discussed in this chapter, process products include, but are not limited to, the following: the project tasks document, the software requirements specification, the software description from the preliminary design phase, the detailed design document, the final source code after development, the test cases, the set of execution cases, the problem reports collected during the testing phase, the test documentation written after testing, and the final user manual for the software. Based on these process products, certain metrics related to software reliability assessment can be proposed, such as the ratio of lines of code to pages in the requirements documentation, which can, to some extent, reflect the level of detail in the software requirements and, in turn, impact the reliability of the software product.

In fact, in the actual software development projects, the collection and control of process data are still in the early stages. Although relevant standards, such as GJB 5000A-2008, have proposed product development models to address software process improvement, outlined the characteristics of effective processes, and provided best practices for process data collection, many engineering units still have a limited and unscientific understanding of these process data. In some cases, process records are even 'fabricated' after development is completed. As a result, the scale, credibility, and effectiveness of process data are compromised.

When conducting software reliability evaluations, the primary departments involved are the software testing department and the software quality department, with their main activities focusing on software testing and validation. Software testing has a well-established theoretical framework that has been refined through long-term practice and improvement. It typically employs a complete set of methods and retains a substantial amount of historical data, which is generally objective, accurate, and reliable. Based on this, we believe that in software process metrics, testing metrics should be given special attention. Therefore, this section introduces software testing metrics from two perspectives: the relationship between testing quality and software quality, and how to measure testing activities.

Many researchers believe that software testing is an essential method for ensuring software quality, and the quality of the software testing process directly impacts the testing quality, which in turn affects the overall software quality to some extent. In software testing, the quality of the testing process influences the quality of the software product. A well-conducted testing process helps improve testing quality, reduce testing costs, and shorten the testing cycle, making the study of the software testing process highly significant. Furthermore, as research in formal verification and program correctness proof is still in its early stages, software testing remains one of the most effective means of ensuring software quality and improving software reliability. Some argue that the software testing process itself is essentially a process of measuring and evaluating the software's quality. By means of this measurement and evaluation, it is possible to verify whether the software meets the user's requirements, which serves as an important criterion for users in product selection.

Different researchers have categorized and summarized software testing metrics from various perspectives. Some researchers divide testing metrics into two categories: testing quality metrics and testing process metrics, which is analogous to measuring software as both a product and a process. For testing quality metrics, the first step is to define testing quality objectives, followed by utilizing extensive testing experience and prior research on testing processes to assess the current testing quality. Commonly evaluated aspects include the quality of intermediate products in testing, the quality of the testing process itself, and the quality of research on optimizing the testing process. These three aspects are then further refined to establish the final measurement system. For testing process metrics, optimization of the testing process can be achieved through factors such as personnel, tools, environment, time, and testing techniques. By further refining these factors, a comprehensive set of measurement indicators can be developed.

Regarding software testing metrics, researchers have classified and summarized them from various perspectives. Some scholars divide testing metrics into two types: test quality metrics and test process metrics. This classification approach is similar to categorizing software into software products and software processes for separate measurement.

For test quality metrics, the first step is to define the test quality objectives, followed by assessing the current test quality based on extensive testing experience and prior research on testing processes. Common areas of focus include the quality of intermediate products, the quality of the testing process itself, the quality of testing techniques, and the quality of research on optimizing the software testing process. These three aspects are further refined to establish the final metric system.

In terms of test process metrics, optimization of the testing process can be achieved through factors such as personnel, tools, environment, time, and testing techniques. Further refinement of these factors leads to the development of a comprehensive set of measurement indicators.

The partial test metrics are listed in Table 5.3.

Table 5.3 Partial test metrics

No.	Metric element name	Description	Collection timing	Collection methods/ Tools used
1	Number of qualified test cases	Number of actual test cases passed by expert review	After test case review is completed	TestDirect
2	Total number of designed test cases	Number of test cases covering all functional requirements	After test case design is completed	TestDirect
3	Number of test cases supported by platform	Number of test cases that the simulation platform can support	After the simulation platform is set up	TestDirect
4	Number of issues found during test case execution	Number of issues found after executing test cases	After all test cases in a round are executed	TestDirect
5	Number of functional points supported by test environment	Functional points that the simulation platform can support	After the simulation platform is set up	Manual
6	Total number of functional points	Sum of functional points obtained by decomposing requirements	After obtaining project requirements in test planning	Manual
7	Number of issues resolved	Number of issues confirmed as resolved after regression testing	After a round of regression	TestDirect
8	Number of issues confirmed by developers	Number of issues confirmed by developers among those found by testing	After stage testing and issue confirmation with developers	TestDirect
9	Number of issues found in documentation review	Number of confirmed issues found in various documentation reviews	After documentation review is completed	Manual
10	Number of issues found in code review	Number of confirmed issues found during code walk-through/review	After code walk-through/review and confirmation with developers	TestDirect
11	Number of issues found in dynamic testing	Number of issues found during dynamic testing when executing test cases	After all test cases are executed	TestDirect
12	Total number of issues found in testing	Total number of issues found throughout the testing process (including documentation review, static/dynamic testing issues)	After the entire testing process is completed	TestDirect

5.2.4 GQM (Goal-Question-Metric): A Method for Analyzing Process Metrics

The process and software testing metrics described in the previous sections are not generated arbitrarily; rather, they require appropriate modeling methods to break down the metric goals in order to derive them. Among the various metric modeling methods, the most well-known and widely used approach is the GQM (Goal-Question-Metric) model, proposed by Professor Victor R. Basili. This model is highly operational and flexible. The implementation steps of the GQM model generally involve defining an overarching goal for the entire software process, which serves as the starting point for the entire metric activity. Based on this goal, and considering the specific circumstances of the project, the goal is then subdivided into several distinct problem elements, which represent the process variables in the final metric results. For each of these problem elements, answers are proposed to address the issues, and these answers collectively form a set of metric indicators. The following is a brief introduction to this method.

5.2.4.1 GQM Overview

Selecting and defining metrics, determining metric goals, and choosing appropriate measurement items are fundamental to the successful implementation of metrics. The widely adopted goal-driven approaches internationally are the GQM and GQ(I) M metric definition models.

GQM is a systematic method for quantifying software and its development processes through measurement. The GQM model, proposed in the mid-1980s by Victor R. Basili and his team at the University of Maryland, is a goal-driven, top-down approach for defining measurements. Its purpose is to guide organizations or institutions in determining which data should be collected. The model is based on the following assumption: for a software organization aiming to measure purposefully, it must first define the goals of both the organization and the project. Then, it tracks these goals to data that operationally defines them, ultimately providing a framework that explains the relationship between the data and the relevant goals. Each organization or project has a set of objectives to achieve. To assess whether each goal has been met, a series of questions must be answered. For each question, a complete and quantifiable answer can be found. The model breaks down and decomposes the organization's goals into measurable indicators, refines these indicators into quantifiable values, and allows for the prediction and control of process performance, enabling quantitative management of software development.

GQM is a goal-oriented measurement method and a logical approach for managers to think about problems scientifically. The GQM model offers a top-down method for defining metrics and a bottom-up approach for data collection and interpretation. First, the goals to be measured are defined, followed by identifying potential problems for each goal in order to define it. The achievement of these goals is measured by answering the corresponding questions. The abstract and vague goals are broken down into specific, measurable problems. Then, for each problem, a set

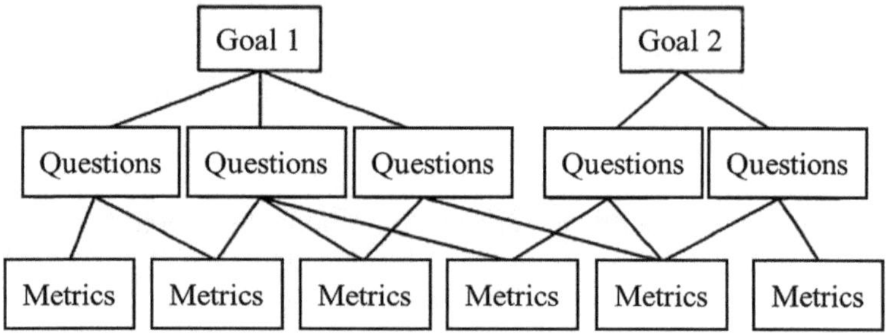

目标1	Goal 1
目标2	Goal 2
问题	Questions
度量	Metrics

Fig. 5.5 The Three-Level GQM Model

of measurement methods is provided, and the data obtained through these methods serve as the answers to the questions. The purpose of the analysis process is to transform conceptual goals into more specific questions, which are then further decomposed into measurable indicators. Thus, this process emphasizes analyzing the hierarchical structure of goals, questions, and indicators, and understanding their interrelationships.

The GQM model is based on a top-down, goal-driven approach and consists of three levels: the conceptual level (goals), the operational level (questions), and the data level (metrics). The three-level GQM model is shown in Fig. 5.5. These three levels form an inheritable structure, where each layer refines the one above it. Through this refinement process, the necessary metrics are derived from the goals.

1. Conceptual Layer (Objectives). Objectives are defined for specific entities and are linked to various quality models for different reasons, based on different perspectives and contextual environments. They describe the purpose of measurement, providing a clear expression of the measurement process through defined objectives. The GQM framework generally includes five elements: measurement object, purpose, attribute, perspective (for whom), and measurement environment. It outlines the measurement goals expected to be achieved through measurement. Each measurement objective consists of three characteristics: focus, object, and perspective, as well as a measurement purpose. The focus defines the specific features of the object that need to be studied, while the perspective outlines the observer's viewpoint regarding the required measurement information. The measurement purpose describes how the measurement results are to be utilized. The definition of project objectives within a specific environment is influenced by various factors, primarily including products, processes, and resources.

2. Operational Layer (Problem). This layer defines a set of questions targeting the objectives of the previous layer, using a series of questions to define the model of the object under study, and subsequently evaluating or achieving specific objectives. These questions describe the measurement objectives from various perspectives. The problem is related to the state model, which provides a more detailed definition of the object of interest. A series of questions used to evaluate established objectives should be based on a model with relevant characteristics, and the selected questions should aim to precisely characterize the measurement objective, rather than selecting data indiscriminately. Otherwise, it may lead to increased workload and data wastage.
3. Quantitative Layer (Measurement). Based on the above model, a series of measurements are associated with each question and are answered in a measurable way. A set of data is defined for each question, providing a quantitative answer and characterizing the features of the problem. The data for the indicators come from both subjective and objective measurements, and careful selection is necessary to portray each question from different perspectives. The same measurement item can also be used to answer different questions within the same objective, and multiple GQM models can share some questions and measurement items. Viewing the same measurement from different perspectives will result in different data values.

5.2.4.2 GQIM Overview

The Software Engineering Measurement and Analysis Group at Carnegie Mellon University's Software Engineering Institute proposed the GQ(I)M model based on the GQM model. The distinction between GQ(I)M and GQM is the addition of a visual indicator layer between the problem layer and the measurement layer in the GQM model, which establishes a connection between the problem and the measurement data. An indicator is a measurement used to evaluate or predict other measurements. It is a synthesis of one or more measurements and reflects a specific aspect of the software product or process. Indicators are typically represented using charts, text, or tables. Model users can define indicators based on the problem and then determine the measurements and data required to construct the indicator. Different measurement objectives require the selection of different indicators. In practice, there are thousands of measurable items, so the most suitable indicators that best reflect the current measurement environment should be selected. These indicators can serve as measurement requirement specifications, guiding the data that needs to be collected, the processing and analysis required, and the planning of these activities.

For example, to determine how many defects were injected and removed at each stage, we can establish an indicator for the defect stage distribution. Figure 5.6 shows the GQ(I)M measurement model.

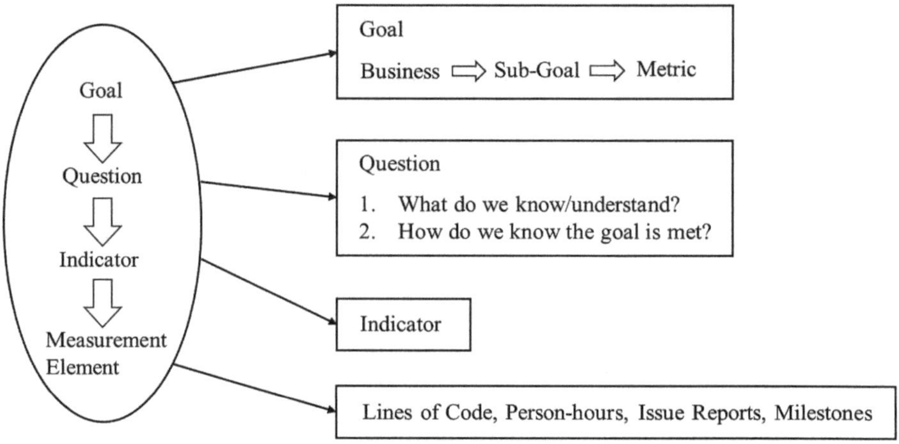

Fig. 5.6 Schematic Diagram of the GQ(I)M Model

5.2.4.3 Implementation Principles and Key Steps of GQM and GQ(I)M

1. Define the goals and objectives of the measurement.
2. Develop a question hierarchy to achieve the goals and objectives.
3. Design a measurement hierarchy according to the question hierarchy.
4. Select appropriate indicators for each measurement.
5. Develop a data collection plan and collect relevant data.
6. Analyze the data and generate reports.
7. Use the reports to improve and optimize the software development process

1. Implementation Principles of GQM and GQ(I)M

   (a) Measurement objectives should be derived from business goals

   Measurement is not about measuring for the sake of it. Rather, it serves the purpose of achieving a specific goal, whether to enhance the understanding of a process or to improve it. Regardless of the goal, it always serves the business objectives of a software company. Therefore, when defining measurement objectives, it is crucial to ensure that they are derived from business goals and remain traceable.

   (b) Define the focus and establish the environment for measurement objectives

   During the implementation of measurement, there can be many measurement objectives. Often, one may feel overwhelmed by the vast amount of data; therefore, it is essential to define the focus and prioritize the measurement objectives. Additionally, to achieve these objectives, it is crucial to create an environment that supports their attainment.

   (c) Use GQM/GQ(I)M to transform informal goals into measurable, actionable structures.

   Once measurement objectives are defined, questions can be formulated based on these objectives. Solving these questions ultimately achieves the measurement objectives. From these questions, intuitive indicators or metrics can be derived, enabling the mapping from measurement objectives to measurement data items.

2. Implementation Steps of GQM

Based on the specific needs of projects and organizations, the GQM model provides a framework to assist software managers in designing a comprehensive software measurement system aimed at achieving management goals. It uses a systematic approach to tailor and integrate various objectives within software processes and product models. GQM effectively addresses the question of 'why' the data being collected is important. This question is vital because it defines how the obtained data can be interpreted, and its answer serves as the foundation for reusing measurement plans and processes in future projects and activities. The application of the GQM method in software development seeks to formalize this process by defining objectives and measurement metrics, enabling them to be easily observed and measured, much like tangible products.

The GQM method offers numerous advantages, including clear conceptualization, ease of understanding, simplicity in training, knowledge reuse, computer management, and adaptability to complex and dynamic application environments.

The measurement activities using the GQM method include the following four stages:

(a) Planning stage: Select the project for which the measurement will be applied, define and characterize it, and develop a plan to generate the project schedule.
(b) Definition stage: Define the measurement process (objectives, questions, metrics, and assumptions) and document it.
(c) Data collection stage: Collect, verify, and analyze data, generate a result set, and take appropriate actions.
(d) Interpretation stage: Convert the collected data into measurement results based on the defined measurement criteria, answer the posed questions, and evaluate the objectives to be achieved. Analyze the data retrospectively to assess whether it aligns with the objectives and provide suggestions for improvement. Offer feedback to stakeholders regarding the measurement results.

The definition stage of GQ(I)M includes the following implementation details:

1. Identify business objectives.
2. Determine what needs to be obtained.
3. Determine sub-objectives. Group related questions and create a series of sub-objectives related to management or execution activities.
4. Determine entities and attributes related to sub-objectives. This process provides information on entities and attributes that must be obtained to achieve sub-objectives.
5. Format measurement objectives. The formatted measurement objectives should include four elements: the object of focus, the purpose of focus, the audience of focus, and the measurement environment.

6. Formulate a measurement plan. When decomposing objectives, process model structure can be used as a reference. Based on the composition of sub-processes in the process, decompose the measurement objectives into sub-measurement objectives for each sub-process.
7. Determine measurable questions and related indicators that contribute to achieving the measurement objectives. Derive quantifiable questions and indicators (including various types of charts) required to achieve these objectives from the formatted measurement objectives.
8. Define the data elements required to construct indicators to solve problems.
9. Define the metrics used and make them feasible. Clearly define the formulas for measurements and the meaning of measurement data within the organization, and use structured methods to ensure that important measurements are not overlooked.
10. Determine the activities for implementing measurements.

These activities include identifying the data sources, defining the methods, frequency, and responsible individuals for collecting and reporting data, determining the users who will utilize the data, specifying how the data will be analyzed and reported, defining tools for process automation and management, and establishing the data collection process.

Prepare a plan to implement the metrics and develop mechanisms for data collection and analysis. This plan should include the objectives of the metrics, the context of the metrics, the scope of the metrics, the relationship with other process improvement activities, the tasks, activities, and resources required for implementing the metrics, the timeline for the metrics, the responsibilities related to the metrics, and the supporting activities.

From the above principles and implementation steps, it is clear that processes 1–5 analyze and decompose the organization's business objectives, generating the 'G' in GQ(I)M. Processes 6 and beyond map the objectives to their corresponding metrics. Throughout the entire process, two traceabilities must be maintained: one is the traceability from the objectives (G) to the business objectives, and the other is the traceability from the metrics (M) to the objectives (G). This ensures that the metrics align with the organization's objectives, prevents unnecessary data collection, and avoids wasting resources.

5.3 Data Training and Validation: Data-Based Evaluation Algorithms

In this section, several data training methods that have been applied in engineering practice will be introduced, including metric data aggregation, metric data normalization, data dimensionality reduction methods, and various data fitting algorithms.

5.3.1 *Metric Data Aggregation*

From the above description, it is clear that software code metrics are defined at the method or class level, making them difficult to directly apply to software quality assessment based on project data. When analysis is required at the project level, code metric data below this level must be aggregated into project-level data for use.

Due to the large scale and inherent complexity of software code, software code metrics that describe aspects such as code size and complexity are more effective in reflecting the characteristics of the software. However, these metrics are defined at the method or class level, making them difficult to directly apply to software quality assessment based on project data. To examine software code metrics from the perspective of the entire project, commonly used aggregation indicators, such as descriptive statistics, distribution parameters, and econometric indicators, are employed to aggregate software code metric data.

5.3.1.1 Characteristics and Aggregated Metrics of Software Code Measurement

Software code metrics within a project are highly skewed, similar to defect distribution following the Pareto principle. Descriptive statistics such as the mean, which are suited for symmetric distributions, are ineffective in capturing this characteristic of software metrics within a project. Skewness and kurtosis, however, are indicators that describe the direction and extent of data skewness. Additionally, power-law distributions, log-normal distributions, and certain econometric indicators that describe the degree of income inequality can effectively characterize this highly asymmetric distribution. Therefore, when considering aggregated metrics, it is essential to identify a combination of indicators that adequately describe the software metrics, rather than relying on a single aggregated metric.

The commonly used aggregated metrics include three main categories: descriptive statistics, statistical distribution parameters, and econometric indicators, as shown in Table 5.4.

Descriptive statistical indicators and econometric indicators are similar in that both methods derive aggregated values by incorporating every data point into a formula, meaning both can be applied to any type of data. Statistical distribution

Table 5.4 Commonly used aggregated metrics for measurement

Data aggregated metrics	Applicable conditions	Aggregated results
Descriptive statistical indicators	Any data	sum, mean, median, variance, skewness, kurtosis, and others
Statistical distribution fitting	Data follows a certain distribution	Distribution parameters
Econometric indicators	Any data	Gini coefficient, Theil index, Hoover index, and others

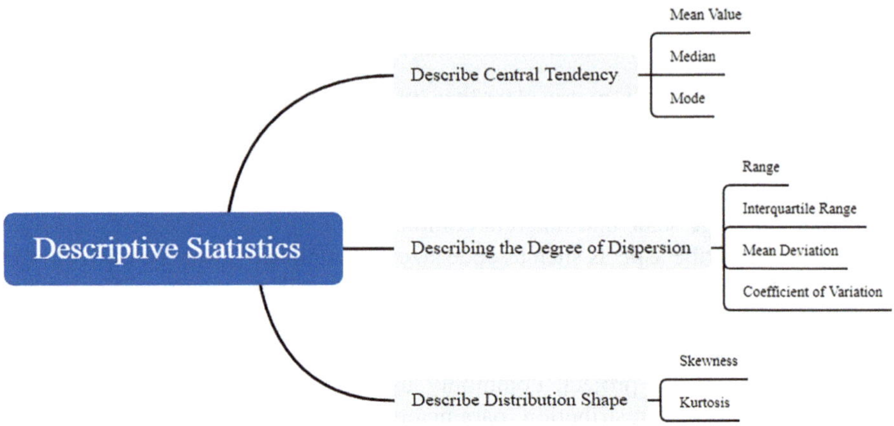

Fig. 5.7 Descriptive Statistical Indicators

parameters, however, can only be used for aggregation when the exact distribution data is known. The advantage of this approach is that a single set of statistical distribution parameters (such as μ and σ to describe a normal distribution) can fully characterize the data, while other methods may require combining multiple indicators to achieve the same level of description. Additionally, research on aggregating software code measurement data has focused on the statistical distribution of software code metrics, with a specific research case to be presented in Sect. 5.4. This section will introduce the concepts and significance of these aggregated metrics.

Descriptive statistical indicators are the most basic, common and widely applicable aggregated metrics, as shown in Fig. 5.7.

1. Measures of Central Tendency of Descriptive Statistics.

   (a) Mean value

$$\overline{x} = \frac{\sum_{i=1}^{n} x_i}{n} \tag{5.1}$$

$$\overline{x} = \frac{\sum_{i=1}^{n} x_i f_i}{\sum_{i=1}^{n} fi} \tag{5.2}$$

The mean calculation formula has two following important mathematical properties:

- The sum of the deviations of all observations from their mean is equal to zero. That is:

$$\sum_{1=1}^{n} (x_i - \overline{x}) = 0 \tag{5.3}$$

- The sum of the squared deviations of all observations from their mean is minimum, i.e.

$$\sum_{i=1}^{n}\left(x_i - \bar{x}\right)^2 \tag{5.4}$$

(b) Median

The key to calculating the median is to determine its position. The formula for determining the position of the median is:

$$\frac{n+1}{2} \tag{5.5}$$

If the number of observations in the data is even, the following formula can be used to calculate the value of the median.

$$m_1 = \frac{1}{2}\left[x_{(n/2)} + x_{(n/2+1)}\right] \tag{5.6}$$

(c) Mode

Mode is also a commonly used method to determine the central position of data distribution. Different observations appear in the sample data with varying frequencies, and the observation with the highest frequency is the mode of the data, denoted as "mo".

Comparison of Mean, Median, and Mode:

The mean is the centroid of all observations;

The mode is the focal point of all observations;

The median is the center of all observations. The relationship between the three is shown in Fig. 5.8.

Experience suggests that when the skewness of the frequency distribution is low, there is generally:

$$\left(m_0 - m_e\right) = 2\left(m_e - \bar{x}\right) \tag{5.7}$$

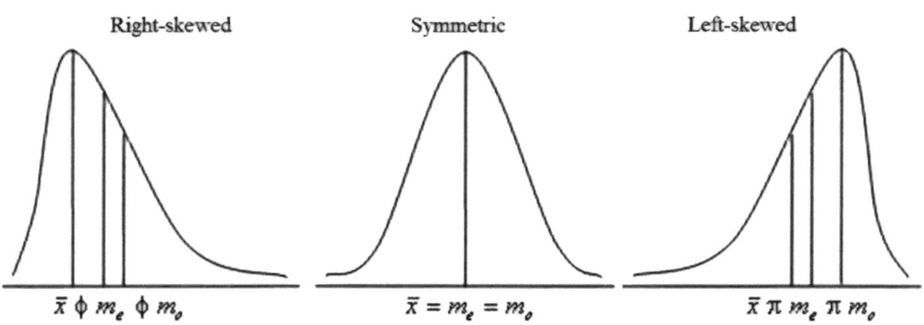

Fig. 5.8 The relationship between the Mean, Median, and Mode

2. Measures of Dispersion of Descriptive Statistics.

   Measures of dispersion in descriptive statistics include range, interquartile range, mean deviation, variance, and standard deviation.

   (a) Range

   The difference between the maximum and minimum observations in the data is called the range, denoted as R. The formula for calculating the range is:

$$R = \max(x_i) - \min(x_i) \tag{5.8}$$

   (b) Interquartile Range

   The difference between the upper quartile and the lower quartile in the data is called the interquartile range. The formula for calculating the interquartile range is:

$$Q_d = Q_u - Q_L \tag{5.9}$$

   (c) Mean Deviation

   The mean deviation is the average of the absolute deviations of each observation from its mean, denoted as M_d. The formula for calculating mean deviation is:

$$M_d = \frac{\sum_{i=1}^{n} |x_i - \overline{x}|}{n} \tag{5.10}$$

   (d) Standard Deviation

   Sample variance is the average of the squared deviations of each observation from its mean, denoted as s^2. The formula for calculating sample variance is:

$$s^2 = \frac{\sum_{i=1}^{n} (x_i - \overline{x})^2}{n-1} \tag{5.11}$$

   The square root of the sample variance is called the sample standard deviation, denoted as s. The formula for calculating the sample standard deviation is:

$$s^2 = \sqrt{\frac{\sum_{i=1}^{n} (x_i - \overline{x})^2}{n-1}} \tag{5.12}$$

   When calculating sample variance or sample standard deviation, it is sometimes for the purpose of estimating the population variance or population standard deviation. Population variance is a parameter used to describe the degree of dispersion in population data. The formula for calculating population variance is:

$$\sigma^2 = \frac{\sum_{i=1}^{n}(x_i - \mu)^2}{N} \tag{5.13}$$

Population standard deviation is the square root of population variance, and its formula for calculating is:

$$\sigma^2 = \sqrt{\frac{\sum_{i=1}^{n}(x_i - \mu)^2}{N}} \tag{5.14}$$

(e) Coefficient of Variation.

Coefficient of variation is the ratio of the standard deviation to the mean, denoted as ν. The formula for calculating coefficient of variation is:

$$\nu = \frac{s}{\overline{x}} \tag{5.15}$$

3. Measures of Distributional Shape of Descriptive Statistics.

Measures of distributional shape of descriptive statistics include skewness and kurtosis.

(a) Skewness

Skewness is a statistical measure of the symmetry of frequency distribution, denoted as SK. The formula for calculating skewness is:

$$SK = \frac{n\sum_{i=1}^{n}(x_i - \overline{x})^3}{(n-1)(n-2)s^3} \tag{5.16}$$

A skewness value of 0 indicates a symmetrical frequency distribution.

If the value is less than 0, the distribution is negatively skewed; If greater than 0, it is positively skewed.

The larger the absolute value of the skewness, the greater the degree of left or right skewness. Particularly, if the absolute value exceeds 1, the distribution is usually considered highly skewed.

(b) Kurtosis

Kurtosis is a statistical measure of the peakedness or sharpness of frequency distribution, denoted as KU. The formula for calculating kurtosis is:

$$kU = \frac{n(n+1)\sum_{i=1}^{n}(x_i - \overline{x})^4 - 3\left[\sum_{i=1}^{n}(x_i - \overline{x})^2\right]^2(n-1)}{(n-1)(n-2)(n-3)S^4} \tag{5.17}$$

A kurtosis value of 0 is referred to as a mesokurtic peak. If it is less than 0, it is called a platykurtic peak, indicating that the speed of concentration of the frequency distribution changes slowly and the distribution shape is relatively flat. If it is greater than 0, it is called a leptokurtic peak, indicating that the speed of concentration of the frequency distribution changes quickly and the distribution shape is relatively peaked or sharp.

5.3.1.2 Distribution Fitting for Software Code Metrics

In mathematical statistics textbooks, parameter tests are typically introduced for situations where the population distribution is known, and unknown parameters need to be tested. These tests are collectively referred to as parametric tests. In practical problems, it is often impossible to precisely determine the population distribution. In such cases, it is necessary to infer the population distribution from sample data in order to identify the distribution the population follows. These statistical tests are known as non-parametric tests. Below, we will primarily introduce the chi-square test and the tests for skewness and kurtosis.

Explanation:

Chi-Squared Test

The chi-squared test is one of the non-parametric methods, primarily used to test whether a population distribution matches a specified distribution type. In practical applications, it may be necessary to determine whether a product's quality characteristics follow a normal distribution or another distribution type. In such cases, the chi-squared test is employed for verification. Specifically, the chi-squared test compares the magnitude of differences between sample observed values and theoretical values for the population.

Skewness and Kurtosis Test

Another method for inferring the population distribution is the skewness and kurtosis test. Given a sufficiently large sample size, the Central Limit Theorem suggests that the sample mean's distribution approximates a normal distribution, and the distributions of sample skewness and kurtosis values approximate their population counterparts. This method compares the differences between the sample's skewness and kurtosis values and the theoretical values for skewness and kurtosis of the normal distribution (or other distributions) to test whether the population distribution is normal or follows another type.

1. Chi-square test

   Test the hypothesis about the population distribution based on the sample x_1, x_2, …, x_n when the distribution of the population is unknown.

   (a) H0: The distribution function of the population x is F(x).

(b) H1: The distribution function of the population x is not F(x).
If the population $\chi 2$ is discrete, hypothesis 1 is equivalent to:
(c) H0: The probability distribution of the population x is $P\{x = ti\} = pi(i = 1, 2, \ldots)$.
If the population x is continuous, hypothesis 1 is equivalent to:
(d) H0: The probability density of the population x is f(x).

When using the chi-square test to test hypothesis H0, if the form of F(x) is known under H0, but the parameter value is unknown, the maximum likelihood estimation method is used to estimate the parameters before conducting the test.

The idea of the chi-square test is to divide the entire set of possible outcomes of a random experiment Ω into k mutually exclusive events. A1, A2, …, Ak(ΣA $i = \Omega, AiAj = \varnothing .,i \neq j, i, j = 1, 2, \ldots, k$). Therefore, under the assumption H0, it is possible to calculate pi = P(Ai) (where $i = 1, 2, \cdots, k$). In n trials, the frequency of event Ai appearing fi/n often differs from pi, but generally speaking, if H0 is true and the number of trials is large, then this difference should not be significant. Based on this idea, Pearson uses the formula:

$$\chi^2 = \sum_{i=1}^{k} \frac{(f_i - np_i)^2}{np_i} \tag{5.18}$$

Or

$$\chi^2 = \sum_{i=1}^{k} \frac{\left(f_i - n\hat{P}_i\right)^2}{n\hat{P}_i} \tag{5.19}$$

As the test statistic for hypothesis testing, and proved the following theorem.

Theorem 5.1 If n is sufficiently large ($n \geq 50$), then when H0 is true (regardless of the distribution within H0), statistic:

$$\chi^2 = \sum_{i=1}^{k} \frac{(f_i - np_i)^2}{np_i} \tag{5.20}$$

The statistic always approximately follows a chi-square distribution with $k - r - 1$ degrees of freedom. Here, r is the number of estimated parameters.

Therefore, under the assumption H0, when calculating 4, we have:

$$\chi^2 \geq \chi_a^{\ 2}(k - r - 1) \tag{5.21}$$

Then, at the significance level α, reject H0; otherwise accept H0.

When using, it is necessary to ensure that n is sufficiently large and npi is not too small. $n \geq 50$, and each npi is not less than 5, and npi is best above 5; otherwise, appropriately combine Ai to meet this requirement.

2. Peak and Skewness Test

The skewness and kurtosis of a random variable refer to the standardized variable x of the third and fourth central moments of x, i.e.

$$\nu_2 = E\left[\left(\frac{x-E(x)}{\sqrt{D(x)}}\right)^3\right] = \frac{E\left[\left(x-E(x)\right)^3\right]}{D(x)^{3/2}} \tag{5.22}$$

$$\nu_2 = E\left[\left(\frac{x-E(x)}{\sqrt{D(x)}}\right)^4\right] = \frac{E\left[\left(x-E(x)\right)^4\right]}{D^2(x)} \tag{5.23}$$

Let x1, x2, ⋯, xn be a sample from the population x, then the central moment estimates of $\nu1$ and $\nu2$ are respectively:

$$g_1 = \frac{B_3}{B_2^{3/2}} \tag{5.24}$$

$$g_2 = \frac{B_4}{B_2^2} \qquad ((5.25)$$

Among them, Bk(k = 2, 3, 4) are the kth central moments of the sample, respectively referred to as g1 and g2 for sample skewness and sample kurtosis.

If the population x is a normal variable, it can be proven that when n is sufficiently large, approximately there is

$$g_1 \sim N\left(0, \frac{6(n-2)}{(n+1)\ln+3}\right) \tag{5.26}$$

$$g_2 \sim N\left(3-\frac{6}{n+1}, \frac{24n(n-2)(n-3)}{(n+1)^2(n+3)(n+5)}\right) \tag{5.27}$$

Let x1, x2, ⋯, xn be a sample from the population x. Now to test the hypothesis H0: x is a normal population.

Note: $\sigma_1 = \sqrt{\frac{6(n-2)}{(n+1)(n+3)}}$, $\sigma_1 = \sqrt{\frac{24n(n-2)(n-3)}{(n+1)^2(n+3)(n+5)}}$, $\mu_2 = 3-\frac{6}{n+1}$, $u_1 = g_{1/\sigma_1}$, $u_2 = (g_2 - \mu_2) / \sigma_2$.

When H0 is true and n is sufficiently large, approximately there is:

$$u_1 \sim N(0,1), u_2 \sim N(0,1).$$

Since the sample g1, g2 converge in probability to the population skewness $\nu1$ and kurtosis $\nu2$, when H0 is true and n is sufficiently large, generally speaking, the deviation of g1 from $\nu 1 = 0$ should not be too large, and the deviation of g2 from $\nu 2 = 3$ should not be too large. Therefore, intuitively, when $|u_1|$or$|u_2|$ is too large, reject H0. Taking the significance level as α, the rejection region for H0 is

$$|u_1| \geq k_1 \text{ OR } |u_2| \geq k_2$$

Among them, k1 and k2 are determined by the formula below, that is, namely

$$P_{H_0}\{|u_1| > k_1\} = \frac{\alpha}{2} \tag{5.28}$$

$$P_{H_0}\{|u_2| > k_2\} = \frac{\alpha}{2} \tag{5.29}$$

Thus, we have $k_1 = Z_{\alpha/4}$, $k_2 = Z_{\alpha/4}$. Therefore, the rejection region is:

$k_1 = Z_{\alpha/4}$, $k_2 = Z_{\alpha/4}$

The content of the image is translated as follows while preserving the original formatting conventions:

Below verifies that when n is sufficiently large, the aforementioned test approximately satisfies the requirement of significance level α. Specifically, when n is sufficiently large:

$$\begin{aligned} p\{\text{reject } H_0|H_0 \text{ is true}\} &= P_{H_0}\left\{\left(|u_1| \geq z_{\alpha/4}\right) \cup \left(|u_2| \geq z_{\alpha/4}\right)\right\} \leq P_{H_0}\left\{|u_1| \geq z_{\alpha/4}\right\} \\ &+ P_{H_0}\left\{|u_2| \geq z_{\alpha/4}\right\} = \frac{\alpha}{2} + \frac{\alpha}{2} = \alpha \end{aligned} \tag{5.30}$$

5.3.2 *Standardization of Metric Data*

Before data analysis, it is usually necessary to standardize the data first, using standardized data for data analysis. Data standardization mainly includes data normalization and non-scaling processing in two aspects. Data normalization mainly solves the problem of different nature data, and direct addition of different indicators cannot correctly reflect the comprehensive results of different forces. It is necessary to consider changing the inverse index data characteristics first, making all indicators have the same trend, and then adding up to get the correct results. Non-scaling processing mainly solves the comparability of data. After the above standardization processing, the original data is converted into dimensionless index evaluation values, that is, and all indicators are on the same scale, allowing for comprehensive evaluation and analysis.

Advantages of standardization: improve the convergence speed of the model; improve the precision of the model, which is particularly significant when involving distance calculations.

There are many methods of data standardization, which can be summarized into linear methods (such as extreme value method, standard deviation method), piecewise methods (such as three-piece method), and curve methods (such as normal distribution). Different standardization methods have different impacts on the evaluation results of the system. There is no general rule that can be followed for the selection of data standardization methods in experiments. The most commonly used are min-max standardization and Z-score standardization.

1. min-max standardization/0-1 standardization/linear function normalization/difference standardization is a linear transformation of the original data, making the results fall into the [0,1] interval, and the conversion function is

$$\mathbf{x}^* = \frac{\mathbf{x} - \mathbf{min}}{\mathrm{max} - \mathbf{min}} \tag{5.31}$$

In the formula:max: Maximum value of the sample data; min: Minimum value of the sample data.

If you want to map the data to the interval [−1,1], replace the formula with:

$$x^* = x^* * 2 - 1 \tag{5.32}$$

The content in the image translates to English as: "or perform an approximation, that is:

$$x^* = \frac{x - x\_mean}{x\_max - x\_min} \tag{5.33}$$

In the formula, x^- represents the mean of the data.

This method has a drawback, which is that when new data is added, it may cause changes in the maximum and minimum values of the sample data, requiring redefinition.

2. Z-score Standardization (zero-mean normalization)

Z-score standardization is one of the most commonly used standardization methods in many statistical tools and is also the default standardization method in SPSS.

Z-score standardization is also known as standard deviation standardization. This method standardizes the data based on the mean and standard deviation of the original data.

The processed data conforms to the standard normal distribution, with a mean of 0 and a standard deviation of 1. The transformation function is

$$x^* = \frac{x - \mu}{\sigma} \tag{5.34}$$

In the formula: μ is the mean of all sample data; σ is the standard deviation of all sample data.

The Z-score standardization method is suitable for situations where the maximum and minimum values of attribute A are unknown, or there are outliers beyond the range of values. This normalization method requires that the distribution of the original data can be approximated by a Gaussian distribution; otherwise, the effect of normalization will be affected.

Usage scenarios for min-max normalization and Z-score standardization, the most commonly used methods:

(a) In classification and clustering algorithms, when distance is needed to measure similarity, or when using PCA technology for dimensionality reduction, the Z-score standardization method performs better.
(b) When not involving distance measurement, covariance calculation, and when the data does not conform to a normal distribution, the min-max normalization method or other normalization methods can be used.

3. Other methods

(a) Logarithmic function transformation. The method of using a logarithmic function with base 10 for transformation can also achieve normalization, the specific method is

$$x^* = \frac{\log_{10}(x)}{\log_{10}(\max)} \tag{5.35}$$

In the formula: max is the maximum value of the sample data. The logarithmic function transformation method requires all data to be no less than 1.
(b) arctan function transformation. The inverse tangent function can also be used to normalize data, that is

$$x^* = \frac{\arctan(x) * 2}{\pi} \tag{5.36}$$

When using the arctan function transformation method, it is important to note that if the desired mapping interval is [0,1], then all data should be no less than 0. Data less than 0 will be mapped to the [−1,0] interval, and not all data standardized will be mapped to the [0,1] interval.
(c) Decimal scaling standardization. This method standardizes data by shifting the decimal point position. The number of decimal places to shift depends on the maximum absolute value of the attribute A's values.

The calculation method for standardizing the original value x of attribute A to x′ using decimal scaling standardization is

$$x^* = \frac{x}{10^j} \tag{5.37}$$

In the formula: j is the smallest integer that meets the condition.

For example, suppose the values of A range from −986 to 917, and the maximum absolute value of A is 986. To perform decimal scaling standardization, divide each value by 1000 (i.e., j = 3), so −986 is normalized to −0.986. Note that decimal scaling standardization will alter the original data, so it is necessary to save the parameters of the standardization method used to standardize subsequent data uniformly.

(d) Sigmoid function standardization. The sigmoid function (also known as the S-shaped curve or logistic function) is commonly used as an activation function in neural networks due to its monotonically increasing nature and the monotonically increasing nature of its inverse function, mapping variables to the interval [0,1].

The sigmoid function, also known as the logistic function, is used for the output of hidden layer neurons, with a value range of (0,1). It can map a real number to the interval (0,1) and can be used for binary classification. It works particularly well when the features are complex or when the differences are not particularly large. The sigmoid function as an activation function has the following advantages and disadvantages.

- Advantages: Smooth and easy to derive.
- Disadvantages: The activation function requires a large amount of computation, and the derivation involves division when calculating the error gradient during backpropagation; backpropagation can easily lead to the vanishing gradient problem, making it impossible to complete the training of deep networks.

The sigmoid function is defined by the following formula, that is

$$S(x) = \frac{1}{1+e^{-x}} \tag{5.38}$$

Its derivative with respect to x can be expressed in terms of itself, that is

$$S'(x) = \frac{e^{-x}}{\left(1+e^{-x}\right)^2} = S(x)\left(1-S(x)\right) \tag{5.39}$$

5.3.3 Dimensionality Reduction Methods

Due to the large number of attribute dimensions formed by the aggregation of software metrics, using these data directly for modeling will affect the performance of the evaluation model. Dimensionality reduction of sample data can eliminate the correlation between features, reduce the workload of feature selection, and ensure the performance of the model. Common dimensionality reduction methods include correlation analysis and principal component analysis, which will be introduced below.

Correlation analysis is a supervised dimensionality reduction method that selects features with significant correlation with label values as inputs to the model. Since the label values of the samples are discrete, the Spearman correlation coefficient, which can measure the degree of correlation between ordered categorical variables, is used.

Principal component analysis is an unsupervised dimensionality reduction method, meaning that dimensionality reduction of data attributes can be performed without considering the label values of the samples. The steps of principal component analysis are as follows: incrementally increase the number of principal components from 1 to (number of samples - 1), and perform principal component analysis on the sample data until the variance explained by the principal components exceeds a set upper limit. At this point, the number of principal components is the number of features after dimensionality reduction.

Both principal component analysis and correlation analysis extract different information by constructing appropriate linear combinations of the original variables. Principal component analysis focuses on considering the "dispersion" information of the variables, while correlation analysis is based on identifying and quantifying the statistical correlation between two sets of variables, which is an extension of the correlation between two random variables to two sets of variables.

5.3.3.1 Correlation Analysis

Correlation analysis is a statistical analysis method that studies the relationships between two or more random variables of equal status. Correlation analysis is the analysis of indicators that have a real connection in the population, and its main subject is the analysis of indicators with causal relationships in the population. It is a process that describes the degree of relationship between objective things and expresses it with appropriate statistical indicators.

To determine the relationship between related variables, one should first collect some paired data and then describe these points on a Cartesian coordinate system. This set of points is called a "scatter plot."

According to the scatter plot, when the independent variable takes a certain value, the dependent variable corresponds to a certain probability distribution. If the probability distribution for all values of the independent variable is the same, it indicates that there is no correlation between the dependent and independent variables; on the contrary, if the values of the independent variable are different, and the distribution of the dependent variable is also different, it indicates that there is a correlation between the two.

The degree of correlation between two variables is represented by the correlation coefficient r. The value of the correlation coefficient r is between -1 and 1, but it can be any value within this range. When positively correlated, the value of r is between 0 and 1, and the scatter plot slopes upward, indicating that when one variable increases, the other variable also increases; when negatively correlated, the value of r is between −1 and 0, and the scatter plot slopes downward, at this time when one variable increases, the other variable will decrease. The closer the absolute value of r is to 1, the stronger the correlation between the two variables; the closer the absolute value of r is to 0, the weaker the correlation between the two variables.

Correlation can be classified according to the following rules:

1. By the degree of correlation, it is divided into perfect correlation, imperfect correlation, and no correlation.
   (a) Two dependent relationships, where the change in the quantity of one indicator is determined by the change in the quantity of another indicator, are called perfectly correlated, also known as a functional relationship.
   (b) Two indicators do not affect each other, and their quantity changes are independent, known as no correlation.
   (c) The relationship between two phenomena, which is between perfect correlation and no correlation, is called imperfect correlation.
2. By the direction of correlation, it is divided into positive correlation and negative correlation.
   (a) Positive correlation refers to the relationship where the quantity changes of the factor indicator and the result indicator are in the same direction.
   (b) Negative correlation refers to the relationship where the quantity changes of the factor indicator and the result indicator are in opposite directions.
3. By the form of correlation, it is divided into linear correlation and nonlinear correlation.

   A value of one phenomenon and the corresponding value of another phenomenon determined on an exponential coordinate system is called linear correlation.
4. By the number of influencing factors, it is divided into simple correlation and multiple correlation.
   (a) If the study is about the correlation between one result indicator and one factor indicator, it is called simple correlation.
   (b) If the analysis involves the impact of several factor indicators on the result indicator, it is called multiple correlation or multivariate correlation.

For different types of data, the methods of correlation analysis are different. This section mainly organizes and summarizes various correlation analysis methods according to different data types.

1. Correlation between discrete and discrete variables
   (a) Chi-square test.

      The chi-square test is a widely used hypothesis testing method for count data. The chi-square test is a statistical measure of the degree of deviation between the actual observed values and the theoretical inferred values of the sample. The degree of deviation between the actual observed values and the theoretical inferred values determines the size of the chi-square value. The larger the chi-square value, the greater the degree of deviation; conversely, the smaller the degree of deviation. If the two values are completely equal, the chi-square value is 0, indicating that the theoretical value completely fits.

The chi-square test belongs to the category of nonparametric tests, mainly comparing the rates (proportions) of two or more samples and the correlation analysis of two categorical variables. Its fundamental idea is to compare the degree of fit or goodness of fit between theoretical frequencies and actual frequencies.

The application of the chi-square test in statistical inference of categorical data includes the chi-square test for comparing two rates or two proportions, the chi-square test for comparing multiple rates or multiple proportions, and the correlation analysis of categorical data, etc.

- The steps of the chi-square test are as follows.
  - State the null hypothesis:

    H0: The distribution function of the population X is F(x).
    If the population distribution is discrete, then the hypothesis is specific to:
    H0: The distribution of the population X is P(X = xi) = pi, i = 1, 2, ⋯.
  - Divide the range of values of the population X into k non-overlapping intervals A1, A2, A3, ⋯, Ak, such as A1 = (a0, a1], A2 = (a1, a2], ⋯, Ak = (ak − 1, ak). Here a0 can be $-\infty$, and ak can be $+\infty$. The division of intervals should be determined based on specific circumstances, but each interval should contain no fewer than 5 sample values, and the number of intervals k should not be too large or too small.
  - The number of sample values in the i-th interval Ai is denoted as fi, known as the group frequency (actual value), and the sum of all group frequencies f1 + f2 + ⋯ + fk equals the sample size n.
  - When H0 is true, based on the assumed theoretical distribution of the population, the probability pi that the value of the population X falls into the i-th interval Ai can be calculated, thus npi is the theoretical frequency (theoretical value) of the sample values falling into the i-th interval Ai.
  - When H0 is true, the frequency fi/n of the sample values falling into the i-th interval Ai in n trials should be close to the probability pi. When H0 is not true, fi/n and pi will differ greatly. Based on this idea, Pearson introduced the following test statistic, that is

$$\chi^2 = \sum_{i=1}^{k} \frac{(f_i - np_i)^2}{np_i} \tag{5.40}$$

Table 5.5 Sample frequency data

Variable	y_1	y_2	Total
x_1	a	b	a + b
x_2	c	d	c + d
Total	a + c	b + d	a + b + c + d

Table 5.6 Inspection data

P ($\chi^2 \geq$ k)	0.5	0.4	0.25	0.15	0.1
k	0.455	0.708	1.323	2.072	2.706
P ($\chi^2 \geq$ k)	0.05	0.025	0.01	0.005	0.001
k	3.841	5.024	6.635	7.879	10.828

– Under the assumption that H0 holds, it follows a chi-square distribution with k − 1 degrees of freedom.

• Test method. Use an independent sample fourfold table.

• With 1 degree of freedom, suppose there are two categorical variables X and Y, their value ranges are $\{x1, x2\}$ and $\{y1, y2\}$, respectively, and their sample frequency contingency table is shown in Table 5.5.

If the hypothesis to be tested is H_1: "X is related to Y", an independence test can be used to examine whether the two variables are related and to provide a more precise measure of the reliability of this judgment. The specific method is: calculate the value of the test statistic $\chi 2$ from the data in Table 5.6. The larger the value of $\chi 2$, the greater the likelihood that "X is related to Y" holds true.

When the values of a, b, c, and d in Table 5.5 are all greater than or equal to 5, the degree of confidence in the conclusion that 'X is related to Y' can be determined by consulting the data in the table.

(b) Information Gain and Information Gain Ratio

Before introducing information gain, let us first introduce two fundamental concepts: information entropy and conditional entropy. Information entropy represents the degree of uncertainty associated with a random variable and is defined as follows:

$$H(X) = -\sum_{i=1}^{k} p_i \log p_i \tag{5.41}$$

Conditional entropy refers to the degree of uncertainty of a random variable given a condition, and is defined as follows:

$$H(X|Y) = -\sum_{j=1}^{n} p(y_j) H(X|y_j) = -\sum_{j=1}^{n} p(y_j) \sum_{i=1}^{k} p(x_i|y_i) \mathit{logp}(x_i|y_i) \tag{5.42}$$

- Information gain is defined as the difference between entropy and conditional entropy, and quantifies the reduction in uncertainty under a given condition

$$\mathrm{Gain}(Y,X) = H(Y) - H(Y|X) \tag{5.43}$$

The larger the information gain, the greater the reduction in impurity achieved by introducing condition X. A higher information gain indicates a stronger correlation between the two variables.
The information gain ratio accounts for the effect of the number of branches when a variable has a large number of distinct values, such as ID. Introducing ID can reduce the impurity of each sub-node to zero, thereby maximizing the reduction in information gain. Therefore, when the number of distinct values for different variables varies significantly, introducing variables with more values tends to result in a higher information gain. Consequently, when using the information gain ratio, the influence of the number of branches must be considered.

$$\mathrm{Gain\_ratio} = \left(H(Y) - H(Y|X)\right) / H(Y|X) \tag{5.44}$$

2. Correlation between two continuous variables.
   (a) Covariance.
   Covariance expresses the co-variation relationship between two random variables. If two variables are uncorrelated, the covariance is 0.

$$\mathrm{Cov}(X,Y) = E\{[X - E(X)],[Y - E(Y)]\} \tag{5.45}$$

   When cov(X, Y) > 0,it indicates a positive correlation between X and Y.
   When cov(X, Y) < 0, it indicates a negative correlation between X and Y.
   When cov(X, Y) = 0, it indicates that X and Y are uncorrelated.
   Covariance can only be used to analyze the correlation between two sets of data. When there are more than two sets of data, a covariance matrix is needed.
   Covariance quantifies the correlation between variables with numerical values, with positive values indicating a positive correlation and negative values indicating a negative correlation. However, it does not measure the strength of the correlation. When dealing with multiple variables, covariance cannot determine which pair of data sets exhibits the strongest correlation. To measure and compare the strength of the correlation, the correlation coefficient method is required.
   (b) Linear correlation coefficient.
   The linear correlation coefficient, also known as the Pearson correlation coefficient, measures the degree of linear relationship between two variables.

$$r = \mathrm{cov}(X,Y)/(D(X)D(Y)) \quad (5.46)$$

The correlation coefficient is obtained by dividing the covariance by the product of the standard deviations of the two random variables. Its value ranges from −1 to 1. The use of the correlation coefficient eliminates the issue of numeric inflation caused by changes in measurement units.

The linear correlation coefficient is meaningful only if there is a linear relationship between the dependent and independent variables; otherwise, it is meaningless.

3. The correlation between continuous and discrete variables.

   (a) Discretization of continuous variables.
   Discretize the continuous variables and then use methods for analyzing the correlation between discrete variables to assess the relationship.

   (b) Boxplot.
   By using boxplots, you can examine the mean, variance, and distribution of continuous variable values when the discrete variable takes on different values
   If the boxplots of the continuous variable corresponding to different values of the discrete variable show little difference, it indicates that the discrete variable has a minimal effect on the continuous variable, and the correlation is low. Conversely, if there is a significant difference, it suggests a high correlation.

5.3.3.2 Principal Component Analysis (PCA)

In multivariate statistical analysis, Principal Component Analysis (PCA) is a method used for statistical analysis and data reduction. It employs orthogonal transformations to linearly transform a set of possibly correlated variable observations into a set of linearly uncorrelated variables, known as principal components. Specifically, principal components can be interpreted as a linear equation, which includes a set of linear coefficients that indicate the direction of the projection. PCA is sensitive to the normalization or preprocessing of the original data (i.e., relative scaling).

1. Basic Contents of PCA
   The basic idea of PCA (Principal Component Analysis) is to transform high-dimensional data into low-dimensional data while retaining the main information of the data and discarding irrelevant or minor information. Through orthogonal transformation, the original data is converted into a set of new uncorrelated variables, known as principal components. Each principal component is a linear combination of the original data. These principal components are arranged according to the size of their variance. The larger the variance, the more a principal component can represent the main information of the original data. By selecting the first few principal components, we can describe most of the

information of the original data. Therefore, PCA can help us reduce the dimensionality of the data, remove noise and redundant features, and extract the essential characteristics of the data

Move the origin of the coordinate axis to the center of the data, and then rotate the axes such that the variance of the data along the C1 axis is maximized, meaning that the projections of all n data points are most widely spread in that direction. This ensures that more information is retained, with C1 becoming the first principal component.

The second principal component, C2: Find a direction C2 such that the covariance (or correlation coefficient) between C2 and C1 is 0 to prevent overlap with C1 information, while maximizing the variance of the data in that direction.

Following this approach, the third, fourth, …, pth principal components are identified. There can be as many principal components as there are random variables in the dataset (p).

Principal Component Analysis (PCA) is commonly used to reduce the dimensionality of datasets while retaining the features that contribute the most to their variance. This is achieved by preserving the lower-order principal components and discarding the higher-order ones. In this way, the lower-order components often capture the most important aspects of the data. However, this approach is not fixed and depends on the specific application. Since PCA is based on the given data, the accuracy of the data has a significant impact on the results of the analysis.

The mathematical definition of Principal Component Analysis (PCA) is an orthogonal linear transformation that projects the data into a new coordinate system, such that the largest variance of any data projection lies on the first coordinate (the first principal component), the second largest variance on the second coordinate (the second principal component), and so on.

Define an $n \times m$ matrix, XT, as the data with mean values subtracted (centered at the origin), where the rows represent data samples and the columns represent data categories (note that XT is defined here, not X). The Singular Value Decomposition (SVD) of X is given by X = WΣVT, where the $m \times m$ matrix W is the eigenvector matrix of XXT, Σ is a non-negative rectangular diagonal matrix of size $m \times n$, and V is the eigenvector matrix of XXT with size $n \times n$. Based on this, we have:

$$Y^T = X^T W = V\Sigma^T W^T W = V\Sigma^T \tag{5.47}$$

When $m < n - 1$, V is generally not uniquely defined, whereas Y is uniquely defined. W is an orthogonal matrix, and, $Y^T W^T = X^T$. The first column of YT corresponds to the first principal component, the second column to the second principal component, and so on.

To obtain an effective method for reducing the dimensionality of the data, we can use WL to map X into a low-dimensional space that only applies the first L vectors, as follows:

$$Y = W_L^T X = \Sigma_L V^T \tag{5.48}$$

where $\Sigma_L = I_{l*m}$, and I_{L*m} is an $l \times m$ identity matrix.
The one-dimensional matrix of X, W, is equivalent to the eigenvector of the covariance matrix C = XXT, as follows:

$$XX^T = W\Sigma\Sigma^T W^T \tag{5.49}$$

"In Euclidean space, given a set of points, the first principal component corresponds to a line passing through the mean point of the multi-dimensional space, while minimizing the sum of squared distances from each point to this line. After removing the first principal component, the second principal component is obtained using the same method, and so on. The singular values in Σ are the square roots of the eigenvalues of the matrix XXT. Each eigenvalue is proportional to the variance associated with its corresponding eigenvector, and the sum of all eigenvalues equals the sum of the squared distances from all points to their multi-dimensional space mean. PCA provides an effective method for dimensionality reduction. Essentially, it uses orthogonal transformations to project as many variables as possible, concentrated around the mean point, into the first dimension. Therefore, dimensionality reduction is the method that minimizes information loss. PCA is characterized by the optimal orthogonal transformation that maximizes the variance of the subspace. However, it requires higher computational costs compared to the discrete cosine transform. Nonlinear dimensionality reduction techniques, in turn, demand even higher computational resources than PCA.

PCA is sensitive to the scaling of variables. If there are only two variables with the same sample variance and a positive correlation, PCA will involve rotating the principal components of both variables. However, if all values of the first variable are multiplied by 100, the first principal component will closely align with this variable, and the second variable will contribute minimally, with the second principal component also closely aligning with the second original variable. This suggests that PCA can be somewhat arbitrary when different variables represent different units (e.g., temperature and mass). However, in Pearson's original paper, 'On Lines and Planes of Closest Fit to Systems of Points in Space,' it was assumed that such considerations were not taken into account in Euclidean space. One way to reduce the arbitrariness of PCA is to scale the variables to have unit variance.

2. Algorithmic steps of PCA:

   (a) Center the data by subtracting the mean value of each dimension from the values in that dimension.
   (b) Calculate the covariance matrix, where the diagonal elements are the variances of the data in each dimension, and the off-diagonal elements are the covariances between different dimensions.
   (c) Perform eigenvalue decomposition on the covariance matrix to obtain the eigenvalues and their corresponding eigenvectors.

(d) Sort the eigenvectors according to the size of their corresponding eigenvalues in descending order, and select the first k eigenvectors as the principal components.
(e) Map the original data onto the principal components to obtain the data after dimensionality reduction

- Organize the dataset.
- Assume a set of observed data with M variables, where the goal is to reduce the data such that each observation is described by L vectors, with L < M. Furthermore, suppose the data is organized into a set of N vectors, each representing a single observation of M variables.
- Arrange the column vectors $x_1, \ldots, x_n$, where each column vector has M rows, into a single matrix X of size M × N.
- Compute the empirical mean.
- The empirical mean of the data matrix X is defined as the mean of each row of X:
  - Compute the empirical mean for each dimension m = 1, 2, …, M.
  - Put the calculated means into an M × 1-dimensional empirical mean vector u:

$$u[m] = \frac{1}{N}\sum_{i=1}^{N} x[m,n] \tag{5.50}$$

- Compute the mean deviation.
- To find a principal component that minimizes the mean squared error of the approximated data, mean subtraction is an essential part of the solution. Therefore, proceed with the following steps.
  - Subtract the empirical mean vector u from each column of the data matrix X.
  - Store the mean-subtracted data in an M × N matrix B:

$$B = X - uh \tag{5.51}$$

  - where h is a 1 × N row vector of all ones, i.e.,

$$h[n] = 1 \tag{5.52}$$

  - where: $n = 1,2,\ldots,N_o$
- Compute the covariance matrix.
- To find the empirical covariance matrix C of size M × M from the matrix B, we have:

$$C = E[B \otimes B] = E\left[B \otimes B^*\right] = \frac{1}{N-1}\Sigma B \cdot B^* \tag{5.53}$$

- In the equation:
  E denotes the expectation value;
  [] denotes the outermost operator;
  ^T denotes the transpose operator;
  * denotes the matrix multiplication operator;
  Find the eigenvalues and eigenvectors of the covariance matrix.
  Compute the eigenvectors of the matrix C. We have:

$$V^{-1}CV = D \tag{5.54}$$

- In the equation,
  - V is a matrix whose columns are the eigenvectors of C;
  - Λ is a diagonal matrix whose diagonal entries are the eigenvalues of C;
  - D is a matrix that contains the eigenvalues on its diagonal, i.e.,D = diag $(\lambda_1, \lambda_2, \ldots, \lambda_M)$;
  - This step typically involves using a computer-based algorithm to compute the eigenvalues and eigenvectors of C;
  - The resulting matrix D is a diagonal matrix of size M × M.

  The eigenvectors and eigenvalues of a covariance matrix are paired, meaning that each eigenvector corresponds to a single eigenvalue, and vice versa. In other words, for a matrix C of size M × M, there are M eigenvectors and M eigenvalues, and each eigenvector corresponds to a unique eigenvalue.

5.3.4 Several Data Fitting Algorithms

For an extended period, many experts believed that achieving human-level artificial intelligence could be accomplished by developing enough rules to process knowledge, a concept known as symbolic AI. While symbolic AI is effective for solving well-defined logical problems, it encounters difficulties in providing explicit rules for addressing more complex and ambiguous problems. As a result, a new paradigm emerged to replace symbolic AI: machine learning. Similar to the development stages of AI, the field of software quality evaluation has also evolved from symbolic AI and expert systems to machine learning. The relationship between software metrics and software quality is inherently complex and ambiguous. This section aims to introduce fitting algorithms that can be employed to model the relationship between software metrics and software quality.

5.3.4.1 Linear Regression

Linear regression is a statistical analysis method that uses regression analysis in mathematical statistics to determine the quantitative relationship between two or more variables that are mutually dependent. It is a widely used analysis tool. The expression of linear regression is y = w′x + e, where e is the error term that follows a normal distribution with a mean of 0.

In general, linear regression can be derived using the least squares method to compute the equation y = bx + a, which gives the slope and intercept of the line. The simplest univariate linear model can be used to illustrate the least squares method. In regression analysis, if there is only one independent variable and one dependent variable, and their relationship can be approximated by a straight line, this type of regression is referred to as simple linear regression. If the regression analysis includes two or more independent variables, and the relationship between the dependent variable and the independent variables is linear, it is called multiple linear regression. In two-dimensional space, linearity corresponds to a straight line; in three-dimensional space, it corresponds to a plane; and in multidimensional space, it corresponds to a hyperplane.

Taking simple linear regression as an example, for a univariate linear regression model, assume that n sets of observation values (X_1, Y_1), (X_2, Y_2), …, (X_n, Y_n) are obtained from the population. For these n points in the plane, an infinite number of curves can be used for fitting. The goal is to obtain the sample regression function that best fits this set of values. In general, this line should be positioned at the center of the sample data. The criterion for selecting the best fitting curve can be defined as minimizing the total fitting error (i.e., the total residual). The following three methods can be applied:

There are three methods that can be used to determine the position of the line in linear regression by minimizing the total fitting error:

1. Minimizing the sum of residuals: This method minimizes the sum of the residuals, but it may suffer from the issue of cancellation, where positive and negative residuals offset each other.
2. Minimizing the sum of absolute residuals: This method minimizes the sum of the absolute values of the residuals. While it is less sensitive to outliers compared to the first method, it is more computationally complex.
3. Least squares method: This method minimizes the sum of squared residuals, which is computationally convenient and results in estimators with desirable properties. However, it is sensitive to outliers.

The most commonly used method is Ordinary Least Squares (OLS), where the chosen regression model minimizes the sum of squared residuals for all observed values. Q represents the sum of squared residuals, which is based on the squared loss function.

The sample regression model can be defined as:

$$Y_i = \hat{\beta}_0 + \hat{\beta}_1 X_i + e_i => e_i = Y_i - \hat{\beta}_0 - \hat{\beta}_1 X_i \tag{5.55}$$

where ei represents the error of the sample(X_i, Y_i).

Define the squared loss function as:

$$Q = \sum_{i=1}^{n} e_i^2 = \sum_{i=1}^{n} \left(Y_i - \hat{\beta}_0 - \hat{\beta}_1 X_i\right)^2 \tag{5.56}$$

By minimizing Q, we can determine the line and the values of β0 and β1. We can regard them as functions of Q, and this becomes an optimization problem where we can obtain the solution by taking the derivative.

Taking the partial derivatives of Q with respect to the two parameters to be estimated, we have:

$$\frac{\partial Q}{\partial \hat{\beta}_0} = 2\sum_{i=1}^{n}\left(Y_i - \hat{\beta}_0 - \hat{\beta}_1 X_i\right)(-1) = 0 \tag{5.57}$$

$$\frac{\partial Q}{\partial \hat{\beta}_1} = 2\sum_{i=1}^{n}\left(Y_i - \hat{\beta}_0 - \hat{\beta}_1 X_i\right)(-X_i) = 0 \tag{5.58}$$

According to mathematical knowledge, the extremum point of a function is where the partial derivatives are equal to zero. We can solve for:

$$\hat{\beta}_1 = \frac{n\Sigma X_i Y_i - \Sigma X_i \Sigma Y_i}{n\Sigma X_i^2 - (\Sigma X_i)^2} \tag{5.59}$$

$$\hat{\beta}_0 = \frac{n\Sigma X_i^2 Y_i - \Sigma X_i \Sigma X_i Y_i}{n\Sigma X_i^2 - (\Sigma X_i)^2} \tag{5.60}$$

The above is the solution to the method of least squares, which is to find the extremum point of the squared loss function.

5.3.4.2 Logistic Regression

Logistic Regression is a generalized linear regression model. For binary classification problems, the occurrence of an event is defined as Y, where the value of Y is either 0 or 1.

$$y = \begin{cases} 1 \text{ it indicates that the event has occurred} \\ 0 \text{ it indicates that the event has not occurred} \end{cases} \tag{5.61}$$

Assuming that p represents the probability of an event occurring, we can consider p as a linear function of the independent variable Xi, i.e.:

$$\theta(p) = \text{logit}(p) = \ln\left(\frac{p}{1-p}\right) \tag{5.62}$$

The change in p with respect to Xi is slow and highly nonlinear near p = 0 or p = 1. Therefore, we need to find a function θ(p) that exhibits significant variation near p = 0 or p = 1, while remaining relatively simple. Thus, we introduce a logistic transformation of p, i.e.:

$$\theta(p) = \ln\left(\frac{p}{1-p}\right) = \beta_0 + \beta_1 X_1 + \beta_2 X_2 + \cdots + \beta_k X_k + \varepsilon \tag{5.63}$$

The variation of θ(p) is large near p = 0 and p = 1, and when p changes from 0 to 1, θ(p) changes from -∞ to +∞. If the relationship between p and Xi is not linear, then the relationship between θ(p) and Xi can be linear. By replacing p with logit(p), we have:

$$p = \frac{e^{\theta}}{1+e^{\theta}} = \frac{e^{\beta_0 + \beta_1 X_1 + \beta_2 X_2 + \cdots + \beta_k X_k + \varepsilon}}{1+e^{\beta_0 + \beta_1 X_1 + \beta_2 X_2 + \cdots + \beta_k X_k + \varepsilon}} \tag{5.64}$$

By expressing p in terms of θ, we have:

$$L_p\left(x_i, x_j\right) = \left(\sum_{l=1}^{n} \left|x_i^{(l)} - x_j^{(l)}\right|^p\right)^{\frac{1}{p}} \tag{5.65}$$

The regression equation that satisfies the above conditions is called a logistic linear regression.

In practical applications, variables may be ordinal multi-class variables, such as evaluation metrics categorized into poor, fair, good, and excellent. This type of logistic regression analysis is typically referred to as a proportional odds model, which requires fitting m – 1 logistic regression models (where m is the number of categories). When working with a dataset, logistic regression is often the first algorithm attempted to become familiar with the classification task. However, logistic regression is sensitive to multicollinearity among the independent variables, so principal component analysis should be applied beforehand to reduce correlations between the features.

5.3.4.3 KNN Algorithm

The core idea of the KNN algorithm is that if the majority of the k nearest neighbors of a sample in the feature space belong to a certain category, then the sample is also classified into this category and exhibits the characteristics of the samples in this category. This method determines the classification of the sample based solely on the categories of the nearest one or a few neighboring samples. The KNN method is only concerned with a small number of neighboring samples when making classification decisions. Since the KNN method relies on a limited number of nearby samples, rather than discriminant methods based on class boundaries, it is more suitable for datasets where the class boundaries are highly overlapped or ambiguous, compared to other methods.

When selecting the similarity between two instances, the commonly used Euclidean distance Lp is defined as follows.

$$L_\infty\left(x_i, x_j\right) = \max_{l} \left|x_i^{(l)} - x_j^{(l)}\right| \tag{5.66}$$

Where xi ∈ Rn, xj ∈ Rn, and L∞ is defined as follows.

$$L_1 = \Sigma k = \ln\left|X_{1k} - X_{2k}\right| \tag{5.67}$$

In the formula, p is a variable parameter.

1. When p = 1, it is the Manhattan distance (corresponding to the L1-norm). Manhattan distance corresponds to the L1-norm, which is the sum of the projection distance of the line segment formed by two points on the axes of the fixed Cartesian coordinate system in Euclidean space. For example, in the plane, the Manhattan distance between point P1 at coordinates (x1, y1) and point P2 at coordinates (x2, y2) is $X_1 - X_2 + Y_1 - Y_2$. Note that Manhattan distance depends on the rotation of the coordinate system, not the translation or mapping of the system on the coordinate axes.

   The Manhattan distance is defined as:

$$L_1 = \sum_{k=1}^{n} \left|x_{1k} - x_{2k}\right| \tag{5.68}$$

The L1-norm is represented as:

$$|x| = \sum_{i=1}^{n} |x_i| \tag{5.69}$$

where

$$x = \begin{bmatrix} x_1 \\ \dots \\ x_n \end{bmatrix} \epsilon R^n$$

2. When p = 2, it is the Euclidean distance (corresponding to the L2-norm).

   The Euclidean distance, also known as Euclidean metric, is the most common way to represent the distance between two or more points in Euclidean space. It is defined as the distance between two points $X_1 = (X_{11}, X_{12}, \ldots, X_{1n})$ *and* $X_2 = (X_{21}, X_{22}, \ldots, X_{2n})$ in n-dimensional space, and is given by the following formula:

$$d_{12} = \sum_{i=1}^{n} \sqrt{\left(x_1 - x_2\right)^2} \tag{5.70}$$

The L2-norm (Euclidean norm) is represented as

$$|x| = \sqrt{\sum_{i=1}^{n} x_i^2} \tag{5.71}$$

Where

$$x = \begin{bmatrix} x_1 \\ \dots \\ x_n \end{bmatrix} \epsilon R^n$$

3. When p → ∞, it is the Chebyshev distance.
   The Chebyshev distance between two points a(x1,y1) and b(x2,y2) in the two-dimensional plane is defined as:

$$\mathrm{d12} = \max\left(\mathrm{x1} - \mathrm{x2|y1} - \mathrm{y2}\right) \tag{5.72}$$

The Chebyshev distance between two points $a = (X_{11}, X_{12}, \ldots, X_{1n})$ and $b = (X_{21}, X_{22}, \ldots, X_{2n})$ in n-dimensional space is defined as:

$$\mathrm{d12} = \max\left(\mathrm{x}_{1i} - \mathrm{x}_{2i}\right) \tag{5.73}$$

Choosing a smaller value of k is equivalent to using a smaller neighborhood of training instances for prediction, which reduces the approximation error of learning. In this case, only the training instances close to the input instance will affect the prediction results. However, the disadvantage of a small k value is that the estimation error increases, making the model more sensitive to nearby instances. If the neighboring instances are noisy, prediction errors may occur. In other words, reducing the k value increases the complexity of the model, making it harder to distinguish between instances and more prone to overfitting.
Choosing a larger k value is equivalent to using a larger neighborhood of training instances for prediction. The advantage is that it reduces the estimation error of learning, but the approximation error increases, leading to less accurate predictions for the input instance. An increase in the k value results in a simpler overall model.

5.4 BP Neural Network

A feedforward neural network, also known as a forward neural network, only has feedback signals during the training process. In the operational phase, signals can only flow from the input layer to the hidden layer, and then from the hidden layer to the output layer. This process is irreversible, which is why it is called a 'feedforward' network. The backpropagation (BP) neural network is a typical example of a feedforward network.

The backpropagation neural network (BP neural network) is primarily used for approximating the input-output relationship of functions. The learning rule of the BP neural network typically employs the steepest descent method. During the backward propagation of the error, from the output layer through the hidden layer to the input layer, the weights of the neurons are adjusted according to the learning rule. The objective is to minimize the sum of squared errors. A schematic diagram of the BP network structure is shown in Fig. 5.9.

The topology of the BP neural network consists of three layers, namely, the input layer, the hidden layer, and the output layer.

Input layer: This layer receives all external input information and performs mathematical operations on it within the neurons, and then transfers the results to other layers of the neural network.

Hidden layer: This layer is responsible for information transformation and is the internal information processing layer of the neural network. The middle layer can be designed as a single layer or multiple layers. Usually, within a certain range, the more layers there are, the better the descriptive performance of the data; the last hidden layer passes information to the neurons of the output layer.

Output layer: The output layer receives signals transmitted from the hidden layer, processes the signals, and then sends the final results to the outside, completing one forward propagation process.

Notice:

The topology of a BP neural network mainly consists of three network layers: the input layer, the hidden layer, and the output layer.

Input Layer: This layer receives all external input information, such as images, text, or other forms of data, and passes it into the neural network. Each input variable corresponds to a neuron in the input layer.

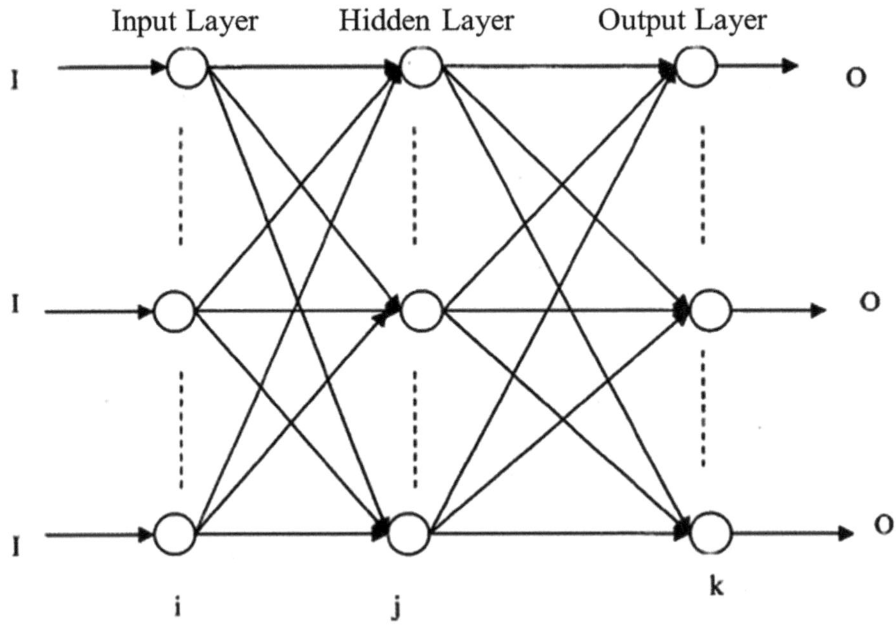

Fig. 5.9 Schematic diagram of the BP network structure

Hidden Layer: This layer is responsible for processing, transforming, and combining the input information, typically using nonlinear functions to handle the data. The number of neurons in the hidden layer can be chosen arbitrarily, and there can be multiple hidden layers. Multiple hidden layers can enhance the network's complexity and expressive power.

Output Layer: This layer processes the output from the hidden layer and provides the final output of the neural network to the outside world. In classification problems, the output layer often uses the softmax function to convert the output into a probability distribution. In regression problems, the output layer usually does not use an activation function.

The BP neural network is trained using the backpropagation algorithm. Specifically, it first performs forward propagation of the input to compute the output and calculate the output error. Then, through error backpropagation, it calculates the error contribution of each neuron in the layers and uses gradient descent to update the network parameters, thereby minimizing the output error. This is a supervised learning algorithm.

The training of the BP neural network in Fig. 5.10 consists of two processes: forward propagation and backpropagation. The training process begins with weight initialization and the input of training samples. Data information then propagates from the input layer to the hidden layer and from the hidden layer to the output layer, which is the forward propagation. After forward propagation, the actual output is compared with the desired output. If the accuracy meets the requirements, another training sample is used to continue the training; otherwise, backpropagation is performed. In backpropagation, the error from the output layer is propagated backward through the hidden layer to the input layer. During this process, the weights of the output and hidden layer neurons are adjusted. New training samples

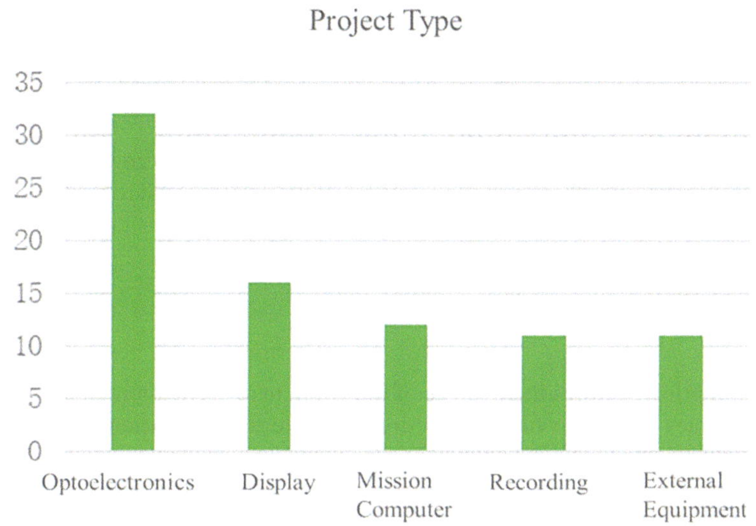

Fig. 5.10 Histogram of Project Type Distribution

are then introduced, and the process continues until the desired accuracy is achieved. After training on a large number of data samples, the weights of each neuron are fixed, and the network enters the operational phase.

Taking the single hidden layer BP neural network shown in Fig. 5.10 as an example, let the input layer have i neurons, the hidden layer have j neurons, and the output layer have k neurons. The mathematical formula for forward propagation is as follows:

The input of the jth neuron in the hidden layer is:

$$\mathrm{net}_j = \sum_i \omega_{ji} o_i \tag{5.74}$$

The output of the jth neuron in the hidden layer is:

$$o_j = g\left(\mathrm{net}_j\right) \tag{5.75}$$

The input of the kth neuron in the output layer is:

$$\mathrm{net}_k = \sum_j \omega_{kj} o_j \tag{5.76}$$

The output of the k_{th} neuron in the output layer is:

$$o_k = g\left(\mathrm{net}_k\right) \tag{5.77}$$

Here, g(·) denotes the activation function.

In the error back propagation process of the BP neural network, the adjustment of neuron weights is accomplished by minimizing the objective function; the most commonly employed method is gradient descent.

Let t_k represent the desired output and ok the actual output. Then, the mean error is defined as:

$$E = 1/2\left(\sum_k \left(t_k - o_k\right)^2\right) \tag{5.78}$$

In this expression, E denotes the objective function. According to the gradient descent algorithm, the change in the weight Δωkj is proportional to the partial derivative of E with respect to ωkj, that is,

$$\Delta\omega_{kj} = \frac{\partial E}{\partial \omega_{kj}} \tag{5.79}$$

This formulation captures the essence of the gradient descent approach, wherein the weight updates are made in the direction opposite to the gradient (i.e., along the negative gradient) to minimize the objective function.

For instance, when the sigmoid function is used as the activation function, defined by:

$$g(x) = \frac{1}{1 + e^{-(x+\theta)}}$$

its derivative with respect to x can be expressed as:

$$g'(x) = g(x)(1 - g(x))$$

Here, g′(x) represents the gradient of the sigmoid function, a result that simplifies the computation of gradients in learning algorithms by enabling the derivative to be directly calculated from the output g(x).

$$\Delta\omega_{kj} = -\eta \frac{\partial E}{\partial \omega_{kj}} = \eta (t_k - o_k) o_k (1 - o_k) o_j \tag{5.80}$$

In this expression, we define the local error term for neuron k as $\delta_k = \left(\frac{-\partial E}{\partial \omega_{ji}}\right) = (t_k - o_k) o_k (1 - o_k)$,

For neurons in the hidden layer, the weight update rule can similarly be expressed as:

$$\Delta\omega_{ji} = -\eta \frac{\partial E}{\partial \omega_{kj}} = -\eta \frac{\partial E}{\partial o_j} (1 - o_j) o_j \tag{5.81}$$

Let $\delta_j = \left(\frac{-\partial E}{\partial o_j}\right)(1 - o_j) o_j$,

then the adjustment for each weight coefficient is derived as follows:

$$\Delta\omega_{kj} = \eta (t_k - o_k) o_k (1 - o_k) o_j \tag{5.82}$$

$$\Delta\omega_{ji} = \eta \delta_j o_i \tag{5.83}$$

Where η denotes the learning rate.

The learning algorithm for a BP (Backpropagation) neural network can be summarized as follows:

1. Initialize the weights: Set the initial values for all connection weights.
2. Sample extraction and input: Retrieve a training sample and input it into the network.
3. Forward propagation: Propagate the input data through the network, from the input layer to the hidden layer, and then to the output layer, to generate the network's output.
4. Error computation: Compare the network's output with the target (desired) output to determine the error.
5. Backpropagation: Initiate the backpropagation process and adjust the neuron weights according to the learning rule.
6. Iteration: Repeat the aforementioned steps for all training samples, monitoring the error throughout the process.
7. Convergence: Conclude the training process once the error falls below a predefined threshold or the desired accuracy is attained.

5.5 A Specific Case Study

In this section, we present a case study in which a data-driven approach to software reliability evaluation was implemented in a specific project.

In the aforementioned project, the author collected comprehensive data from 82 historical software projects conducted by a research institution. The dataset comprised software code metrics, testing metrics, and survey as well as scoring data related to software quality evaluation objectives. Utilizing this multifaceted dataset, the research involved the aggregation of software code metrics, rigorous data preprocessing, and the development of a software quality model. Moreover, associated supporting tools were created to facilitate the process. This case study is presented as a practical reference for readers.

5.5.1 Data Collection

5.5.1.1 Data Collection Plan

A preliminary data collection plan was developed for a research institution, using the previously presented software quality/reliability evaluation framework, relevant standards, and the organization's actual conditions as references. The data collection process was divided into the following three steps.

Step 1: Determine Software Quality Evaluation Objectives

1. A questionnaire for software quality evaluation objectives (see Appendix 2) was designed based on the "GB/T 25000-10—2016 System and Software Engineering—System and Software Quality Requirements and Evaluation" standard. The purpose of this questionnaire is to ascertain the software quality sub-characteristics that are prioritized by the research institution.
2. A survey was conducted among developers, testers, and departmental managers within the organization.
3. The results of the questionnaire were compiled and analyzed to calculate the scores for each software quality sub-characteristic.
4. Software quality evaluation objectives were then established based on these scores and the practical requirements of the organization.

Step 2: Scoring of Software Quality Evaluation Objectives

1. Based on the standards "GB/T 25000-10—2016 System and Software Engineering—System and Software Quality Requirements and Evaluation" and "GB/T 30961—2014 Embedded Software Quality Metrics", the 10 sub-characteristics designated as evaluation objectives were further subdivided. This subdivision led to the development of a scoring table for software quality evaluation objectives (see Appendix 3).

2. Software testers evaluated these objectives using the scoring table, wherein each sub-item was assigned a score of 5, 4, 3, or 2, or an option for "unable to judge". The average score for each sub-characteristic was computed by averaging the scores assigned to its sub-items.
3. Although the resulting average scores were non-integer values, the high degree of dispersion rendered continuous function fitting methods unsuitable. Consequently, the scores for the software quality evaluation objectives were converted into categorical labels to facilitate sample classification, mapping to quality grades of "excellent", "good", "fair", and "poor" in accordance with the rules specified in Table 5.7.

Step 3: Software Metric Collection

1. Collect software code metric data for each project, including all 33 function-level code metrics provided by the static code analysis function of the Helix QAC testing tool, as well as 5 comment line metrics and 4 nesting depth metrics for functions provided by LARA Testbed. These 42 software code metrics are listed in Table 5.8.

Table 5.7 Rules for scoring and quality level correspondence

Score	Level
$4.5 < \text{score} \leq 5.5$	Excellent
$3.5 < \text{score} \leq 4.5$	Good
$2.5 < \text{score} \leq 3.5$	Fair
$2.0 < \text{score} \leq 2.5$	Poor

Table 5.8 Collecting code matrix

Abbreviation	Thole name	Tool
Metric	Description	Testbed
TC	Total comments	LARA tested
CH	Comments in headers	LARA tested
CD	Comments in declarations	LARA tested
CE	Comments in executable code	LARA tested
BC	Blank comments	LARA tested
NL	Number of loops	LARA tested
ND	Nesting depth	LARA tested
IN	Order 1 intervals	LARA tested
MN	Max Int nesting	LARA tested
AKI	Akiyama's criterion	helix QAC
AV1	Average size of statement in function (variant 1)	helix QAC
AV2	Average size of statement in function (variant 2)	helix QAC
AV3	Average size of statement in function (variant 3)	helix QAC
BAK	Number of backward jumps	helix QAC

(continued)

Table 5.8 (continued)

Abbreviation	Thole name	Tool
Metric	Description	Testbed
CAL	Number of functions called from function	helix QAC
CYC	Cyclomatic complexity	helix QAC
ELF	Number of dangling else-ifs	helix QAC
FN1	Number of operator occurrences in function	helix QAC
FN2	Number of operand occurrences in function	helix QAC
GTO	Number of goto statements	helix QAC
KDN	Knot density	helix QAC
KNT	Knot count	helix QAC
LCT	Number of local variables declared	helix QAC
LIN	Number of code lines	helix QAC
LOP	Number of logical operators	helix QAC
M07	Essential cyclomatic complexity	helix QAC
M19	Number of exit points	helix QAC
M29	Number of functions calling this function	helix QAC
MCC	Myer's interval	helix QAC
MIF	Deepest level of nesting	helix QAC
PAR	Number of function parameters	helix QAC
PBG	Residual bugs (PTH-based est.)	helix QAC
PDN	Path density	helix QAC
PTH	Estimated static program paths	helix QAC
RET	Number of return points in function	helix QAC
ST1	Number of statements in function (variant 1)	helix QAC
ST2	Number of statements in function (variant 2)	helix QAC
ST3	Number of statements in function (variant 3)	helix QAC
SUB	Number of function calls	helix QAC
UNR	Number of unreachable statements	helix QAC
UNV	Unused or non-reused variables	helix QAC
XLN	Number of executable lines	helix QAC

2. Collect testing metrics for each project, including 16 groups of metrics related to test cases and 9 metrics related to defects. The 25 testing metrics are shown in Table 5.9.

Table 5.9 Collecting testing metrics

Testing metric (Test case related)	Testing metric (Bug related)
Total number of test cases executed	Total number of problems
Total number of failed test cases	Key issues
Number of functional test cases executed	Important issues
Number of functional test cases failed	General issues
Number of performance test cases executed	Recommended improvements
Number of performance test cases failed	User requirement issues
Number of interface test cases executed	Software requirement issues
Number of interface test cases failed	Design issues
Number of regression test cases executed	Code issues
Number of regression test cases failed	
Number of security test cases executed	
Number of security test cases failed	
Number of boundary test cases executed	
Number of boundary test cases failed	
Number of recovery test cases executed	
Number of recovery test cases failed	

5.6 Actual Data Collection

Step 1: Software Quality Evaluation Objective Survey

The software quality evaluation objective questionnaire was distributed within the defined scope, ultimately resulting in completed responses from 5 department managers, 5 developers, and 13 testers. Following an analysis of the overall score rankings and subsequent confirmation by relevant personnel, 10 software quality sub-characteristics were selected as the evaluation objectives. Their respective weights were calculated based on the scores, as presented in Table 5.10.

Step 2: Project Data Collection

The research institution provided data for 82 aviation on-board embedded software projects. These projects covered five types of software: optoelectronic, display, mission computer, recording, and external equipment. The project sizes ranged from 7 functions (208 lines of code) to 403 functions (25,545 lines of code). The distribution histograms for project types and sizes are shown in Figs. 5.10 and 5.11.

All of these projects underwent document review, code review, and configuration testing by the software testing department. Each project was tested by a single tester. The following data were collected for each project:

Table 5.10 Weights of software quality sub-characteristics

Characteristic	Sub-characteristic	Weight	Description
Reliability	Availability	0.15	The degree to which a system, product, or component can be operational and accessible when required for use
Functionality	Functional correctness	0.13	The degree to which a product or system provides the correct results with the required precision
Reliability	Maturity	0.12	The degree to which a system, product, or component meets reliability requirements during normal operation
Reliability	Fault tolerance	0.11	The degree to which a system, product, or component operates as expected despite the presence of hardware or software faults
Functionality	Functional completeness	0.10	The degree to which a set of functions covers specified tasks and user objectives
Maintainability	Modifiability	0.09	The degree to which a product or system can be effectively and efficiently modified without introducing defects or reducing the quality of the existing product
Maintainability	Analyzability	0.09	The degree to which it is possible to evaluate the impact of an expected change (to one or more parts of the product or system), diagnose defects or failures, and identify the effectiveness and efficiency of the parts to be modified
Maintainability	Testability	0.08	The degree to which test criteria can be established for a system, product, or component and the effectiveness and efficiency of evaluating whether the criteria have been met through testing
Reliability	Recoverability	0.07	The degree to which a product or system can recover the data directly affected by an interruption or failure and restore the desired system state
Performance efficiency	Timeliness	0.06	The degree to which a product or system meets the response time, processing time, and throughput requirements while performing its functions

1. Data were collected for 42 software code metrics, which included all 33 function-related code metrics provided by QAC, as well as 5 comment line metrics and 4 nesting depth metrics for functions provided by the Testbed tool.
2. Test metrics were collected, including 9 metrics related to defects and 16 metrics related to test cases.
3. Software quality evaluation objective ratings were collected, with the tester for each project rating the 10 software quality sub-characteristics based on the evaluation table.

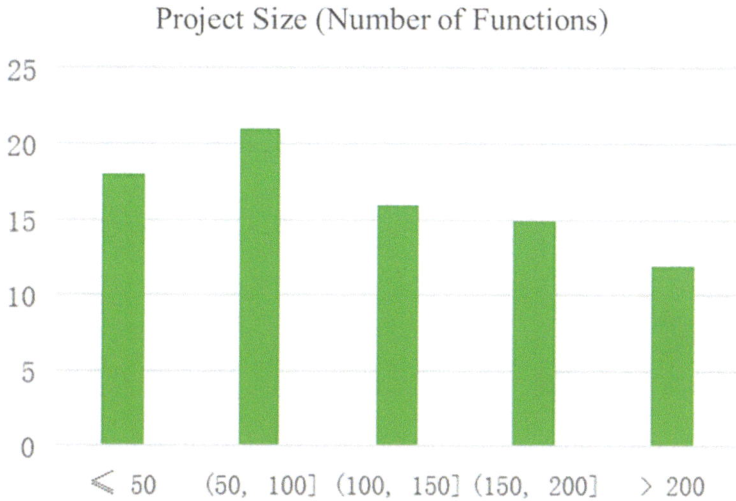

Fig. 5.11 Histogram of Project Size Distribution

5.6.1 Aggregation of Software Code Metric Data

To examine software code metrics from the perspective of the overall project, common aggregation measures such as descriptive statistics, distribution parameters, and econometric indicators will be employed to aggregate the software code metric data.

5.6.1.1 Characteristics of Software Code Metrics

Figure 5.12 shows the distribution of cyclomatic complexity for a project in this study.

In practice, distribution parameters, descriptive statistical values, and econometric coefficients were used to aggregate the code metric data. The following sections provide a description of each of these aggregation methods.

5.6.1.2 Distribution Fitting of Code Metrics

Distribution fitting is a method used to determine the distribution that best describes the data. Since the software code metric data is highly skewed, the lognormal distribution and power-law distribution were selected as the candidate distributions for fitting.

1. Pre-processing for Distribution Fitting

   For both the lognormal distribution and power-law distribution, the samples must be greater than zero. First, the proportion of zero-value samples in the 42

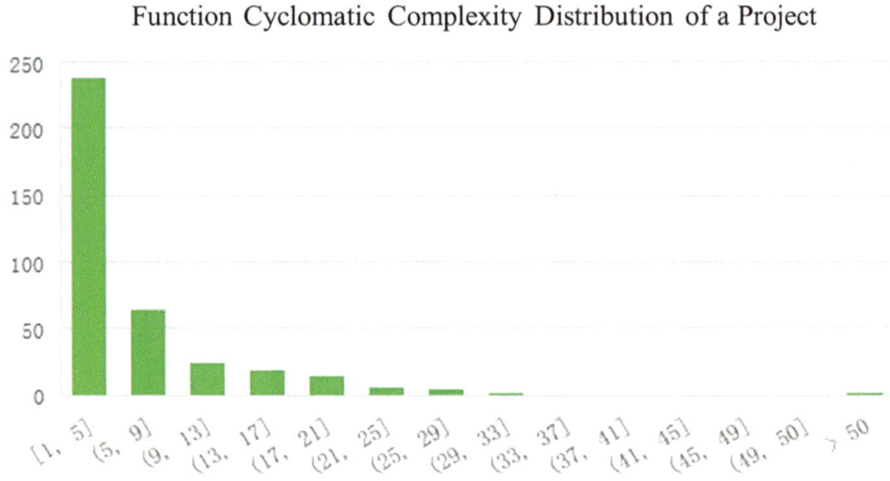

Fig. 5.12 Histogram of Cyclomatic Complexity Distribution for a Project

Table 5.11 Non-zero code metrics

No.	Abbreviation	Name
1	AKI	Akiyama's criterion
2	CYC	Cyclomatic complexity
3	FN1	Number of operator occurrences in function
4	LIN	Number of code lines
5	M07	Essential cyclomatic complexity
6	PTH	Estimated static program paths

code metric datasets was analyzed. The following 6 code metrics, all consisting of non-zero samples, were selected (two projects contained approximately 1% of units with STLIN equal to 0, which were excluded as outliers), as shown in Table 5.11.

2. Lognormal Distribution Fitting

   The lognormal distribution is defined as follows: Let X be a continuous random variable that takes positive values. If ln(X) follows a normal distribution with mean μ and variance σ^2, then X is said to follow a lognormal distribution. The logarithms of the sample data for each of the 6 code metrics were taken, and normality tests were conducted. In this study, the normality test was performed using the scipy.stats.normaltest() method from the Python library, which employs the D'Agostino-Pearson omnibus test. This test combines skewness and kurtosis to provide a comprehensive assessment of normality. The proportion of projects with normality test p-values greater than 0.05 out of 40 projects is shown in Table 5.12.

3. Power-law Distribution Fitting

   The power-law distribution is characterized by the property that, in a log-log plot, it appears as a straight line with a negative slope equal to the power-law

Table 5.12 Result of power-law distribution fitting

No.	Code metric	Lognormal distribution pass rate/%	Power law distribution pass rate/%	Result
1	AKI	50	90	Power law distribution
2	CYC	27.5	90	Power law distribution
3	FN1	82.5	55	Lognormal distribution
4	LIN	42.5	12.5	Failed
5	M07	5	92.5	Failed
6	PTH	2.5	92.5	Power law distribution

exponent. For each of the 6 code metrics, the sample data were sorted in descending order. The sample values were treated as the dependent variable, and the rank of the sample values was treated as the independent variable. A linear regression was performed on the log-log plot. The proportion of projects with a coefficient of determination (R^2) greater than 0.8 from the 40 projects is shown in Table 5.12.

4. Distribution Parameter Estimation

   For the code metric FN1 (the number of operators in a function), which passed the lognormal distribution fitting test, maximum likelihood estimation was employed to estimate the parameters μ and σ. These estimated values were then used as the aggregation results, as follows:

$$\mu = \frac{1}{n}\sum_{i=1}^{n} \ln y_i, \sigma^2 = \frac{1}{n}\sum_{i=1}^{n}\left(\ln y_i - \mu\right)^2 \tag{5.84}$$

For the 3 code metrics that passed the power-law distribution fitting test, maximum likelihood estimation was applied to estimate the parameter α. The estimated value was then used as the aggregation result, as shown in the following equation:

$$\alpha = 1 + \frac{1}{n}\sum_{i=1}^{n} \ln \frac{y_i}{y_{min}} \tag{5.85}$$

5. Aggregation of Software Code Metrics

   Four code metrics have been determined through fitting statistical distributions. The results of maximum likelihood estimation for the distribution parameters are used as the aggregation values for these metrics. For other metrics, exact distributions could not be found, so descriptive statistical values and econometric indicators are used as approximations for the aggregation values. The selection of these aggregation values considers both the comprehensiveness of the sample description and the explanatory power of the input variables. The selected aggregation values are listed in Table 5.13.

   After aggregation, the four code metrics with known distributions produced six aggregated values, while the remaining 38 code metrics resulted in 304 aggregated values. In total, 310 aggregated software metrics were obtained from the 42 software code metrics data.

Table 5.13 Selected aggregation values

Aggregate value	Characteristics
Maximum value	Used to describe extreme values
Sum	Used to describe the overall scale of the project
Arithmetic mean	Used to describe the average scale of functions
Standard deviation	Used to describe the degree of dispersion
Skewness	Used to describe the degree of asymmetry in data distribution
Kurtosis	Used to describe the degree of data aggregation
Gini coefficient	Used to describe the degree of data inequality
Hoover coefficient	Used to describe the degree of data inequality

5.6.2 *Data Pre-processing*

Data pre-processing methods transform the collected data into a format suitable for modeling. Data pre-processing includes the following 3 components:

1. Since the granularity of software code metrics is finer than that of other data types, aggregation of the software code metric data is required.
2. Due to the varying dimensions and magnitudes of software metric data, standardization procedures such as data trend harmonization and dimensionless normalization are necessary to normalize the data.
3. Given the large number of attribute dimensions generated from the aggregation of software code metrics, dimensionality reduction is essential to ensure model performance.

5.6.2.1 Normalization of Software Code Metrics

If data with large scale differences is used for analysis, the metrics with higher values will dominate the influence on software quality, thus diminishing the impact of metrics with lower values. Since the non-zero values in the software metric data, formed after aggregation, range from 10,910^9109 to 10−1610^{−16}10−16, directly proceeding with subsequent calculations may lead to a loss of accuracy. Therefore, it is essential to normalize the sample data.

Z-score normalization is preferred in this study because it is more suitable for measuring similarity using distance-based methods (e.g., K-nearest neighbor algorithm) and for dimensionality reduction using PCA (Principal Component Analysis). In contrast, min-max normalization can scale the covariance of the data values by a multiple after linear transformation. As a result, Z-score normalization is applied to process the data in this experiment.

5.6.2.2 Dimensionality Reduction of Software Code Metrics

In this experiment, each sample has 310 attributes, while the total number of samples is 82. Since the maximum number of linearly independent vectors in a matrix cannot exceed the number of rows or columns in the matrix, at most 82 attributes are useful. Therefore, dimensionality reduction of the sample data is required to eliminate irrelevant data and ensure the model's performance. Common dimensionality reduction methods include correlation analysis and principal component analysis (PCA). The following sections introduce these two methods and present the results after processing.

1. Correlation Analysis

   Dimensionality reduction of software code metrics is performed step-by-step using correlation analysis. The steps are as follows:

   Step 1 removes irrelevant features. Use Spearman correlation analysis to select features that are significantly correlated with each label ($p < 0.05$).

   Step 2 removes duplicate features. These features are sorted in descending order of the absolute value of the correlation coefficient, and starting from the first feature, sequentially scan and remove features that are significantly correlated ($p < 0.05$) and highly correlated ($r > 0.9$) with this feature.

   The results of the correlation analysis are shown in Appendix 4.
2. Principal Component Analysis

   Principal component analysis (PCA) was performed on the sample data until the variance explained by the principal components exceeded 80%. The number of principal components at this point represents the number of features after dimensionality reduction.

   As shown in Fig. 5.13, PCA was conducted with different numbers of principal components, while simultaneously plotting the curve of the variance percentage as it changes with the number of principal components. When the number of principal components reached 18, the variance percentage had reached 80%, meaning that 18 principal components could explain 80.91% of the information in the 310 attributes.

   Figure 5.14 shows the variance explained by adding new principal components. It can be observed that the variance explained by the 18th principal component has decreased to 1.33%. When the number of principal components reaches 81, the variance explained totals 100%, resulting in an average variance explained rate of 100%/81 = 1.23%. Therefore, the gain from adding more principal components (1.27%) is close to the average value, making the selection of 18 principal components a reasonable choice.

5.6.3 Comparison of Data Training Methods

This section primarily compares the impact of three classification algorithms—logistic regression, the KNN algorithm, and the BP neural network—on the performance of software quality evaluation models. The data preprocessing method

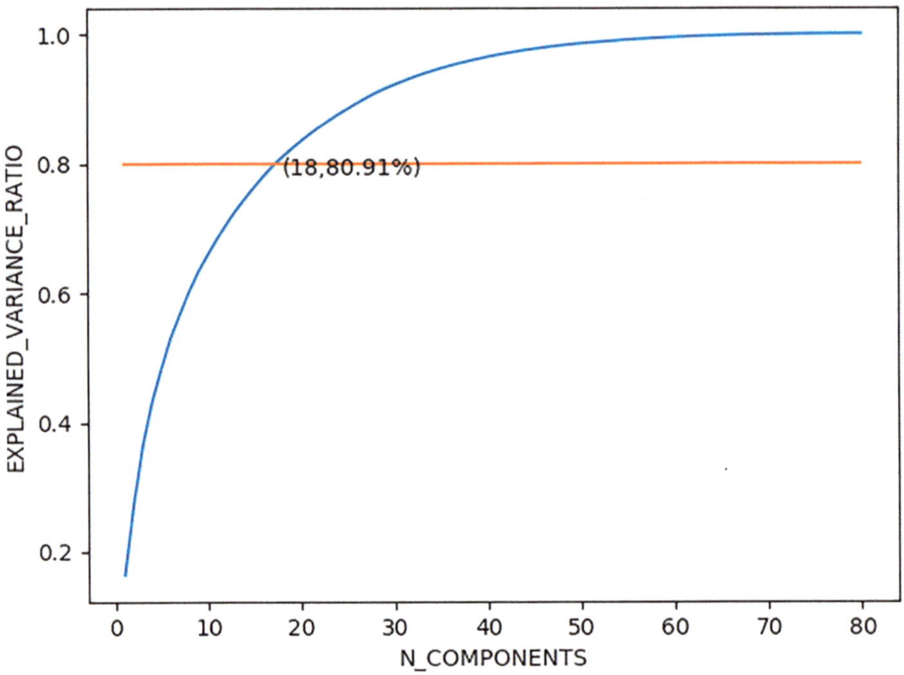

Fig. 5.13 Percentage of Variance Changes

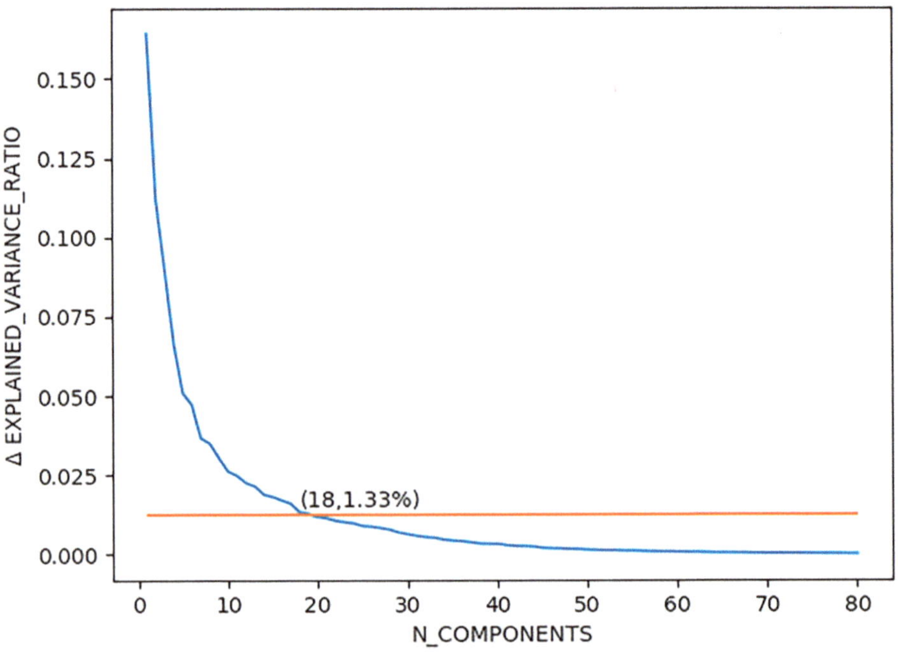

Fig. 5.14 Explained Variance Ratio of Added Principal Components

transforms the collected data into 82 sets of training samples. For the training data of the model, each project represents a sample, where the software code metric aggregation value of each project forms the feature of the sample, and the software quality target rating level of each project constitutes the label. All data analysis and modeling processes were carried out using the third-party machine learning library Scikit-learn (sklearn), which was implemented in Python. The definitions of the three algorithms are provided below.

1. Logistic regression is a linear machine learning algorithm used for classification.
2. The k-nearest neighbor (KNN) algorithm is a straightforward nonlinear machine learning classification method.
3. The backpropagation neural network (BPNN) classification algorithm is a prototypical nonlinear classification method.

5.6.3.1 Cross-validation

The basic concept of cross-validation is to partition the original dataset into distinct subsets, where one subset is used as the training set and another as the testing set. Initially, the classifier is trained with the training set, and subsequently, the trained model is evaluated on the testing set to assess the classifier's performance.

Due to the inherent randomness in the division of the training and testing sets—which may influence the model's performance metrics—the "ten-fold cross-validation" approach is employed to mitigate this variability. In ten-fold cross-validation, the dataset is divided into 10 equal parts; in each iteration, 9 parts are used for training and the remaining part for testing, with the overall performance estimated as the mean of the 10 results. To further minimize the impact of random data partitioning, this process is repeated 10 times using independent ten-fold cross-validations, thereby achieving a more precise estimation. The specific procedures are outlined as follows:

1. Randomly partition the dataset into ten equal folds.
2. For each fold, designate nine folds as the training set and the remaining fold as the validation set.
3. Train the classification model on the training set.
4. Evaluate the model's classification accuracy on the validation set.
5. Repeat steps 2–4 until each fold has served once as the validation set.
6. Repeat the entire ten-fold cross-validation procedure ten independent times to further reduce variability.

This method is known as k-fold cross-validation and is commonly used to compare different data training methods. By using multiple iterations of training and testing on different subsets of the data, it helps to reduce the bias and variance of the model, and provides a more robust estimate of the model's performance.

Table 5.14 Results of logistic regression

Software quality evaluation target	Logistic regression	
	Feature count	Accuracy/%
Functional completeness	18	66.77
Functional correctness	18	57.67
Usability	18	56.92
Fault tolerance	18	63.91
Maturity	18	72.69
Time characteristic	15	62.56
Modifiability	18	42.85
Analyzability	17	49.91
Recoverability	18	50.25
Testability	18	47.36
Average	17.6	57.09

5.6.3.2 Logistic Regression Classification Model

Due to the lower data requirement of simpler models, a linear classifier is preferred as the software quality evaluation model. Linear classifiers classify by linearly combining features. Common linear classifiers include perceptron (PLA), linear discriminant analysis (LDA), logistic regression (LR), and support vector machine (SVM). In this case, we adopt logistic regression as our linear classifier. Logistic regression is a generalized linear regression analysis model that uses machine learning to implement classification or prediction by constructing a regression function. The modeling steps are as follows:

1. Randomly partition the dataset into ten equal folds.
2. For each fold, designate nine folds as the training set and the remaining fold as the validation set.
3. Train the classification model on the training set.
4. Evaluate the model's classification accuracy on the validation set.
5. Repeat steps 2–4 until each fold has served once as the validation set.
6. Repeat the entire ten-fold cross-validation procedure ten independent times to further reduce variability.

The results of the logistic regression are shown in Table 5.14.

5.6.3.3 K-Nearest Neighbors Algorithm (KNN)

The KNN algorithm is a simple machine learning algorithm. Using $k = 5$ and the 10x10 cross-validation method, the following steps were taken to model the software quality evaluation model based on the K-nearest neighbors algorithm. The results are shown in Table 5.15.

1. Randomly divide the dataset into 10 equal parts.
2. Rotate the selection of 9 parts as the training set and 1 part as the validation set.

Table 5.15 Results of the KNN algorithm

Software quality evaluation target	KNN algorithm	
	Feature count	KNN accuracy/%
Functional completeness	18	66.08
Functional correctness	18	58.75
Usability	18	58.25
Fault tolerance	18	64.82
Maturity	18	63.08
Time characteristic	15	67
Modifiability	18	49.92
Analyzability	17	44.73
Recoverability	18	54.42
Testability	18	48.82
Average	17.6	57.59

3. Perform dimensionality reduction on the feature data of the training set using correlation analysis/principal component analysis, and apply the same transformation to the test set as was applied to the training set.
4. Build the KNN classification model using the training set.
5. Evaluate the classification accuracy of the model using the test set.
6. Repeat steps 2–5 until each data subset has been used as the validation set.
7. Repeat steps 1–6 for a total of 10 iterations.

It can be observed that the accuracy of the KNN classification model is almost the same as that of the linear model. This may be due to the fact that the KNN algorithm calculates the Euclidean distance of sample features, and after standardization processing, the sample features have lost their original dimensions. In other words, each feature contributes equally to the calculation of the Euclidean distance, but in reality, the impact of each feature on the result is not equal. Therefore, it is necessary to multiply the feature data by a suitable weight before calculating the Euclidean distance. To find the optimal weight, genetic algorithm is used. The general process of genetic algorithm is as follows.

1. Randomly initialize a population of N individuals, where each individual represents a set of weights.
2. Evaluate the fitness of each individual corresponding to a chromosome in the population and select the top k individuals as parents based on their fitness.
3. Apply the crossover operator to generate offspring by selecting chromosomes from the parents according to the crossover rate.
4. Apply the mutation operator to mutate the offspring's chromosomes based on the mutation rate.
5. Repeat steps 3 and 4 until the new population contains N individuals.
6. Return to step 2 until the specified number of genetic generations, M, is reached.

Taking functional completeness as an example, each modeling process still performs ten-fold cross-validation on the sample data, with each modeling process as the objective function of the genetic algorithm. The average accuracy returned is set

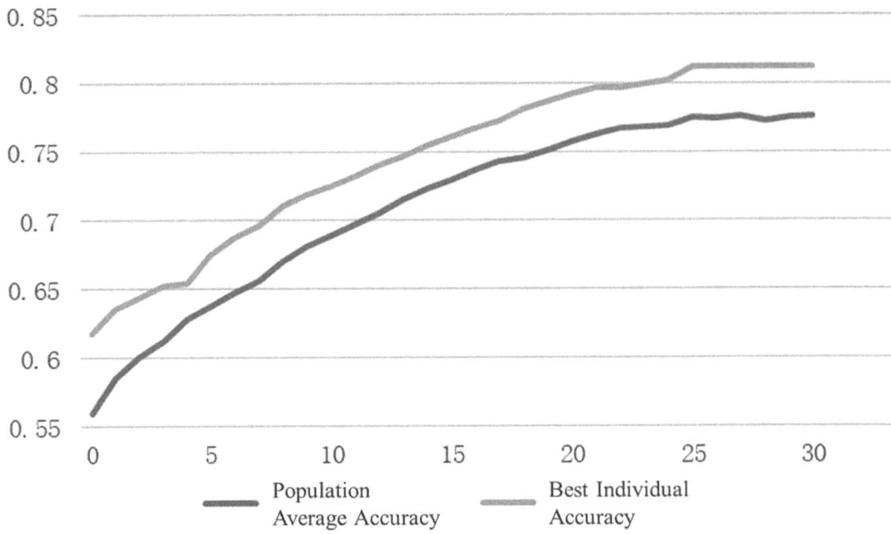

Fig. 5.15 Population Evolution Process

as the objective value, and the genetic algorithm is used to search for feature weights that can increase the accuracy. The population size is set to 100, the crossover rate is 75%, the mutation rate is 5%, and the genetic evolution process ends after 30 generations. The evolutionary process is shown in Fig. 5.15.

The optimal individual of the randomly initialized initial population can easily achieve an accuracy of around 61.77%, while the average accuracy of the population is only approximately 55.95%. After 25 generations of evolution, the accuracy of the optimal individual stabilizes at around 81.18%, and the average accuracy of the population stabilizes at approximately 77.51%. This indicates that the population has likely reached a point near the optimal value. While it is uncertain whether this is a local or global optimum, expanding the population to verify whether it is a local or global optimum would be time-consuming and costly, as the fitness of each individual has undergone ten-fold cross-validation. Therefore, the current results from the genetic algorithm are used as the feature weights. For comparison, the accuracy of the functional completeness evaluation model established using the logistic regression and KNN classifiers is only around 67%.

Using ten-fold cross-validation, the software quality evaluation model based on the KNN algorithm was built again. Before establishing the KNN classification model for each evaluation objective, the feature data was multiplied by the optimized weights. The average accuracy obtained from cross-validation is shown in Table 5.16. It can be observed that after optimizing the features with the genetic algorithm, the average accuracy has increased.

Table 5.16 Results of KNN algorithm and genetic algorithm

Software quality evaluation target	KNN algorithm + Genetic algorithm	
	Feature count	Accuracy/%
Functional completeness	18	79.38
Functional correctness	18	72.05
Usability	18	71.55
Fault tolerance	18	78.12
Maturity	18	76.38
Time characteristic	15	80.3
Modifiability	18	63.22
Analyzability	17	58.03
Recoverability	18	67.72
Testability	18	62.12
Average	17.6	70.89

Table 5.17 Structure of BP neural network

Neural network	Number of input layer units	Number of hidden layer units	Number of output layer units
NN_1	18	17	1
NN_2	18	18	1
NN_3	18	17	1
NN_4	18	16	1
NN_5	18	14	1
NN_6	15	15	1
NN_7	18	15	1
NN_8	17	16	1
NN_9	18	15	1
NN_{10}	18	12	1

5.6.3.4 BP Neural Network Classification Model

The BP neural network requires a relatively large amount of data, so the simplest 3-layer BP neural network structure is initially selected. The first step is to determine the topology of the BP neural network. For m software quality evaluation objectives, m independent 3-layer BP neural network models need to be established. For the i-th BP neural network, denoted as NNi, the number of neurons in the first layer (input layer) corresponds to the number of elements in the related independent variable set {IV'1, IV'2, …, IV'Ti} of the software quality evaluation objective Ti; the number of neurons in the third layer (output layer) is 1; and the number of neurons in the second layer (hidden layer) is determined experimentally. The final results for each neural network are shown in Table 5.17.

The Sigmoid function is selected as the activation function for the hidden layer, with the error squared cost function, dynamic learning rate, and random weight initialization set. After determining the structure and parameters of the neural network, the software quality evaluation model based on the BP neural network is built using the ten-fold cross-validation method, following the steps outlined below:

1. Randomly divide the dataset into 10 equal parts.
2. Use each of the 10 parts in turn as a validation set, while the other 9 parts are used as training sets.
3. Use principal component analysis (PCA) to reduce the dimensionality of the feature data in the training set, and apply the same transformation to the test set.
4. Use the training set to establish a BP neural network classification model.
5. Use the test set to evaluate the model's classification accuracy.
6. Repeat steps 2–5 until each part of the data has been used as a validation set.
7. Repeat steps 1–6 for a total of 10 times.

The average number of iterations, average convergence time, and classification accuracy of multiple BP neural networks trained multiple times are shown in Table 5.18.

It can be seen that compared with the KNN algorithm optimized by genetic algorithm, the performance of the BP neural network model is not ideal, which may be due to overfitting caused by too few sample data and too many iterations. In order to improve the accuracy of the neural network model, the following two optimization measures are adopted.

1. In the case of limited data, data augmentation can be used to increase the diversity of training samples and avoid overfitting. By using random noise methods to add new samples to the original dataset, the original 82 sets of data are expanded to 164. This can help to increase the size of the dataset and improve the accuracy of the BP neural network model."
2. Because the neural network based on gradient descent method is very slow at the beginning of learning but fastest at the end, while the genetic algorithm simulat-

Table 5.18 Result of BP neural network

	BP neural network			
Software quality evaluation target	Features count	Iteration count	Convergence time/ms	Accuracy/%
Functional completeness	18	1588	60	67.01
Functional correctness	18	1783	83	66.26
Usability	18	1503	55	59.44
Fault tolerance	18	1389	40	72.48
Maturity	18	1624	58	75.45
Time characteristic	15	1332	27	71.17
Modifiability	18	1307	28	55.08
Analyzability	17	1383	33	54.44
Recoverability	18	1485	50	64.5
Testability	18	1357	32	54.66
Average	17.6	1475	46.6	64.05

Table 5.19 Result of optimized BP neural network

Software quality evaluation target	Optimized BP neural network			
	Features count	Iteration count	Convergence time/ ms	Accuracy/%
Functional completeness	18	1265	33	86.34
Functional correctness	18	1269	22	85.44
Usability	18	808	20	61.99
Fault tolerance	18	702	22	81.25
Maturity	18	848	24	76.49
Time characteristic	15	1073	20	77.37
Modifiability	18	712	12	62.91
Analyzability	17	737	13	71.65
Recoverability	18	828	21	69.2
Testability	18	788	16	71.29
Average	17.6	903	20	74.39

ing the theory of biological evolution is very fast at the beginning but slow at the end, genetic algorithm is used to optimize the BP neural network model. Similar to Sect. 5.3.2, each individual in the population contains all weights and biases of the neural network, and the individual corresponding to the optimal fitness value is the one with the optimal initialization weights and biases.

After increasing the sample size through data augmentation and optimizing the initialization weights and thresholds of the BP neural network using a genetic algorithm, the training results of the BP neural network are shown in Table 5.19. It is evident that both data augmentation and the genetic algorithm have a significant impact on optimizing the BP neural network. The data augmentation method helps reduce the risk of overfitting in the BP neural network. Although the sample size has increased, the genetic algorithm notably accelerates the convergence speed of the BP neural network.

5.6.3.5 Conclusion

A comparison of the performance of software quality evaluation models built using logistic regression, the KNN algorithm, and the BP neural network reveals that although logistic regression and the KNN algorithm are better suited for small-sample learning, their actual performance is suboptimal. This can primarily be attributed to the fact that simpler models often focus on basic relationships, whereas the relationship between software metrics and software quality is considerably more complex. While both the KNN algorithm and BP neural network models initially performed inadequately, their performance was significantly enhanced through optimization using genetic algorithms and data augmentation techniques. In conclusion, the optimized BP neural network model delivered the best results, achieving

Table 5.20 Comparison result of three models

Software quality evaluation target	Logistic regression accuracy/%	Optimized KNN algorithm accuracy/%	Optimized BP neural network accuracy/%
Functional completeness	66.77	79.38	86.34
Functional correctness	57.67	72.05	85.44
Usability	56.92	71.55	61.99
Fault tolerance	63.91	78.12	81.25
Maturity	72.69	76.38	76.49
Time characteristic	62.56	80.3	77.37
Modifiability	42.85	63.22	62.91
Analyzability	49.91	58.03	71.65
Recoverability	50.25	67.72	69.2
Testability	47.36	62.12	71.29
Average	57.09	70.89	74.39

an average classification accuracy of 74.39%, owing to its robust ability for nonlinear fitting. The comparison of the results from the three models is presented in Table 5.20.

5.6.4 Case Outcomes

The comprehensive software quality evaluation system is a hierarchical framework that progresses from software metrics to quality characteristics and ultimately to overall software quality. In the book's example, the sample data consist of pre-testing software measurements along with ratings provided by testers. The evaluation model derived from these data is optimally applied during the pre-testing phase.

For newly developed example software, the software code measurement data are collected prior to testing, and the comprehensive quality of the software is then computed using the model illustrated in Fig. 5.16. Each software quality evaluation target is predicted by an associated BP neural network model, and, following the grading rule where scores of 5, 4, 3, and 2 correspond to "excellent", "good", "fair", and "poor" respectively, the overall quality score is determined by the formula Ti × ωi. In this instance, a computed score of 3.9 yields a quality rating of "good", as shown in Table 5.21.

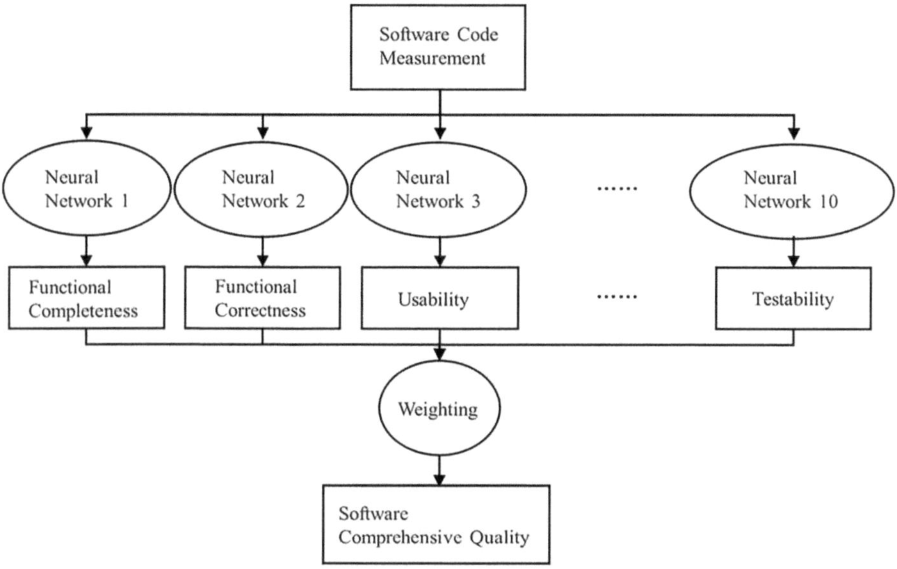

Fig. 5.16 The Comprehensive Software Quality Evaluation Model

Table 5.21 Weights of software quality sub-characteristics

Feature	Sub-feature	Classification result	Weight ω_i
Functionality	Functional completeness	Good	0.1
Functionality	Functional correctness	Excellent	0.13
Performance efficiency	Time characteristic	Excellent	0.06
Reliability	Maturity	Good	0.12
Reliability	Fault tolerance	Medium	0.11
Reliability	Usability	Excellent	0.15
Reliability	Recoverability	Medium	0.07
Maintainability	Modifiability	Good	0.09
Maintainability	Analyzability	Medium	0.09
Maintainability	Testability	Medium	0.08

5.7 Chapter Summary

As software size and complexity continue to increase, managing, evaluating, and predicting the quality and reliability of complex software systems become increasingly challenging. Software reliability assessment plays a vital role in ensuring software quality, while software metrics are essential for evaluating, managing, and improving software product quality. Establishing a robust software reliability evaluation framework is crucial for optimizing resource allocation throughout the software lifecycle and identifying areas where software quality may be lacking. This chapter presents a framework and associated techniques for data-driven software reliability assessment.

Furthermore, this chapter presents a case study in which a neural network learns the relationship between software metrics and software quality from historical software data, thereby establishing a customized and automated comprehensive evaluation system for assessing software quality through software metrics. The analysis of the case study reveals that, after augmenting the sample data through data augmentation techniques, the software quality evaluation model built using the BP neural network optimized by a genetic algorithm performs the best, achieving an average classification accuracy of 74.39%. This performance can be attributed to the neural network's powerful ability to learn complex relationships, which enables it to outperform relatively simpler models in capturing the relationship between software metrics and software quality.

Both related research and examples have demonstrated the feasibility of using data-driven approaches in the evaluation of complex software quality and reliability. By leveraging historical software data, machine learning techniques such as neural networks have been shown to effectively model the relationship between software metrics and software quality. This approach can significantly improve the accuracy and efficiency of software quality evaluation and reliability assessment, and it has the potential to play an important role in the development of high-quality and reliable software.

Chapter 6
Model-Based Testing Techniques

Before a software system is delivered to users, designers must take certain verification measures to ensure the system's reliability. Software testing is one of the most important and effective methods frequently used for this purpose. As more standards related to software testing are introduced, the maturity of the software testing process has also continuously improved. However, in today's environment, where software is becoming increasingly complex and requirements are more diverse, traditional software testing methods are struggling to keep up with these changes. The inadequacies of traditional software testing are summarized as follows:

1. The formalization of software testing is not high;
2. The time for starting software testing is relatively late;
3. The efficiency of software testing is relatively low, and subjective factors have a greater impact;
4. Software testing is not easy to adapt to software requirement changes and operating platforms;
5. Traditional methods are difficult to adapt to new technologies.

Currently, there is still no highly efficient and high-quality testing implementation method for complex software systems.

In the field of software development, the Model-Driven Architecture (MDA) proposed by the Object Management Group (OMG) has become increasingly mature. Model-driven approaches have effectively enhanced the efficiency and quality of software development, while overcoming challenges such as interoperability issues between intermediate components and difficulties in platform migration.

With the growing adoption of model-driven development, model-based testing (MBT) has also gained increasing attention. Compared to traditional software testing, MBT presents the entities and activities in the testing process in the form of models. This approach can help testers better understand the system, uncover test-related information, and provide more structured and formalized expressions of the testing process. MBT facilitates the automation of testing and improves the overall testing process.

Y. Wang et al., *Reliability Technology for Complex Software Systems*,
https://doi.org/10.1007/978-981-95-3889-8_6

Introducing model-based testing into the field of complex software testing can effectively address some of the challenges associated with testing such systems. This chapter discusses MBT technologies and explores potential solutions that MBT may offer for complex software verification and testing.

6.1 Introduction to Model-Based Testing

6.1.1 MDA and MBT

Before introducing model-based testing (MBT) in detail, it is important to understand the Model-Driven Architecture (MDA) and the relationship between MDA and model-based testing. This will help in comprehending the technologies and concepts discussed in the subsequent chapters.

6.1.1.1 The Fundamental Elements of MDA

During the 1990s, software middleware experienced rapid development, leading to the emergence and widespread use of several platform technologies. Over the course of more than a decade, middleware products supporting different technology standards coexisted, and no single "ultimate winner" emerged. The differentiation of middleware created interoperability challenges between different middleware platforms, which led to expensive and time-consuming integration efforts. Additionally, the "binding" of enterprise business logic to specific platform technologies increased the difficulty of migrating software systems to new platforms and raised the risk of business operations becoming dependent on specific platform technologies.

In response to these challenges, the Object Management Group (OMG) officially introduced the Model-Driven Architecture (MDA) framework specification in 2001. MDA is a methodology and standard system for software system development based on the use of models.

The core of MDA consists of the OMG's modeling standards, including UML, CWM, XMI, and MOF. The following is an introduction to these four standards:

1. UML (Unified Modeling Language): UML is a standard visual modeling language used for object-oriented analysis and design. It is used to visualize, define, construct, and document software systems. All models within MDA are described using UML.
2. CWM (Common Warehouse Metamodel): CWM is a metamodel used to define the representation of business and technical metadata, particularly in the database and business analysis domains. It provides a model-based approach to enable metadata exchange between heterogeneous software systems.
3. XMI (XML Metadata Interchange): XMI is a standard XML-based format for exchanging metadata between software tools and systems. It defines an XML

data exchange format for various models using standardized XML document formats and Document Type Definitions (DTDs).

4. MOF (Meta Object Facility): MOF is a modeling standard introduced by OMG for defining and managing metadata. It serves as a meta-meta-model, meaning it is the model of models. The UML and CWM metamodels within MOF are based on MOF itself. The establishment of the MOF standard ensures interoperability between different metamodels.

MDA, based on various OMG-established standards, separates application logic from platform technology. It enables the creation of platform-independent models for enterprise applications. MDA provides a comprehensive set of specifications and methodologies that guarantee interoperability, portability, and reusability throughout the software system lifecycle.

According to the relationships and definitions of the models, MDA's models are divided into a four-layer structure, as shown in Fig. 6.1.

From M3 to M0: Each higher level serves as the foundational definition for the level below it, with the next level being the application of the level above, i.e., the description of objects. The M3 level is the meta-meta-model level, which includes only MOF. The meta-meta-model provides a common standard for the definition of modeling languages. The M2 level is the meta-model level, which represents the various modeling languages built on top of MOF. The M1 level is the application model layer, where modeling languages are used to create descriptive models for enterprise applications. The M0 level represents the real-world systems of enterprise computing, information, or conceptual and cognitive models of the real world as understood in human cognition, serving as the ultimate goal of model description.

As shown in Fig. 6.1, the MDA application model layer consists of three types of models.

1. The Computation Independent Model (CIM) is a system model established by MDA based on the Computation Independent View (CIV). It is used to describe the system's requirements, functions, behavior, and operating environment, and is also referred to as the business model. The term "computation independent" is used because CIM focuses on representing the external behavior and operating environment of the system, without delving into the internal structure and implementation details. CIM serves as a bridge for communication and collaboration between domain experts and system designers regarding system requirements,

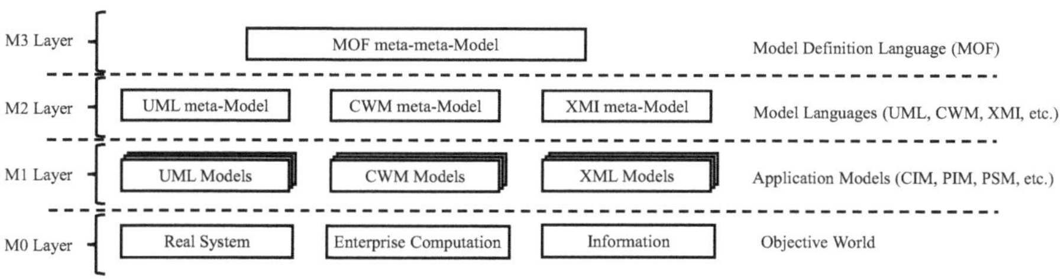

Fig. 6.1 The Four-Layered Meta-Model Architecture of MDA

and directly supports the construction and implementation of the Platform Independent Model (PIM) and Platform Specific Model (PSM).

2. The Platform Independent Model (PIM) is a system model established by MDA based on the Platform Independent View (PIV). PIM represents abstract business logic, and it is termed "platform independent" because it does not contain any information specific to the implementation platform or technology. The platform independence of PIM enables it to be implemented on any technology platform.
3. The Platform Specific Model (PSM) is a system model established by MDA based on the Platform Specific View (PSV). PSM is derived from the corresponding Platform Independent Model (PIM) and includes both the business logic specifications defined in the PIM and the specific implementation details related to the selected platform and technology.

6.1.1.2 Model Transformation

If we compare the MDA framework to a building, the "building materials" of the MDA framework would include: high-level models; one or more precisely defined languages used to create high-level models; definitions for converting PIM to PSM; languages for writing these definitions that can be executed by transformation tools; tools capable of executing transformation definitions; and tools that can perform PSM-to-code transformations.

Therefore, within the MDA framework, its main components include models, PIM, PSM, languages, transformations, transformation definitions, and transformation tools. Among these, model transformation is a crucial technology.

MDA separates the software system model into a Platform-Independent Model (PIM) and a Platform-Specific Model (PSM), unifying them through model transformation rules in an effort to address the challenges posed by requirement changes. In the MDA framework, a platform-independent modeling language is initially used to construct the Platform-Independent Model (PIM). The PIM is then transformed into the Platform-Specific Model (PSM) based on the mapping rules of the specific platform and implementation language. Finally, application code and testing frameworks are generated from the PSM.

Models are formal and must be described using a language with a strict syntax and unambiguous semantics. Although metamodels have rigorous syntax, their semantics are not formalized and cannot be "understood" by computers, which makes it difficult to ensure the accuracy of transformation results during MDA model conversion. Therefore, the model transformation work in MDA primarily focuses on the application model layer, M1.

Model transformation is the process of applying model mapping techniques to convert one model of a system into another. In MDA, the core of model transformation is the conversion from PIM to PSM.

MDA provides two methods of model mapping: type mapping and instance mapping. Type mapping establishes the mapping from the model language types used in PIM to those used in PSM, while instance mapping involves marking PIM model elements to indicate how they are transformed into PSM model elements in a specific way.

Corresponding to these two model mapping methods, MDA provides two basic model transformation methods: type-based model transformation and instance-based model transformation.

In software development, there are several types of model transformations as follows:

1. Transformation from CIM to PIM: This transformation is the conversion from a business model to an analysis model, and it is primarily implemented manually.
2. Transformation from PIM to PIM: This occurs when a model is improved, filtered, or specialized during the development lifecycle, but no platform-specific information is required. Typically, this transformation involves converting an analysis model into a high-level design model. PIM to PIM transformation is typically related to model refinement and is a form of forward engineering.
3. Transformation from PIM to PSM: In the practical application of MDA, transformation tools can convert a PIM into multiple different PSMs based on different mapping rules. An independent PSM is generated for each specific technology platform, and bridges are created between these PSMs to establish connections.
4. Transformation from PSM to PSM: Transformation from PSM to PSM is related to the refinement of platform-specific models. It involves refining a PSM. When component implementation and deployment are involved, transformation from PSM to PSM is required.
5. Transformation from PSM to PIM: This transformation is used to extract an abstract, platform-independent model from an existing implementation model. From a software engineering perspective, this is a form of reverse engineering and is often difficult to fully automate. Ideally, the result of transforming from PSM to PIM should correspond to the result of transforming from PIM to PSM.
6. Transformation from PSM to Code: This transformation generates the final code from the PSM model.

Table 6.1 lists some common model transformation techniques along with their advantages and disadvantages. In MDA, the establishment of accurate models and the development of model transformation techniques will significantly improve the quality and efficiency of the software development process.

6.1.1.3 Model-Driven Software Testing

Model-driven software testing introduces the principles of model-driven development into the software testing process. It is an innovative and promising automated testing method that provides testers with a convenient and efficient way to achieve automated testing. Similar to model-driven software development, model-driven software testing treats the testing process as a collection of models, allowing testers to focus on modeling the software testing process and defining the transformations

Table 6.1 Common model transformation techniques and their advantages and disadvantages

Model transformation technology	Advantages	Disadvantages
Based on XMI model transformation technology	Based on text transformation technology, relatively simple syntax	Not intuitive, more steps in transformation, high overhead, easy to cause inconsistency between pre and post models
Based on QVT model transformation technology	Can define relationships between model elements using both declarative and imperative languages	The visualization function of models is not yet resolved in QVT specifications
Based on OCL model transformation technology	The static semantics of OCL is clear and can be used for model constraints, element queries, and transformations, as well as defining modeling languages	OCL dynamic semantics description capability is relatively weak, which is essential for complete model transformations. The OCL metamodel must be extended to support imperative language features
Based on UML profile model transformation technology	Fully compatible with UML, can be processed with UML modeling tools, especially suitable for PSM modeling	Limited extension capabilities, not as effective as directly using MOF to define a new modeling language
Using XSLT for model transformation and code generation	Convenient for code generation and reverse engineering	High overhead and complexity in transformation
Graphical model transformation technology	Uses UML class diagrams and text template languages to describe models, and graphically indicates matching relationships between model elements	Limited functionality, and not sufficiently formalized

of these models. The tasks are then executed automatically by computers, reducing manual effort and improving testing efficiency and accuracy.

During the software lifecycle, different models are created at various stages, such as software requirement models, software design models, and testing models. As carriers of business information and specific implementation technologies, models serve as communication tools between customers, software developers, and testers. Furthermore, models, as an essential concept in MDA, are a prerequisite for both model-driven software development and model-driven software testing.

Figure 6.2 illustrates a model-driven software testing process integrated with the model-driven software development process. The left part of the figure describes an MDA-based software development process, starting with the construction of the PIM. The PIM represents business functions and behaviors and serves as a logical application representation. It is typically created by business experts and modeling specialists. Model-driven software development focuses on constructing the PIM, and tools can be employed to automatically perform the mapping from PIM to PSM and the transformation from PSM to code. The right part of the figure represents the

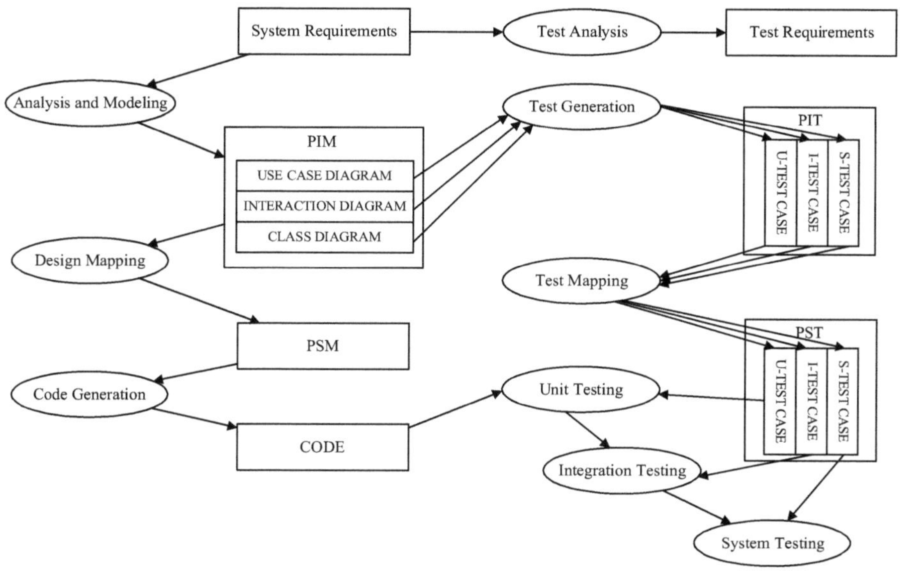

Fig. 6.2 Model-Driven Software Development Integrated with Model-Driven Software Testing

model-driven software testing process, corresponding to the PIM in the development phase. Model-based testing methods are applied, allowing for the generation of test cases for unit, integration, and system testing, including test execution conditions and expected results.

Similar to the PIM-to-code transformation, a set of transformation rules is defined to implement the conversion of the test cases in the platform-independent testing model (PIT) to the test cases in the platform-specific testing model (PST). This allows PST to directly execute corresponding unit testing, integration testing, and system testing on the respective platforms.

The model-driven software testing process integrates the advantages of model-based software testing and model-driven software development, moving the focus of testing work to the construction of PIT in advance. Model transformation work can be supported by tools, reducing the involvement of testers and avoiding the introduction of new errors.

The model-driven software testing process integrates the benefits of model-based software testing and model-driven software development, shifting the focus of testing work to the construction of PIT. Model transformation tasks can be supported by tools, reducing the involvement of testers and minimizing the introduction of new errors.

Model-driven testing technology has two key technical aspects: (1) constructing a testing model that can be used to generate test cases; (2) generating test cases using specific methods.

The testing model can be developed using the following approaches:

1. Building a testing model based on the system design model using a modeling language. The PIM (platform-independent model) from the system design model can be directly used for test generation. This model is implemented during the development process and can be transformed from the design model (PIM) to the testing model for generating test data. This approach primarily involves model transformation techniques
2. Directly starting from analyzing requirements to building a testing requirement model. Testing requirement modeling is a crucial step in the test case generation process and a prerequisite for automatic test case generation. Several researchers have made significant improvements in testing requirement modeling and proposed a visual testing requirement metamodel, as shown in Fig. 6.3.

The core of software testing lies in the generation of test cases. In model-driven testing, test cases can be generated based on the PIM, testing requirement models, and testing models derived from PIM transformations. When testers are provided with a limited set of correct inputs, it is unlikely that all possible test cases can be generated. Consequently, appropriate testing strategies are required to determine which test cases should be included in the test suite. Different test case generation strategies must be employed for different testing models.

1. Testing models generally have traceable paths, such as directed graphs, finite state machines (FSMs), and others. The generation of test cases primarily involves traversing these paths using specific algorithms. Depending on the direction of the search path, the traversal can be categorized into breadth-first search (BFS) and depth-first search (DFS). Additionally, based on different algorithmic strategies, the traversal can be classified into non-recursive and recursive traversal algorithms.
2. Generating Test Cases Using Intelligent Algorithms

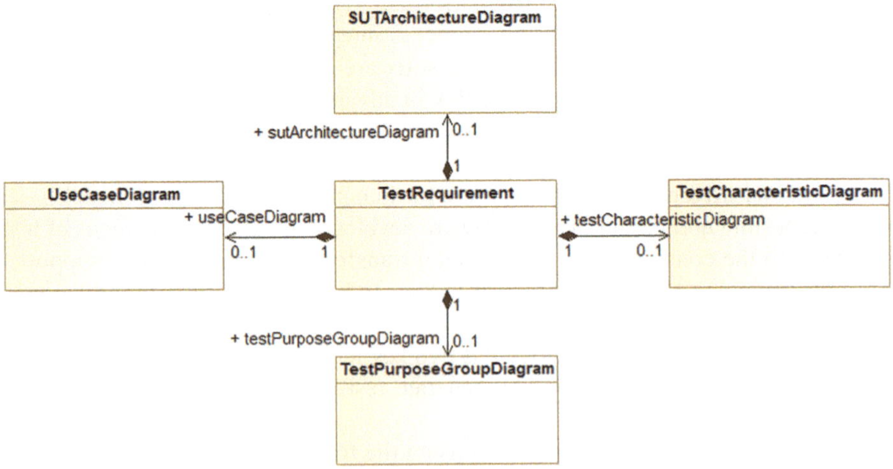

Fig. 6.3 A Visual Metamodel for Testing Requirements

Generating test cases through intelligent algorithms is a widely used approach in contemporary testing practices. Popular algorithms for test case generation include random, ant colony optimization, genetic algorithms, tabu search, and particle swarm optimization. Each of these algorithms has its unique advantages and limitations, with no algorithm inherently superior to another. The key distinction lies in the suitability of each algorithm for different application scenarios. The choice of algorithm should be made based on the specific characteristics of the software model under test and the testing objectives.

3. Using Model Checking to Generate Test Cases

   In model-driven testing, models are used to represent the expected behavior of the software, and the primary goal of testing is to identify discrepancies between the actual implementation of the software and the behavior described by the model. The foundation of model checking lies in the requirements, and the input for checking the behavior primarily comes from historical failure data and requirement analysis. The counterexamples discovered during this process, which deviate from the model, serve as valuable test cases and can be used in the execution of test cases.

6.1.1.4 Model-Based Testing

It should be noted that model-based testing should not be confused with model-driven testing. Model-based testing belongs to the category of specification-based software testing. Initially applied in hardware testing, it has been widely used in telecommunications switching system testing. Recently, it has also been applied to software testing, garnering increasing attention in both academic and industrial circles.

Characteristics of model-based testing include the fact that both test case generation and test result evaluation are based on the system model under test and the derived models. In model-based testing, after constructing the test model, further tasks are required, such as utilizing a specific coverage criterion to guide test case generation and determining when to terminate testing. A coverage criterion defines which specific elements of the software need to be satisfied or covered during testing.

Currently, there is limited research on the automatic transformation of design models into test models in model-based testing. Most research focuses on generating test cases or test code directly from the design models. The advantage of this approach is the rapid generation of test code, but its disadvantage is the lack of a detailed description of the testing process, making it difficult to reuse, evolve, and customize the test code according to user requirements.

Model-based testing provides the foundation for model-driven software testing and is a necessary step toward model-driven testing (MBT). It can be considered that model-based testing is less automated and formalized than model-driven testing.

6.1.2 MBT Framework

After the previous introduction, we now have a preliminary understanding of Model-Based Testing (MBT). Next, we will focus on the MBT framework and the related technologies. The content in this section primarily refers to the Model-Based Testing Engineer certification standard from the ISTQB Foundation Level certification.

There are specific activities involved in the MBT process that are generally not part of traditional test analysis and design. The two main activities are as follows:

1. MBT Modeling Activities: This includes not only the crucial task of constructing the system under test model but also activities related to the management of MBT models, the development and integration of modeling methods, the definition of modeling guidelines, the development of the MBT model structure, and the creation of modeling elements. Examples of these activities include generating specific charts for MBT and working with tools relevant to MBT modeling.
2. Generation of Testing Artifacts Based on MBT and Coverage Criteria: This involves the creation of test artifacts, such as test cases, based on the MBT approach and the selected coverage criteria.

One of the key impacts of MBT on the testing process is automation and the generation of test artifacts. Compared to traditional test design, MBT moves the test design phase to earlier stages of the project. When graphical models are used, MBT can serve as an early requirement validation technique that fosters and enhances communication. Model-based testing often complements traditional testing, offering testing teams an opportunity to verify the consistency between the test cases automatically generated by MBT and the test cases created manually.

The following four points help us understand the role of MBT technology in the software development lifecycle, making it applicable to both sequential and iterative software development lifecycles:

1. Due to MBT's ability to abstract complex functional requirements and expected behaviors, it is typically applied at higher levels of testing, such as integration testing, system testing, and acceptance testing.
2. MBT can also be used, albeit occasionally, for component (unit) testing. For instance, certain MBT methods, based on techniques that parse code annotations, can be effectively utilized in component (unit) testing.
3. MBT models can represent both the expected behavior of the system and/or its environment. As such, MBT is primarily employed in functional testing.
4. Enhanced or specialized MBT models can be applied to non-functional testing, such as security testing, load/stress testing, and reliability testing.

The primary techniques involved in model-based testing include modeling techniques, model checking, and test case generation. Currently, model-based testing is still in its nascent stages in many organizations within China. The following sections will introduce these techniques in detail, one by one.

6.2 Modeling Language and Test Model

6.2.1 Modeling Language

In model-based testing, the first step is to select an appropriate modeling language and construct the test model for the system under test. The following sections will introduce several modeling languages with a focus on their specific characteristics.

Modeling languages can be broadly categorized into three types: informal, semi-formal, and formal. The higher the level of formalization in a test model, the greater the degree of automation achievable in software testing. However, the most commonly used modeling languages are generally informal or semi-formal. These include UML and its extension, MARTE, for real-time embedded systems, Petri nets, Markov chains, and AADL. Markov chains are used to calculate the probability of transitioning to future states, independent of the current state; AADL (Architecture Analysis and Design Language) is primarily used for modeling system architecture, including both software and hardware components, with a focus on real-time embedded systems; Petri nets are employed for modeling real-time systems, addressing issues of uncertainty and concurrency between events. Extended versions of Petri nets can also be used to model UML sequence diagrams.

UML (Unified Modeling Language) is renowned for its strong object-oriented features and rich modeling capabilities. It can be used to describe both the static structure and dynamic behavior of a system. Some models within UML, such as activity diagrams, exhibit formal characteristics. Due to these features, UML is frequently utilized for modeling software development and testing processes. UML is particularly well-suited for functional testing. However, in many systems—such as embedded software systems—non-functional attributes are equally important as functional requirements. UML has certain limitations in expressing real-time and non-functional attributes, which is a key shortcoming when applied to these systems.

To address this limitation, OMG (Object Management Group) introduced MARTE (Modeling and Analysis of Real-Time and Embedded Systems), a modeling and analysis language specifically designed for real-time embedded systems. MARTE significantly enhances the ability to describe embedded systems. As an extension of UML, MARTE expands UML's modeling capabilities by incorporating three elements: constructs, tag values, and constraints. It also utilizes the Value Specification Language (VSL) to describe non-functional attributes, either qualitatively or quantitatively.

6.2.1.1 Real-Time Embedded System Modeling and Analysis (MARTE)

Due to its well-defined structure, ease of expression, powerful functionality, and broad applicability, UML has been widely adopted in the field of software engineering. It enjoys strong versatility and is supported by a wide range of tools. Extensions of UML can be applied to various domains. However, these extensions often suffer from poor applicability and a lack of standardization, significantly limiting their practical use. In the embedded systems domain, non-functional attributes are critical quality aspects alongside functional requirements. UML lacks standardization for modeling these non-functional attributes, making its direct application to embedded system modeling a labor-intensive process, with quality control being difficult to ensure.

The Systems Modeling Language (SysML), adopted by OMG, is capable of modeling complex systems, including software, hardware, devices, and processes. While SysML offers more robust modeling capabilities than UML for embedded systems, it still falls short in supporting the modeling of critical features such as hard real-time constraints and real-time resource management. This limitation considerably impacts SysML's applicability in the embedded systems domain.

To address the needs of real-time and embedded systems, OMG introduced the UML Profile for Schedulability, Performance, and Time (SPT), which aims to provide an annotation mechanism for modeling scheduling analysis and performance analysis using a limited set of identifiers. However, practical experience indicates that SPT requires significant improvements and refinement in several areas:

1. Specification for software and hardware.
2. Compatibility with UML 2.0.
3. Definition of platform models, application models, and allocation models.
4. Modeling of non-functional attributes.

In response to these challenges, OMG proposed the Modeling and Analysis Language for Real-Time and Embedded Systems (MARTE), with the primary objective of enhancing UML's modeling capabilities in the embedded systems field.

MARTE consists of three main components: the Foundation Model Package, the Design Model Package, and the Analysis Model Package. The Foundation Model Package provides the basic models needed to represent non-functional attributes, time and time-related concepts, allocation mechanisms, and general resources. Further refinement of these foundational models enables the design and analysis of real-time embedded systems. The Design Model Package focuses on model-based design. The models it provides can describe not only the real-time embedded characteristics of application programs but also the software and hardware execution platforms of embedded systems. The Analysis Model Package explains model-based analysis and offers fundamental elements for quantitative analysis. These elements can be further refined to support scheduling and performance analysis.

6.2.1.2 Comparing MARTE and AADL

MARTE and AADL are two widely used modeling languages for real-time embedded systems. Today, embedded systems are extensively applied in both industrial production and daily life, playing a crucial role. Therefore, it is necessary to discuss the respective advantages and disadvantages of MARTE and AADL. The similarities between MARTE and AADL can be summarized in the following aspects:

1. Common Objective: Both are designed for the modeling and analysis of real-time embedded systems.
2. Similar Functions: Both can be used for the design of hardware and software for embedded systems, as well as for defining and analyzing allocation mechanisms.
3. Semantic Equivalence: The concepts in MARTE and AADL exhibit semantic equivalence. The correspondence between the concepts in MARTE and AADL is illustrated in Table 6.2

Although MARTE and AADL share many similarities in modeling real-time embedded systems, they each retain distinct characteristics. The differences between the two are primarily reflected in the following six aspects.

1. Concepts and Modeling Capabilities
   (a) MARTE inherits the fundamental concepts of UML, offering a rich set of concepts and modeling capabilities that cover all phases from requirements modeling to code generation.
   (b) AADL's concepts are derived from the Digital Simulation Language (DSL), which contains fewer concepts and is limited to modeling specific abstract layers.
2. Semantic Rigor
   (a) Like UML, MARTE has a more flexible semantic structure, which is not conducive to automatic code generation and simulation. Instead, it is suitable for modeling platform-independent systems.
   (b) AADL, on the other hand, features a well-defined, formal, and executable semantic structure.

Table 6.2 The concept relationship between MARTE and AADL

MARTE construct	AADL construct
HwProcessor','HwASIC'','HwPLD','HwComputingResource', 'ComputingResource','ProcessingResource'	processor
'DeviceResource','HwActuator','HwArbiter','HwBridge','HwDevice','HwDMA', 'HwCLock','HwCoolingSupply','HwComponent','HwI_O','HwISA','HwMedia', 'HwMMU','HwPowerSupply','HwResource','HwSensor','HwStorageManager', 'HwSupport','HwTimer','HwTimingResource','HwWatchDog'	device
'HwBus'	bus
No stereotype,SysML::Block	system
'StorageResource','HwMemory','HwCache','HwRAMI,'HwROM','HwDrive'	memory
'MemoryPartition','RtUnit','DeviceBroker','MemoryBroker', 'SynchronizationResource','MutualExclusionResource', 'SwMutualExclusionResource','SwTimerResource'	process
'SwSchedulableResource','SchedulableResource','ConcurrencyResource'	thread
none	threadGroup
'Type','PpUnit','Alarm','InterruptResource','MessageComResource', 'SwCommunicationResource','NotificationResource','SharedDataComResource', UML2!DataType or UML2!Signal	data
An operation(class method)in a thread or a subProgram	subProgram

3. Usability
   (a) MARTE is applicable to general modeling and analysis of embedded system applications.
   (b) AADL is specifically designed for modeling and analysis of applications involving synchronized data flow and event flow.
4. Model Representation
   (a) Similar to UML, MARTE offers multiple views to model embedded systems from different perspectives, making the models easier to understand. Consistency checking among these views helps identify design flaws in the system.
   (b) AADL can model embedded systems using text, XML, and graphical representations. However, these three forms are essentially equivalent and represent the same model from different perspectives.
5. Modeling Non-Functional Features
   (a) MARTE integrates necessary non-functional features within the model, covering aspects such as schedulability, performance, and timing.
   (b) AADL lacks this level of integration for non-functional features.
6. Supporting Tools
   (a) UML is supported by a wide range of mature tools. As MARTE is an extension of UML for embedded systems, it also benefits from extensive tool support.
   (b) AADL, being a relatively new technology, is still in the early stages of research, especially in China. It has been predominantly applied in fields such as aerospace and military applications and has not been widely adopted. Consequently, the tools supporting AADL are still underdeveloped.

Although both MARTE and AADL retain distinct characteristics in embedded system modeling, they are not mutually exclusive. In fact, due to the many similarities between the two in the embedded modeling domain, the integration of MARTE and AADL has become a viable area of research. Numerous scholars, both domestically and internationally, have already begun studying this integration. By combining MARTE for platform-independent system modeling with AADL's precise and well-defined executable semantics, it is possible to use conversion tools or conversion rules to translate MARTE models into AADL models. This approach allows for simulation and formal verification of embedded system models, enabling the early detection of design flaws in the development of embedded systems.

6.2.1.3 UML Testing Modeling Language: UML 2.0 Testing Profile

In response to the limitations of UML's testing modeling capabilities, OMG proposed the UML 2.0 Testing Profile (U2TP) testing modeling standard in June 2001. U2TP provides precise definitions for testing the system structure and behavior of systems modeled using UML. It supports various testing-related activities, including design, visualization, specification, analysis, construction, and documentation.

U2TP provided following logic concepts to support software testing

1. Test Architecture
2. Test Behavior
3. Test Data
4. Test Time

These concepts are defined as follows: Test Architecture specifies the structure of U2TP;Test Behavior defines the behavior of U2TP;Test Data represents the data required throughout the execution of U2TP testing behavior; Test Time quantifies the duration of the test behavior in U2TP.The definitions of these four logical concepts are discussed below.

Test Architecture defines a set of concepts to describe the static structure of the testing system, including Test Context, Test Result Arbiter, Scheduler, System Under Test (SUT), and Test Components. The Test Context manages the entire testing process, encompassing test components, test cases, and test configurations. The Test Result Arbiter determines whether a test case has passed or failed. The Scheduler is responsible for starting, stopping, and dynamically creating test components. The SUT represents the system or the behavior of the system being tested. Test Components are the execution environments where test cases run, and the behavior of a test case is realized through the interaction between the SUT and the test components.

Test Behavior defines a set of concepts to describe the behavior of the testing system, including Test Case, Test Target, Test Log, and Test Result Category; Test Case defines a specific testing behavior; Test Target allows test designers to express the goals of the test; Test Log records entities during the execution of the test case to facilitate deeper analysis.

The Test Result Category includes four types: Pass (test behavior is consistent with the expected result), ail (test behavior is inconsistent with the expected result), Error (an error or exception occurs within the testing system), Inconclusive (unable to determine the consistency between the test behavior and expected results).These categories can be set or retrieved through the Arbiter in the Test Case model.

Test Data defines a set of concepts to describe the test data, including Data Pool, Data Partition, Data Selector, and Coding Rules. Test Data defines a set of concepts to describe the test data, including Data Pool, Data Partition, Data Selector, and Coding Rules. Data Pool is related to the Test Context and contains Data Partition and specific data values. Data Partition is used to define a set of rules to separate the test data. Data Selector provides different strategies for selecting and verifying data. Coding Rules allow users to define encoding and decoding rules for test data.

Test Time Control defines two concepts: Timers and Time Zones. Time Zone is used to handle coordination and synchronization issues for various test components, especially in distributed testing, and specifies that each test component belongs to at most one time zone. Test components in the same time zone have the same time characteristics, that is, they are synchronized. Timer is used to manipulate and control test behavior to ensure the termination of the test case. The U2TP concepts are listed in Table 6.3.

U2TP (UML 2.0 Test Profile) is a testing modeling language based on UML 2.0. It can be used independently of UML and supports testing modeling at various levels, ranging from unit testing to system testing. U2TP can also be extended and applied to various domains, including remote communications, IT, aerospace, and others.

Similar to MARTE, the testing models described by U2TP are platform-independent. This feature makes U2TP well-suited for constructing platform-independent testing models (PIT) in model-driven embedded software testing processes, significantly enhancing the adaptability of the testing models.

6.2.1.4 Other Test Modeling Languages

1. Testing and Test Control Notation (TTCN-3)

   TTCN-3 is a standardized testing language developed by the European Telecommunications Standards Institute (ETSI). It has been adopted by ISO as an international standard (Z.140 series). The goal of TTCN-3 is to adapt to the constantly evolving testing requirements and to provide new testing concepts, architectures, and powerful specification methods for emerging software architectures such as ODP, CORBA, TINA, DCE, and next-generation network protocols. TTCN-3 has been widely applied in fields such as mobile communications, wireless LANs, broadband technologies, and more. In China, research and tool development related to TTCN-3 have also been initiated, including at institutions such as the University of Science and Technology of China.

   TTCN-3 is divided into several usage forms, including Core Language, Tabular Presentation, and Message Sequence Chart (MSC) Presentation. The Core Language is the foundation for other formats; it is complete, independent,

Table 6.3 Concepts in U2TP

Testing system structure	Test behavior	Test data	Test time
Tested Software	Test Purpose	Communication	Timer
Test Component	Test Case	Data Partition	Time Zone
Test Environment	Default	Data Pool	
Test Configuration	Judgment	Data Selector	
Test Control	Confirmation		
Judger	Test Log		
Part of Utility Program	Log Operation		

and serves as the standard interchange format between TTCN tools, providing the semantic foundation for the other forms.

The Core Language of TTCN-3 consists of four parts.

(a) Data Types: Provides basic data types and customizable general data types for users.
(b) Test Data: Used for sending and receiving data during the test process.
(c) Test Configuration: Defines the components of the test and establishes communication ports for different test configurations.
(d) Test Behavior: Defines the dynamic behaviors of the test.

Using TTCN-3 to build test models offers several characteristics:

(a) International Standard: TTCN-3 is a standardized language with high modularity, allowing for reusable structures and distributed development.
(b) Multiple Test Case Executions: It supports executing a single test case multiple times and allows the user to freely choose which test case to run.
(c) Separation of Configuration and Testing Process: The separation of configuration files and the test process enhances the portability of test scripts.
(d) Support for Multiple Data Types: TTCN-3 supports a variety of data types.

2. Currently, Test-Driven Development (TDD) has become a mainstream approach in software development. The fundamental idea of TDD is to write the test code before the functional code, then write only the code that passes the tests. This approach drives the entire development process through tests. TDD helps in writing concise, usable, and high-quality code with excellent flexibility and robustness, allowing the code to quickly adapt to changes and accelerating the development process.

   A key concept in Test-Driven Development is the testing framework, and JUnit is a unit testing framework for Java. Developed by Kent Beck and Erich Gamma, JUnit has become the most successful framework in the xUnit family, which originated from Kent Beck's sUnit.

   The JUnit framework consists of six packages. The relationship between these packages is shown in Fig. 6.4.

   There are seven core classes in the JUnit framework. The names and functions of these classes are listed in Table 6.4.

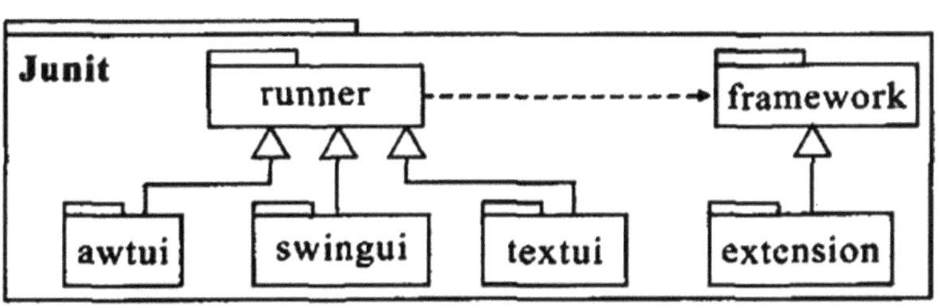

Fig. 6.4 JUnit Packages and Their Relationship

Table 6.4 Core classes in the JUnit framework

Name	Description
Assert	Verifies expected values against actual values, and indicates errors when assertions fail
TestResult	Mainly used to collect test results, including all error information
Test	An interface, implemented by TestCase and TestSuite, results are passed to TestResult after execution
TestListener	An interface, events such as start, result, and errors during testing are notified to TestListener
TestCase	Defines the test environment, tests generally start by inheriting from the TestCase class
TestSuite	Implements running multiple TestCases.
BaseTestRunner	In the JUnit framework, there are 3 TestRunners to launch the test interface (including text control, Swing, AWT), BaseTestRunner is the parent class of these 3 classes

JUnit is easy to use and extend, making unit testing highly accessible. Based on JUnit, several new unit testing frameworks have emerged for specific applications, such as StrutsTest for testing Struts applications and GroovyTestCase for testing Groovy programs.

The JUnit framework standardizes and enhances the effectiveness of Java unit testing, making it more conducive to integration testing. Currently, most Java development tools provide direct support for JUnit unit testing.

3. JUnit is easy to use and extend, making unit testing highly accessible. Based on JUnit, several new unit testing frameworks have emerged for specific applications, such as StrutsTest for testing Struts applications and GroovyTestCase for testing Groovy programs.

   The JUnit framework standardizes and enhances the effectiveness of Java unit testing, making it more conducive to integration testing. Currently, most Java development tools provide direct support for JUnit unit testing.

The XML Metadata Interchange (XMI), based on XML, has become a substandard of the Model-Driven Architecture (MDA). The primary purpose of XMI is to facilitate seamless data exchange between modeling tools and metadata repositories in distributed heterogeneous environments. XML plays a crucial role in connecting various analysis and design models by utilizing its capabilities.

In traditional software testing methods, test programs were written specifically for a particular programming language, making the development process cumbersome and time-consuming. Testers had to create different test programs for each language. However, most modern programming languages share similar features, such as scope, functions, loops, etc. Additionally, almost all object-oriented programming languages include concepts like classes, packages, or interfaces. In this context, testing programs essentially call the functional modules of the program under test and verify whether the return values match the expected values.

This similarity across programming languages allows modern testing frameworks and tools to utilize generic, cross-language test scripts for testing programs written in different languages. The test scripts need only call the functions of the program under test and check if the return values align with the expected ones. This approach significantly simplifies the testing process, enhances testing efficiency, and increases the repeatability of tests.

By abstracting the characteristics of various programming languages, it is possible to define a language-independent, generic XML schema to describe the format for generating test cases. This schema can then be used with XSLT tools to perform text transformations, generating test scripts suitable for specific platforms. Therefore, by incorporating XML into the testing process and describing platform-independent test model information, the applicability of the test models can be significantly improved.

Earlier, a detailed introduction was provided for several commonly used software testing modeling languages. In the following section, a comparison of these modeling languages will be presented.

1. Supporting Range of Test Types

   (a) U2TP can support the construction of test models at various levels, ranging from unit testing to system testing.
   (b) TTCN-3 is particularly well-suited for black-box testing, with consistency testing being one of its key strengths.
   (c) JUnit is specifically designed for unit testing programs developed in Java.

2. Dependency of the Model on the Platform

   Since U2TP is not associated with any specific implementation technology, the test models it describes are platform-independent. In contrast, test scripts generated by frameworks such as JUnit are platform-dependent and can only run on specific platforms, such as Eclipse.
3. Ease of Model Maintenance

   (a) U2TP is correlated with UML models, and the test models constructed by U2TP are relatively easy to adapt to changes in requirements or design defined by UML. Therefore, model maintenance is relatively straightforward.
   (b) TTCN-3 lacks a corresponding development model, meaning that if requirements change, the test models need to be recreated, making maintenance relatively difficult.
   (c) The test models established by JUnit are tied to specific platform implementations and can only accommodate changes to the low-level implementation of those platforms.

4. Level of Abstraction in the Model

   (a) The test models created by U2TP cannot be executed directly and need to be further transformed into test models with a lower level of abstraction.
   (b) Test models built with TTCN-3 and JUnit have precise executable semantics and can be directly executed on their supporting platforms for testing.

Although U2TP fully leverages the advantages of platform-independent modeling, this very advantage means that U2TP models lack precise executable semantics. As a result, U2TP models cannot be directly used for testing and must undergo further transformation to be executable for testing purposes.

6.2.2 Test Model

A model refers to a physical, mathematical, or other logical representation of a system, entity, phenomenon, or process. It serves as an approximation or simulation of the object or process being studied, expressed in a particular form. Objectivity and validity are fundamental requirements in modeling. A model must encapsulate all the information that reflects the essential characteristics of the prototype, so that through studying the model, one can comprehend the primary features of the prototype.

In model-based testing, the development of the test model is a crucial step. Due to the variation in software characteristics and testing objectives, a variety of models have been proposed in the industry for model-based testing. Some researchers have summarized state models used in model-based testing, while others have classified and provided explanations of models based on their notations, as outlined in Table 6.5.

In scientific research and practical engineering, commonly used testing models can generally be classified into several categories: finite state machine models, UML models, meta-modeling language models, MARTE models, and AADL models.

1. Finite State Machine (FSM), also known as a finite state automaton, is a mathematical model that represents a finite set of states, as well as the transitions and actions between these states. An FSM can describe the sequence of states an object undergoes during its lifecycle, as well as how it responds to various external events.

   A finite state machine (FSM) is a directed graph in which nodes are states and edges are transitions. Source and sink nodes become initial and terminal states,

Table 6.5 Classification of models

Markers	Models
Pre/Post (or State-based) Markers	B, OCL, JML
Transition-based Markers	FSM, State Diagrams, I/O Automation, Simulink Stateflow
History-based Markers	Message Sequence Charts
Functional Symbols	Algebraic Models
Operational Symbols	Petri Nets
Statistical Symbols	Markov Chains
Data Flow Symbols	Simulink Models

paths in the finite state machine are modeled as directed graph paths, and so on. Most finite state machine notations add information to the edges (transitions) to indicate the cause of the transition and actions that occur as a result of the transition. In some circles, this is referred to as an Extended Finite State Machine (EFSM). Here is a formal definition.

Definition: A Finite State Machine is a four-tuple (S, T, In, Out)
where:
S is a set of states
T is a set of transitions
In is a set of inputs that cause transitions
Out is a set of outputs that can occur on transitions.
The quadruple of the state machine depicted in Fig. 6.5 is as follows:
S={s1,s2,s3,s4,s5}
T={<s1,s2>,<s1,s3>,<s2,s3>,<s2,s4>,<s3,s5>,<s4,s5>}
In={e1,e2,e3,e4}
Out={a1,a2,a3}

The finite state machine model commonly refers to both the finite state machine (FSM) and the extended finite state machine (EFSM). The extended finite state machine (EFSM) builds upon the FSM model by incorporating variables, predicates that trigger transitions, and operations triggered by transitions. This extension enables the description of control processes and data, providing a precise representation of the dynamic behavior of software systems. Due to the

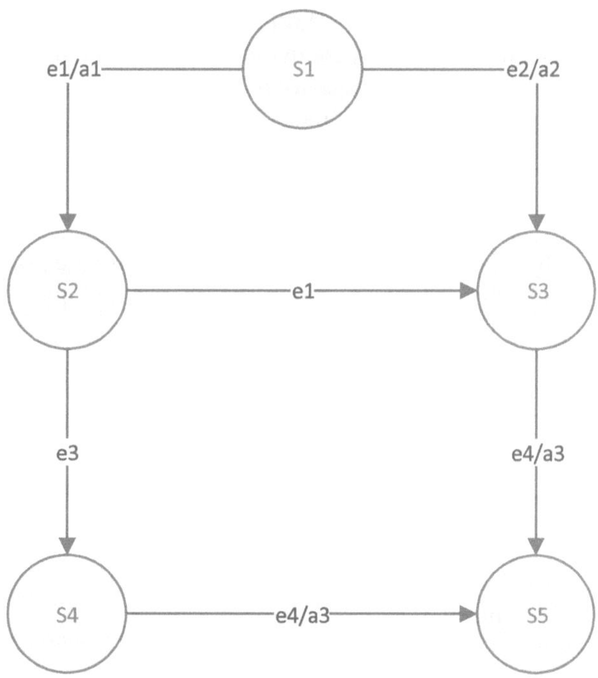

Fig. 6.5 Example Finite State Machine

precision, deductibility, and verifiability of finite state machines, they have been widely studied and applied. In practical engineering, finite state machines are particularly well-suited for modeling and testing control systems, event-driven systems, and, most notably, protocols. The extended finite state machine further allows for the description of control processes and data. A substantial amount of research has been focused on state machine modeling, test case generation, and optimization

2. UML (Unified Modeling Language), also known as the Standard Modeling Language, is a language used for visually modeling software-intensive systems. The definition of UML includes two components: UML semantics and UML notation. Thanks to the strong support and promotion from the Object Management Group (OMG), UML has been widely adopted in both industry and academia, and testing based on UML models has become one of the primary research directions in model-based testing.

   UML is a development methodology used to specify, visualize, construct, and document object-oriented software-intensive systems during the development phase. In the context of modeling large-scale, complex systems—particularly at the software architecture level—UML has proven to be an effective best practice. UML is a modeling language, and models are primarily represented in diagrammatic form. A typical modeling diagram typically consists of several blocks or boxes, lines, and text, which serve as supplementary information for the model. Although these elements may appear simple, they are crucial as they can be interconnected and extended according to UML rules.

   As of UML 2.0, there are 13 types of UML diagrams. UML 1.5 defined nine of these, while UML 2.0 introduced four additional types. The nine UML 1.5 diagrams are: use case diagram, class diagram, object diagram, state diagram, activity diagram, sequence diagram, collaboration diagram, component diagram, and deployment diagram. The four UML 2.0 diagrams are: package diagram, timing diagram, composite structure diagram, and interaction overview diagram.

   (a) Use case diagram: describes the system functionality from the user's perspective.
   (b) Class diagram: describes the static structure of the classes in the system.
   (c) Object diagram: shows the state of multiple objects in the system at a certain moment.
   (d) State diagram: describes the control flow from state to state and is commonly used for dynamic modeling.
   (e) Activity diagram: describes the workflow of business implementation use cases.
   (f) Sequence diagram: describes the dynamic collaboration between objects, emphasizing the order in which objects send messages and showing the interaction between objects.
   (g) Collaboration diagram: describes the cooperation relationship between objects.

(h) Component diagram: a special type of UML diagram used to describe the static implementation view of the system.
(i) Deployment diagram: defines the physical architecture of software and hardware in the system.
(j) Package diagram: organizes and groups the model elements that make up the system.

Timing diagram: represents a graph of lifeline state changes.

Composite structure diagram: represents a diagram of the internal structure of a class or component.

Interaction overview diagram: represents a diagram that uses activity diagrams to show the control relationship between multiple interactions.

Based on practice and literature research, state diagram, class diagram, timing diagram, and activity diagram are commonly used in UML.

3. A meta-model defines the standards for describing a particular model, specifically the elements that make up the model and the relationships between those elements. The meta-model can be viewed as a formal language, where the model is an article written in that language, with the elements of the meta-model acting as the vocabulary of the language, and the relationships between elements forming the syntax. Every model is explained by a meta-model, although this meta-model may not always be explicitly visible. Models and meta-models are relative concepts: the meta-model is a model of the meta-meta-model, and the interaction between models and meta-models forms an infinite loop, with each successive level becoming more abstract.

   The concept of a meta-model is relative to that of a model; without a model, a meta-model would lose its meaning. Some typical meta-models proposed by OMG include UML, SysML, SPEM, and CWM. The International Organization for Standardization (ISO) has also published a standard for meta-models, ISO/IEC 24744.

   Meta modeling is the technique of creating a meta-model that describes a modeling language and provides the modeling tools that support that language. Meta modeling enables the definition of custom meta-models and establishes a meta-model by defining a modeling language suitable for a specific domain. This makes the meta-model more versatile and allows for flexible customization and extension to meet the unique needs of a domain.

   Meta models are general and can be flexibly customized and extended to cater to the specific needs of a domain. They also address the issue of tool development being project-specific. Currently, a significant amount of research focuses on applying meta-modeling techniques to define and build a test requirements meta-model, which is then used to generate test cases in conjunction with the system under test model.

4. MARTE Model

   Modeling and Analysis of Real-Time and Embedded Systems (MARTE) is a standard proposed by OMG aimed at extending UML for embedded systems, addressing the limitations of UML in modeling and expressing the complexities of real-time and embedded systems.

MARTE consists of three main components: the Basic Model Package, the Design Model Package, and the Analysis Model Package, as illustrated in Fig. 6.6.

MARTE extends UML and logically inherits its syntax and semantics. It supports modeling across all stages, from software requirements to code implementation, and creates platform-independent models that align with the needs of MDA for platform-independent model development. The introduction of MARTE significantly enhances UML's modeling capabilities in the embedded systems domain and offers the following advantages:

(a) MARTE supports the modeling of non-functional properties.
(b) MARTE enhances UML by incorporating comprehensive timing and resource models.
(c) MARTE defines concepts for modeling both software and hardware platforms.
(d) MARTE introduces concepts for associating applications with platforms.
(e) MARTE provides support for quantitative analyses, including scheduling and performance.

5. AADL model

The Architecture Analysis and Design Language (AADL) was proposed by the Society of Automotive Engineers (SAE) in 2004, building upon MetaH and UML. It was published as the SAE AS5506 standard, with the objective of

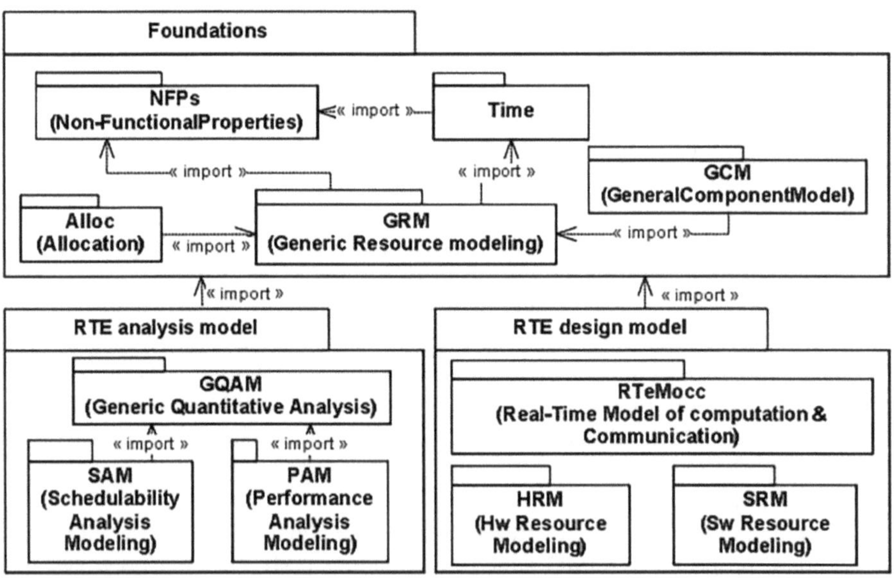

Fig. 6.6 Structure of UML MARTE

providing a standardized and sufficiently precise approach for the design and analysis of both the software and hardware architectures, as well as the functional and non-functional properties of embedded real-time systems. AADL employs a unified model with various analysis methods to integrate critical aspects such as system design, analysis, verification, and automatic code generation within a single framework.

Components serve as the fundamental elements within an AADL model, representing either hardware or software entities within the system. Each component is associated with a specific component type, which defines its functional interface. A component type functions as a specification that dictates how other components interact with it, and it is characterized by features, flows, and property associations.

AADL defines three types of components. Table 6.6 provides a detailed description of the types and content of these components.

AADL utilizes concepts such as components and connections to model the software and hardware architecture of a system. Functional and non-functional properties are described through features and properties. Pattern transformations are employed to capture the evolution of the runtime architecture. User-defined properties and attachments provide support for system extensibility. For complex systems, AADL organizes and models the system using packages.

AADL offers three modeling methods—textual, XML-based, and graphical—which can be employed to describe the aforementioned processes.

AADL is characterized by its simple syntax, powerful functionality, and extensibility. It has been extensively applied in the development of embedded systems across aerospace, mechanical, and electronics sectors. In China, numerous research institutions have conducted in-depth studies and extensions of AADL.

It is crucial to emphasize that designing test constraints prior to generating test cases from a model is an essential step. Test constraint design involves addressing the interactions and constraints among various parameters. By managing these constraint relationships, additional test information can be incorporated into the model, facilitating the generation of a test case set that better aligns with test requirements. This approach also mitigates the state explosion problem. In most research, test constraint languages are used for the design of test constraints.

Table 6.6 Types and contents of AADL components

Component type	Content
Software Component	Data, Processes, Threads, Thread Groups, Subprograms, etc.
Execution Platform Component	Processor, Storage Device, Bus, Peripheral Devices, etc.
Integrated Component	System Components

6.2.3 Modeling Case Study

This section presents a modeling case study using a simple garage door control system to illustrate model-based testing (MBT) techniques. A state machine is employed to represent the behavior of the garage door control system, thereby characterizing the system under test. This state machine model serves as the foundation for subsequent testing phases.

The case is derived from the book "Model-Based Testing: A Software Craftsman's Approach." Below is the requirement specification for the garage door control system.

The garage door control system consists of the following components:

1. A drive motor, which provides power to the system, allowing it to move in both forward and reverse directions.
2. Garage door tracks equipped with sensors, positioned at both the fully open and fully closed states, to detect whether the door is open or closed.
3. A control device that sends control signals to manage the system's operations.
4. A laser beam sensor located near the floor, which only operates when the garage door is closing.
5. The garage door itself.

If the laser beam is interrupted (possibly by a pet) or if an obstacle is encountered while the door is closing, the door will immediately stop and reverse direction. While the garage door is in motion—either opening or closing—if the control device sends a signal to stop the door, it will halt. When the door is restarted, the control signal will direct the door to resume in the same direction it was moving prior to stopping. The door will stop moving once it reaches the fully open or closed position.

The process of constructing a state machine for the system under test typically involves the following five steps:

1. Identifying the system states.
2. Identifying the events and actions that trigger state transitions.

   Through an analysis of the garage door control system's requirement specifications, we can identify the system states and the events and actions that cause state transitions, as outlined in Table 6.7. Events are predicate conditions within

Table 6.7 The states, events, and actions of garage controller system

Input events	Output events (Actions)	States
e1: control signal	a1: start drive motor down	s1: Door Up
e2: end of down track hit	a2: start drive motor up	s2: Door Down
e3: end of up track hit	a3: stop drive motor	s3: Door stop going down
e4: laser beam crossed	a4: reverse motor down to up	s4: door stop opening
		s5: Door closing
		s6: Door opening

Table 6.8 State transition table

State	S1	S2	S3
S1	×	√	√
S2	√	×	√
S3	√	√	×

Table 6.9 State-event transition table

	State		
Event	S1	S2	S3
e1	S2	S2	S1
e2	S3	_	S1
e3	_	S3	S2

the extended finite state machine, while actions are operations triggered after a transition.

3. Determine the transition between states.

   To analyze the transition relationships between each state, a state transition table can be used to determine the complex relationships between each state, as shown in Table 6.8. Assuming that three states (S1, S2, and S3) have been identified in the system, we can create a state orthogonality table to determine whether each state can transit to another state. The state transition table can help us analyze the existence of transitions between states, ensuring completeness and accuracy. Additionally, the state transition table can help us identify implicit requirements or error information, thus improving the model or requirements.

4. Identify the events and actions that trigger state transitions

   Similarly, to analyze the conditions and actions between state transitions, we can use state/event table to determine the complex relationships between states. Here is an example of a state/event orthogonal table, where each event is analyzed to see if it can cause a state transition. These tables can help us analyze the complex relationships of the system, as shown in Table 6.9.

   Finally, after identifying the states, events, actions, state transitions, and the events and actions that trigger state transitions in the garage door control system, we can employ a state machine to model the system. The resulting model is presented in Fig. 6.7.

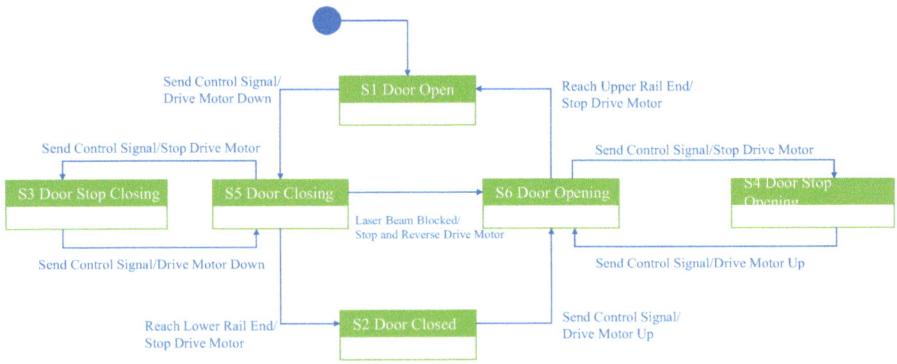

Fig. 6.7 State Machine Model of the Garage Door Control System

6.3 Model Checking Techniques

After developing a model of the system under test, it is generally necessary to validate the model before advancing to the next phase. In engineering, the most straightforward method is to conduct a model review. While graphical models facilitate communication among stakeholders, the review process still faces challenges related to low efficiency and limited effectiveness in validation.

Formal verification, a method grounded in rigorous mathematical theory, can effectively address the challenges in model validation. Among the various formal techniques, model checking is one of the most successful approaches. Model checking validates finite-state systems through the use of theorems and logical proofs, which not only significantly reduces manual effort and enhances automation but also effectively conserves resources and improves the accuracy of the testing process.

6.3.1 Introduction to Model Checking

Model checking is a formal verification technique that first formalizes both the software under test and its properties. For systems with a finite number of states, a state transition system (S) is used to model the system's behavior, while a rigorously defined temporal logic formula (F) is employed to specify the system's properties. Through this formalization, the question 'Does the system satisfy the given properties?' is transformed into the mathematically provable problem of 'Does the system model S satisfy the system property formula F?', denoted as $S \models F$. The finiteness of the system's states ensures that this problem is decidable, meaning a computer can provide a solution within a finite amount of time. The key components of a model checking problem are the system model S and the corresponding formula F that describes the property to be verified. Using specific algorithms, it can be determined

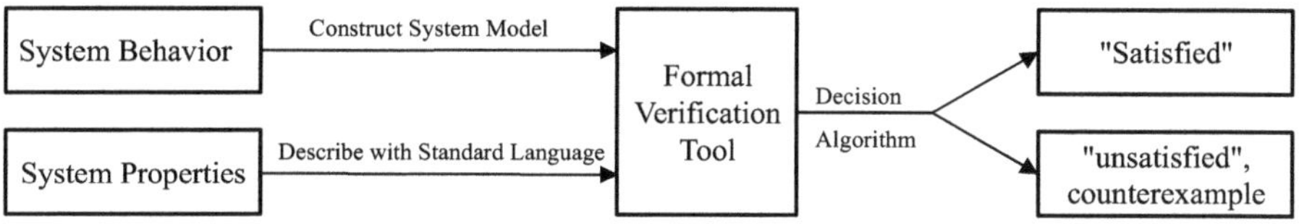

Fig. 6.8 The Basic Process of Model Checking

whether S ⊨ F holds. If it does, the algorithm returns 'yes'; if not, it returns 'no' and provides a counterexample that demonstrates the model's failure to meet the property, as shown in Fig. 6.8. Based on this counterexample, testers can analyze the underlying issues in the system and take steps to improve it.

Model checking generally involves three steps.

1. **Model Creation**: To perform model checking, the system under verification must be transformed into a formal model that is compatible with the model checking program, such as a state machine. During the modeling process, certain abstractions are often necessary based on the verification goals and the constraints of temporal and spatial resources. Elements that do not impact the verification results can be omitted.
2. **Property Specification**: First, it is essential to define the properties that the system is expected to satisfy. These properties are then translated into logical formulas in a specific form, in accordance with the model and the requirements of the model checking program established in step (1). Temporal logic is commonly employed to express the time-dependent behaviors of the software during execution. While model checking can verify whether a model satisfies a given property, it cannot ensure that the specified set of properties fully encompasses all the requirements that the software must meet.
3. **Verification**: After inputting the model and property specifications into the model checking program, the verification process is executed by algorithms, thereby eliminating the potential for human error. However, the verification results still require manual analysis. When the model fails to satisfy a property, it is crucial to examine the counterexample path provided by the program in order to trace the source of the error. Errors typically arise from incorrect system design, faulty models, or improper specifications. In some cases, the model may exceed the capabilities of the model checking program, leading to issues such as memory exhaustion. In such cases, adjustments to the model or modifications to the input parameters of the model checking program may be required

6.3.2 Building the System Model

Software typically interacts with the external environment. For example, in ATM software, processes such as data changes and state transitions are driven by events like user input or responses from remote servers. However, during model checking, the program does not interact with the external world. Therefore, when establishing a system model, two key parts are typically considered: modeling the software and modeling the external environment to simulate input and output data and provide stimulus events.

There are 3 approaches to model creation, listed in increasing order of automation:

1. Manual Model Creation
2. Automatic Model Generation
3. Building Models from Software Code

Manual model creation requires an analysis of the software's logic to abstract states and transitions, representing these as a model automaton. This process can also be automated using specific algorithms, known as automatic model generation. However, the input for model checking is not always a state machine. Some model checking programs directly use code written in common programming languages like C or Java, or an intermediate language styled like C code, as input for the model. This approach streamlines the model checking process, significantly reducing the time required to analyze the software and build the model. Additionally, changes to the software can be rapidly reflected in the model by updating the relevant parts.

Model checking is typically performed by representing the system as a state machine. This state machine can be modeled as a six-tuple (state set, initial state, state transition set, output function, input set, and the set of supported output tools). The model transitions between states based on input and generates output, a process known as a transition. The state machine can represent the lifecycle of a software system or subsystem and can also describe the lifecycle of an object in object-oriented software. It clearly depicts the states the model contains, how external events influence the model's state transitions, and how the transition conditions constrain these transitions. The model also shows how state transitions affect various variables within the system. This state machine model is typically equivalent to a UML state diagram created during the software development process, and a direct mapping can be established between the two.

The state machine used in model checking differs from the traditional state machine in certain respects. To more accurately reflect software logic, the concept of the Extended Finite State Machine (EFSM) is introduced. Unlike the traditional Finite State Machine (FSM), the EFSM adds preconditions to transitions. A transition along a given path can only occur when the precondition of that path is satisfied, and additional operations, such as variable assignments, may be included during the transition process.

Furthermore, traditional state machines are insufficient for describing complex control systems, which is why hierarchical state machines, concurrent state machines, and signals are introduced. In a hierarchical state machine, state nodes may be composite nodes that contain sub-state machines. The states within these sub-state machines are referred to as sub-states, reflecting an OR decomposition of the state: when an object is in a composite node, it is actually located in one of the sub-states within the sub-state machine.

A concurrent state machine may consist of multiple sub-state machines that are either local or overall. Each sub-state machine has its own current state, and the overall system state is formed by combining the states of these sub-state machines, representing an AND decomposition of the state. These concurrent state machines are not completely independent and can synchronize via signals. For instance, an output event from a transition in one sub-state machine can be sent as a signal and received as input by another sub-state machine, triggering further transitions.

States that do not contain sub-state machines are called basic states. Hierarchical and concurrent states are collectively known as composite states. Additionally, special states may be used to preserve the historical state of a sub-state machine after it has exited a particular state.

6.3.3 Property Specification

The behavior of software is represented as state transitions within a state machine model. For an entire software system, property specifications based on state transitions are more meaningful than traditional testing techniques that only focus on program inputs and outputs. This is because the internal workings of a software system are closely tied to state transitions, whereas program inputs and outputs only reflect the start and end of execution. Furthermore, many software systems, such as operating systems, are designed to run continuously, without a clear, finite sequence of computations, making traditional testing methods unsuitable. Temporal logic describes the expected properties of a system using logical formulas composed of atomic propositions and Boolean operators. By combining temporal operators with traditional Boolean logic, temporal logic formulas can express temporal properties of a model. In temporal logic, time is not a conventional notion of time, but refers to an infinite, discrete sequence. One direction of this sequence represents the future, while the other represents the past, with each position in the sequence referred to as a moment. Temporal logic imposes constraints on this sequence, such as 'at the next moment in time,' 'eventually,' and 'forever in the future,' making it well-suited for modeling state transition sequences. These temporal operators can be combined with Boolean expressions and operators or nested within themselves.

Temporal logic primarily encompasses Computation Tree Logic (CTL) and Linear Temporal Logic (LTL).

1. The computation tree is a tree structure that unfolds the model M, with its root node representing a state of M. As shown in Fig. 6.9, the state machine model M is unfolded with S5 as the root. The formal representation of Computation Tree Logic (CTL) for the properties of the computation tree includes temporal operators and path quantifiers, such as 'All' and 'Exist.' Computation Tree Logic effectively captures the branching behavior of states, as demonstrated in Fig. 6.10.

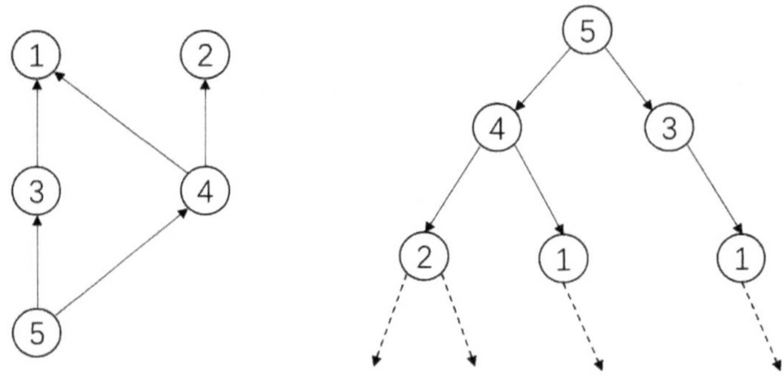

(a) State Machine Model M (b) Computation Tree After M Expansion

Fig. 6.9 Computation Tree Transformation

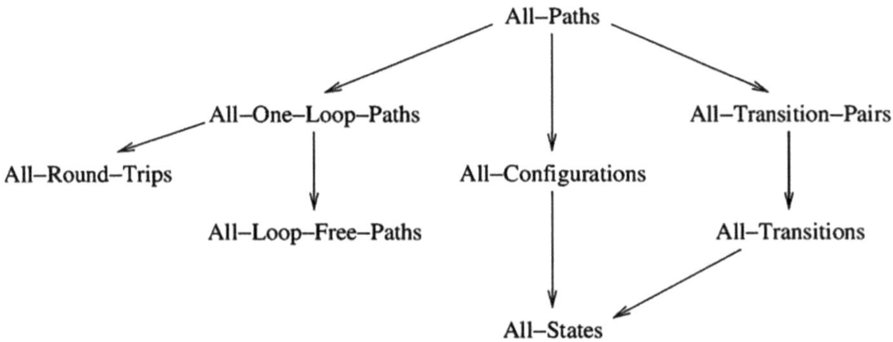

Fig. 6.10 Strong and Weak Relationships Between Coverage Criteria for Finite State Machines

2. LTL (Linear Temporal Logic) is a type of temporal logic that ensures the validity of property specifications across all moments in the system. The logical operators of LTL include "and," "or," "not," logical equivalence (←→), and logical implication, among others. The temporal operators in LTL include "G" (globally), "U" (until), "F" (future), and "X" (next time). When determining whether the state transition model of the system satisfies a property expressed in LTL, satisfiability proofs are used to check if any path violating the LTL formula can be generated starting from the initial state. If every path satisfies a particular LTL formula, then no path can satisfy the negation of that formula. A limitation of LTL is that it cannot utilize existential quantifiers. When property specifications involve both universal and existential quantifiers, LTL is generally not suitable for verification.

   Both CTL (Computation Tree Logic) and LTL are capable of accurately describing temporal properties, but they differ in several ways. LTL allows for the analysis of a subset of paths based on priorities defined by the tester, and it has only one successor state at any given moment. In contrast, CTL may have multiple successor states for each state. For more complex properties, such as state transitions where returning to the initial state is always possible, CTL can express these properties, whereas LTL cannot. LTL defines time in relation to paths, while CTL defines time in relation to states. The behavioral assertions in CTL are achieved by decomposing paths starting from a given state. Both CTL and LTL use the path quantifier "A," with identical descriptions. However, only CTL supports the use of the existential path quantifier "E." Consequently, for properties involving both universal and existential quantifiers, CTL is typically required for verification.

6.3.4 Algorithms and Implementation of Model Checking

The mainstream model checking algorithms can be divided into two main types. The first is the logic rewriting algorithm based on the on-the-fly method. This algorithm separates temporal logic formulas into two components: the part (f1) that must be satisfied in the current state, and the part (f2) that must be satisfied in the succeeding state. During the checking process, if the current state satisfies f1, the process continues by verifying whether f2 holds in the subsequent state. This procedure is repeated until f2 reaches a state that is guaranteed to be true.

The second type is the state machine-based model checking algorithm, which constructs the system model state machine and the property state machine based on the system model and the property to be checked, respectively. Initially, the synchronous product algorithm is employed to compute the intersection of these two state machines, i.e., to calculate their synchronous product. Next, the emptiness checking algorithm is used to determine whether the set of acceptable languages of the synchronous product is empty. If it is empty, the model satisfies the property under verification. If not, it indicates that the model does not satisfy the property, and a counterexample exists.

The input to the synchronous product algorithm consists of the model state machine and the property automaton. While the model state machine is constructed during the modeling phase, the property automaton is derived by transforming the LTL statement. The LTL statement essentially forms a set, where each element represents a path consisting of a series of state machine transitions, with each transition's variables satisfying the corresponding LTL specification. A special automaton, known as the Büchi automaton (property automaton), can then be constructed whose set of acceptable paths exactly matches this set. There are well-established algorithms for transforming LTL statements into Büchi automata, and we will utilize these tools to perform the conversion.

If the model satisfies the property, the behavior of the model automaton S will be entirely contained within the behavior set of the property automaton P. Formally, this is expressed as $L(S) \subseteq L(P)$. The negation of P, denoted as P′, represents the behavior set of the automaton corresponding to paths that violate property P. Thus, $L(S) \subseteq L(P)$ is equivalent to $L(S) \cap L(\neg P) = \varnothing$ $L(S) \cap L^Z(P) = \varnothing$. By generating the synchronous product of S and P, an automaton with the behavior set $L(S) \cap L(\neg P)$ can be obtained.

To implement this process in Python, two data structures—FSM (Finite State Machine) and AccFSM (Accepting Finite State Machine)—should be designed to represent ordinary state machines and state machines with accepting nodes, respectively.

An ordinary state machine contains a set of nodes, "states," where each node corresponds to a state, and a set of transitions, "trans," where each transition connects a starting node to an ending node. The model is input into the program as an FSM object. The state machine with accepting nodes inherits the properties of an ordinary state machine but adds a set of accepting states, "accStates," which is a subset of "states." The property automaton, in addition to the elements of an ordinary state machine, also needs to mark specific accepting states, allowing it to be represented as an AccFSM object. The following getSynProduct() function takes an FSM object and an AccFSM object as inputs and returns the synchronous product as an AccFSM class.

```python
def getSynProduct(model, Büchi):
    """
    Construct the Cartesian Product (synchronization
product)
    of the Model and the Büchi Automaton.
    """
    # Cartesian product of states
    newStates = [(m, b) for b in Büchi.states for m in
model.states]
    # Check if a transition is valid in the
synchronization product
    # transM: (start, end) from model
    # transB: (start, end, condition) from Büchi
    def isTransValid(transM, transB):
        func = transB[2]
        if func == 1:  # unconditional transition
            return True
        return func(transM[1])
        # Here transM[1] is the target state index
        # model.states[transM[1]] is the corresponding
ending state
    # Cartesian product of transitions
    newTrans = [
        [(m[0], b[0]), (m[1], b[1])]
        for b in Büchi.trans
        for m in model.trans
        if isTransValid(m, b)
    ]
    # Cartesian product of accepting states
    newAccStates = [(m, b) for b in Büchi.accStates for m in
model.states]
    # Build the synchronized product FSM
    synProduct = AccFSM(newStates, newAccStates, newTrans)
    return synProduct
```

Next, the emptiness checking problem for the synchronous product of automata can be addressed using graph theory. The algorithm begins by searching from the initial state to identify a cycle in the graph that contains at least one accepting node. Tarjan's algorithm, a well-known graph theory algorithm, is used to detect whether a directed graph contains any strongly connected components, which inherently contain cycles. By applying Tarjan's algorithm to the synchronous product state machine, the presence of a strongly connected component indicates that the model does not satisfy the property. In this case, the path from the initial state to the cycle, along with the cycle itself, forms a counterexample path. If no such cycle is found, it confirms that the model satisfies the property. The implementation of this algorithm is as follows:

````
```python
class AccFSM:
    # ... other methods not shown ...
    def checkEmpty(self):
        """
        Check if the automaton (AccFSM) has an
accepting loop.
        Return the two stacks [s1, s2] if a loop is found,
otherwise None.
        """
        # First stack: initialized with the initial state
        s1 = [self.states[0]]
        # Second stack: initialized empty
        s2 = []
        # Two maps to track visited states in two nested DFS
searches
        m1 = dict().fromkeys(self.states, 0)
        m2 = dict().fromkeys(self.states, 0)
        # First-level DFS
        while s1 != []:
            x = s1[-1]
            succX = self.succ(x)  # get successors of x
            found = 0
            for state in succX:  # scan successors
                if m1[state] == 0:  # if state has not
been visited
                    found = 1
                    m1[state] = 1
                    s1.append(state)  # push to stack s1
                    break
            # If no new successor, backtrack
            if found == 0:
                s1.pop()
                # If x is an accepting state, enter
````
````

```
second DFS
                if x in self.accStates:
                    s2.append(x)
                    # Second-level DFS: search for loops
                    while s2 != []:
                        v = s2[-1]
                        # If we find a path back to x,
loop exists
                        if x in self.succ(v):
                            return [s1, s2]
                        all_done = 1
                        for w in self.succ(v):
                            if m2[w] != 1:
                                all_done = 0
                                m2[w] = 1
                                s2.append(w)   # push to
stack s2
                                break
                        if all_done == 1:
                            s2.pop()
        # If no loop found
        return None
```

6.3.5 *Model Checking Tools*

Currently, there are dozens of model checking tools available on the market, which reflect the evolution of model checking technology. The characteristics of these tools highlight the differences in model checking technology across various periods and directions, both in theoretical and practical applications. Based on the differences in modeling languages, these tools can be classified into the following two categories.

6.3.5.1 Model Checking Tools Oriented Toward Formal Specification Languages

1. SMV (Symbolic Model Verifier) is a model checking tool that verifies the reachability of states and whether they satisfy CTL formulas by computing fixed points. The fundamental unit of an SMV model is a module, which can establish both synchronous relationships and execute concurrently in an asynchronous manner with other modules.

2. NuSMV (New Symbolic Model Verifier) is a model checking tool that integrates bounded model checking technology based on SAT. It is a refactored version of SMV and supports both CTL and LTL formulas.
3. STeP (Stanford Temporal Prover) is a verification tool that uses a model checker to handle verification problems of subsystems and consolidates the results using a theorem prover. STeP is not limited to finite state systems; by combining model checking technology with deduction, it can be applied to a broader range of systems, including programs with infinite data domains.
4. CWB (Concurrency Workbench) is an automated tool designed for the operation and analysis of concurrent systems. It can detect equivalence among models. Its modeling languages are CCS (Calculus of Communicating Systems) and LOTOS (Language Of Temporal Ordering Specification), enabling the analysis of the state space of given programs and detecting various semantic equivalences and sequentiality. It can also determine whether a model satisfies the property specifications expressed in μ-calculus.

6.3.5.2 Model Checking Tools Oriented Toward Source Code Languages

1. SPIN (Simple Promela Interpreter) is an explicit model checking tool that uses PROMELA (Process Meta Language) to describe models. This language bears resemblance to C, but is based on an asynchronous composition of process structures. In addition to checking whether a model satisfies LTL formulas, SPIN can also verify certain safety properties and assertions. The core approach is to represent each process and LTL formula as an automaton and determine whether the model satisfies the properties by checking if the automaton accepts a language.
2. BLAST (Berkeley Lazy Abstraction Software Verification Tool) is an automated verification tool for temporal safety properties of C programs. It employs lazy abstraction techniques to reduce solving complexity, and optimizes verification performance by leveraging the properties of counterexamples during the verification process.
3. SLAM (Software LAnguage Modeling) is a model checking tool for C programs. It extracts a process model containing only boolean variables from the source code of a C program and verifies the abstracted model. SLAM is built on three tools: C2bp for code extraction, Newton for model simplification, and Bebop for model checking.

Model checking tools can also be classified into several categories, including those that combine model checking with theorem proving, symbolic model checking tools, and bounded model checking tools. Symbolic model checking tools use Ordered Binary Decision Diagrams (OBDDs) to represent state transition graphs and employ Boolean logic formulas to specify system properties. Bounded model checking relies on the Boolean satisfiability problem (SAT) to verify whether the system satisfies the property within a bounded number of steps, k. If the property cannot be verified, the value of k is increased, and the verification process is repeated.

Model checking tools can also be classified based on their underlying theoretical foundations, into tools based on automata theory and those based on fixed-point theory. When classified by the type of system being verified, model checking tools can be categorized into those designed for real-time systems, concurrent systems, and hybrid systems.

6.4 Coverage Criteria

6.4.1 *Test Coverage Criteria Based on Models*

Test adequacy criteria, also known as coverage criteria or testing criteria, are a set of rules that guide test engineers in developing effective test cases that can improve the reliability and correctness of software systems. J.B. Goodenough and S.L. Gerhart defined test adequacy criteria in their literature on testing theory.

In model-based testing (MBT), coverage criteria are a crucial element. On one hand, they reflect the test requirements; on the other hand, they provide test engineers with a means to measure test adequacy.

It should be noted that the classification and comparison of coverage criteria are meaningful only within the same system. Different classification frameworks may lead to different coverage criteria. One set of coverage criteria can be based on transitions in the control flow, and the strong and weak relationships between each coverage criterion can be analyzed. Coverage criteria can also be categorized into two main types: control flow-based and data flow-based. Some researchers classify coverage criteria into five categories: structural model coverage criteria, data coverage criteria, fault-based coverage criteria, requirement-based coverage criteria, and statistical coverage criteria.

Here, we will introduce several types of coverage criteria used in model-based testing (MBT), including coverage criteria for finite state machines, UML models, and metamodels.

1. The coverage criteria for finite state machines primarily include all-states coverage, all-transitions coverage, all-transition-pairs coverage, all-loop-free-paths coverage, all-one-loop-paths coverage, all-round-trips coverage, and all-paths coverage. The relationships between these coverage criteria are shown in Fig. 6.10. To understand the strong and weak relationships between coverage criteria, consider the following example: when coverage criterion 1 is satisfied, it necessarily satisfies coverage criterion 2. In this case, we can say that coverage criterion 1 is stronger than coverage criterion 2. Among the coverage criteria for finite state machines, all-paths coverage is the strongest, while all-states coverage is the weakest. Among the coverage criteria for finite state machines, all-

transitions coverage, all-paths coverage, and all-states coverage are the most commonly used.
2. The coverage criteria for UML models are divided into five categories: node coverage, transition coverage, all-predicates coverage, transition-pair coverage, and complete sequence coverage. For different UML models, the categories of coverage criteria correspond to specific modeling elements. For example, in UML state diagrams, node coverage refers to state coverage, all-predicates coverage (i.e., condition coverage), and complete sequence coverage (i.e., path coverage). In UML sequence diagrams, node coverage corresponds to object coverage, transition coverage refers to control focus coverage, all-predicates coverage includes message coverage, and complete sequence coverage corresponds to lifeline coverage. Among the coverage criteria for UML models, transition coverage, complete sequence coverage, and node coverage are the three most commonly used.
3. The coverage criteria used in meta-models, like those in other models, are primarily based on covering modeling elements. Among them, the most commonly used coverage criterion is all-paths coverage after converting the meta-model into a graph. Due to the specific domain of meta-models, it is also possible to define meta-model-specific coverage criteria. For example, some researchers have proposed coverage criteria based on meta-classes, meta-associations, and attributes for specific designed meta-models.

In model-based testing, in addition to the commonly used coverage criteria based on modeling elements, logical coverage criteria and data coverage criteria, which focus on logic and data, are also frequently employed in various models. Some researchers have also defined and designed new coverage criteria from different perspectives. For example, from the perspective of balancing testing cost and efficiency, the minimum testing cost transition coverage criterion for FSM has been developed. Meanwhile, mutation testing has gradually become one of the widely recognized coverage criteria in the industry.

6.4.2 Evaluation of Coverage Criteria

Currently, research on coverage criteria for model-based testing primarily focuses on exploring the relationships between various coverage criteria and evaluating test adequacy based on these criteria.

Common coverage criteria for model-based testing include typical state machine coverage criteria and UML control flow coverage criteria, among others. The typical state machine model coverage criteria are defined based on the model's characteristics and serve as a set of rules to evaluate how thoroughly the test case set covers the model. When applied to a specific test model, these model coverage criteria provide a set of test results for model elements (such as states, transitions, events, etc.) or combinations of model elements.

There is already substantial research in the industry on evaluating traditional code-structure and function-based testing coverage criteria. Most of these studies use axiomatic systems to compare software testing coverage criteria based on program code structure, assessing the relative strengths and weaknesses of each coverage criterion. However, there is currently a lack of in-depth research on the relationships between various coverage criteria in model-based testing.

Test adequacy analysis is a measure of the completeness of software testing, typically quantified by test coverage. Based on the classification of software testing, test adequacy analysis can be divided into code-based and function-based test adequacy analysis. Currently, research on test adequacy analysis primarily focuses on the evaluation objects and methods used. For example, some researchers have proposed analyzing test adequacy at different levels to obtain an overall measure of system test adequacy. However, these results generally evaluate test adequacy from different perspectives in a localized manner.

At the seventh International Conference on Software Quality in 2007, Mayer et al. mentioned the establishment of the first International Workshop on Software Test Evaluation Committees (STEV2007) and pointed out that software test evaluation has long been neglected. Test coverage criteria are one of the key means of quantitatively measuring and controlling the software testing process. However, there are still some issues with testing based on coverage criteria in model-based testing, which are outlined as follows:

1. Measurement methods based on coverage criteria mainly involve separately quantifying various coverage rates to measure the completeness of software testing, with the goal of satisfying corresponding test coverage criteria. However, the current measurement methods lack feedback. If the dynamic characteristics of various test coverage rates could be comprehensively utilized to dynamically track and quantitatively evaluate the effectiveness and weaknesses of different testing methods and test cases, this would provide substantial feedback for the measurement results.
2. Currently, more than ten testing coverage measurement metrics have been proposed, including statement coverage, branch coverage, and condition coverage. Each of these metrics reflects different aspects of test completeness, with its own advantages and disadvantages. However, in practical engineering applications, due to time and cost constraints, only a few coverage criteria are typically used to independently evaluate the adequacy of software testing by separately calculating corresponding coverage rates. Thus, it is worth considering how to fully utilize all available coverage data and their characteristics to develop a comprehensive measurement method for test adequacy in software testing.

6.5 The Generation of Test Cases Based on Models

Test cases are the core product of software testing, and the quality of the test case set directly affects the final outcomes of the software testing process. Generating high-quality test case sets has been a challenge that researchers have long sought to

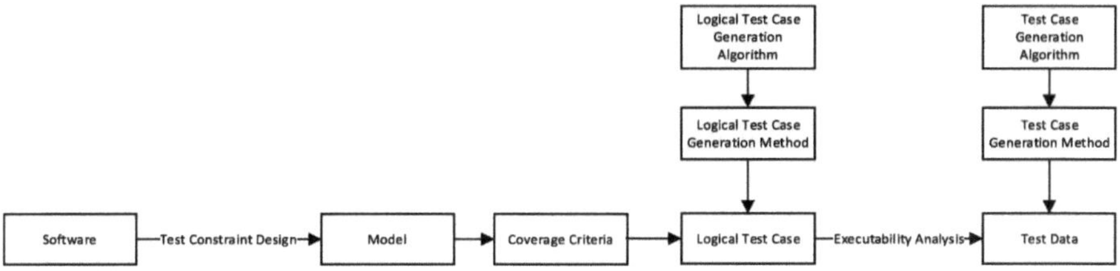

Fig. 6.11 Model-Based Test Case Generation Framework

address. The model-based test case generation framework, as shown in Fig. 6.11, begins by modeling entities such as test requirements in the software testing process. During this process, it is necessary to consider test constraint information, design test constraints, and establish a test model. For different test models, coverage criteria are defined, and test cases are generated based on these criteria. The process consists of 3 steps: (1) generating logical test cases; (2) performing executability analysis; and (3) generating test data. By applying different methods and corresponding algorithms to implement these steps, a test case set that meets the coverage requirements is generated. The test data refers to the specific values of variables in the logical test cases, which can be used as input to execute the program for testing.

In this section, we will introduce model-based test case generation techniques, discussing methods for generating logical test cases, analyzing the executability of test cases, and generating test data. We will also analyze the algorithms used in test case generation, evaluate the performance of each algorithm, and explore ways to optimize these algorithms.

It should be noted that a test case consists of three components: test data, constraint conditions, and expected output. In this section, the focus is on the consideration of constraint conditions in test case generation, as well as the generation of logical test cases and test data. The generation of expected output will not be discussed in detail. Therefore, the test case generation in this section refers specifically to the generation of test data and constraint conditions.

6.5.1 Method for Generating Logical Test Cases Based on Models

As mentioned earlier, model-based test case generation consists of three steps: logical test case generation, executability analysis, and test data generation. Logical test case generation, being the first step, is crucial and directly influences the quality of the generated test cases.

The classification of model-based logical test case generation methods varies among researchers due to different criteria. Some categorize the methods into three types: random, path-oriented, and goal-oriented. Path-oriented methods identify a set of paths in the software based on its control flow and generate logical test cases based on these paths. Goal-oriented methods focus on selecting targets, such as statements or branches, without considering the paths. Other researchers classify the methods based on their generation techniques, including randomization, interval arithmetic, symbolic execution, Korel method, and intelligent algorithms.

Based on the implementation strategy of test case generation, logical test case generation methods can be categorized into five types: randomization, category partitioning, graph theory algorithms, heuristic search, and model checking. A brief overview of each of these methods is provided below.

1. Randomization, as the name suggests, generates test cases randomly within the test case space until the coverage criteria are met. Randomization is simple and easy to implement; however, it often generates a large number of redundant test cases and may introduce loops in the logical test cases. As a result, its fault detection capability is weak, and the efficiency of the generated test cases is low. Currently, randomization is often used as a comparison algorithm in experimental studies.
2. Category partitioning is a testing method based on specification analysis that generates a set of logical test cases based on the analysis of inputs and outputs. The method defines categories based on the input domain and then divides these categories into multiple choices. The combinations of these choices correspond to the inputs for the logical test cases. This method generates logical test cases by determining the behavior equivalence classes within the SUT structure.
3. Test models can also be converted into graphs or trees, and graph theory algorithms can be applied to search or traverse these "intermediate models" to generate logical test cases. The advantage of this approach is that it leverages well-established graph theory algorithms, and during the conversion of test models into "intermediate models," relevant test information can be added to produce better logical test cases. However, the generated logical test cases may still contain redundancy and may include non-executable cases. If the graph is too complex, the problem of "state explosion" may also arise.
4. When the solution space is too large and a smaller search scope is desired to generate a relatively small number of logical test cases that meet the coverage criteria, heuristic search methods can be used. Heuristic search algorithms explore each position in the solution space, evaluate the best position, and continue the search from that position until the goal is reached.

   Compared to other methods, test case generation methods based on heuristic search are among the best-performing approaches. Their main advantages are as follows:

   (a) When the solution space is too large, it is impractical to traverse or search all test case sets that meet the coverage criteria. The goal is to generate as few test cases as possible while satisfying the coverage criteria. Research has shown that this problem is equivalent to the set cover problem, which is NP-complete. For NP-complete problems, most of the solutions provided by the industry are based on heuristic search algorithms. This is the primary reason why heuristic search-based test case generation methods are widely used in many studies.

(b) When using certain graph theory algorithms, there are often prerequisite assumptions made about the graph or tree, some of which may not align with the actual situation, leading to decreased testing effectiveness. Additionally, test case sets generated by some graph theory algorithms often contain a large number of redundant or ineffective test cases. Heuristic search algorithms, by evaluating each search position, reduce the generation of redundant and ineffective test cases.

(c) Heuristic search-based test case generation methods offer significant advantages in terms of both testing efficiency and effectiveness. They possess strong search capabilities, enabling them to test more complex systems effectively.

(d) Using heuristic search algorithms to generate logical test cases can be viewed as part of the search-based software testing field. Currently, search-based software testing is a highly active research area. It frames traditional software testing activities as optimization problems and employs various search algorithms to alter the solutions to these activities.

5. Model checking is a formal verification technique that involves formally representing both model requirements and property specifications, followed by the use of verification tools or algorithms to ensure their consistency. In model-based testing, model checking can also be utilized to generate logical test cases. This approach treats model checking as a coverage criterion and generates logical test cases by verifying the consistency between the model and the specification.

6.5.2 The Executability Analysis Method for Test Cases

Because test models contain variables and predicate conditions, conflicts in the variable data may lead to contradictions in the predicate conditions of logical test cases, making them non-executable. Therefore, before generating test data, executability analysis must be performed on logical test cases to prevent the generation of non-executable test cases and avoid resource waste.

Executability analysis methods can be broadly classified into two types: direct analysis and indirect analysis.

1. The direct analysis method involves using symbolic execution to directly assess the executability of test cases by analyzing variable data and preconditions. This approach takes into account the impact of variables and preconditions on logical test cases, and generates test cases through executability analysis. However, such methods often require model simplification, the omission or simplification of external variables, and are limited in the types of variables they can handle.

2. The indirect method bypasses forward analysis and deduction by constructing or searching for executable paths. It typically generates executable logical test cases through path construction or by utilizing search techniques, including heuristic search.

The traditional logic test case generation method based on the executable analysis tree is an indirect approach. It involves constructing an executable analysis tree and using search algorithms to traverse the tree in order to generate executable test cases. When applying this method, certain strategies are employed to transform initial variables in the model, which helps eliminate conflicts to some extent. Subsequently, a directed graph is constructed, and graph theory algorithms are used to traverse it, resulting in a transformed executable analysis tree. This tree is then converted into executable logical test cases.

Heuristic search algorithms for executability analysis of test cases generally require converting the model into an intermediate model. This intermediate model is better suited for executability analysis and allows for the quantification of feedback information, which, combined with heuristic search, guides the generation of executable test cases.

6.5.3 *Test Data Generation Based on Models*

While a few researchers have attempted to generate test data directly by defining test inputs and outputs within the model, most research and practice involve a process of generating test cases based on the model that includes both logical test case generation and the determination of specific variable values within those test cases, i.e., test data generation. Section 6.5.2 has covered the generation of logical test cases, and here we will explore methods for generating test input data.

Commonly used test data generation methods can be classified into five types based on the implementation strategy: manual method, category partitioning, graph theory algorithms, heuristic search, and symbolic execution.

1. The manual method involves testers manually identifying and selecting test data based on test constraints and logical test cases. By analyzing the test constraints, testers determine the variable values within the logical test cases. Although this method is time-consuming and increases testing costs, it is still widely used in research due to different focuses. The manual method can be effective when test cases have specific requirements or the testing scenario is particularly complex. However, for large and complex software systems, the manual method often fails to meet the test data generation requirements, necessitating the use of automated or semi-automated methods.
2. The category partitioning method is a specification-based testing technique that generates a test data set by analyzing input and output. The method identifies categories based on the characteristics of the input domain and divides these categories into multiple choices, which can be understood as input equivalence

classes. Test data is generated by determining the value combinations for these choices.

3. One approach to test data generation involves constructing a graph or tree based on logical test cases and constraints, and using graph theory algorithms to traverse the model and generate test data. Test data can be generated by identifying the intersection points of the test case table and tree, where each intersection represents an applicable condition that is evaluated as either true or false, thus providing test conditions for the test case. Another method is to use segmented gradient optimization to generate test data.
4. The process of generating test data using heuristic search can be described as follows: after selecting a logical test case, initial values for the test case are randomly selected from the input domain. The model is then iteratively explored multiple times, with the test data being determined heuristically.

   Although the test case generation method based on heuristic search algorithms demonstrates better performance compared to other methods, and a substantial body of research exists on this topic, the quality and efficiency of model test case generation using heuristic search still require improvement, as outlined in the following aspects.

   (a) The issues arising from constraints between variables

   Most models contain variables that interact with each other and are subject to constraints. If test constraints are not carefully designed before using the model to generate test cases, some crucial test information may be lost. This can lead to test cases that do not meet testing requirements, exhibit redundancy, and are of low quality. Additionally, this issue exacerbates the problem of state explosion.

   (b) The issues arising from the limitations of heuristic search algorithms

   Among various heuristic search algorithms, each algorithm has its own advantages and limitations due to different strategies. When using heuristic search algorithms for test case generation, problems may arise because of these limitations. For example, the hill-climbing algorithm, due to its inherent nature, cannot escape local optima; thus, once it reaches a local optimum, the search terminates. The genetic algorithm improves the quality of the test case population rapidly in the early iterations through the crossover operator. However, as the diversity of the population decreases, the rate of evolution toward the optimal solution slows down. The particle swarm algorithm suffers from the issue of "premature convergence," and when the search space does not contain the optimal solution, the algorithm becomes trapped in a local optimum.

   (c) Issues arising from the design of the fitness function

   The fitness function is a key component of heuristic search algorithms. Its design reflects both the testing requirements and the quality expectations. Only by correctly evaluating the test cases can the heuristic search algorithm effectively perform optimization.

Currently, numerous fitness functions have been designed in research to meet various testing objectives. How to design and optimize these fitness functions accurately and comprehensively remains a key focus of related research.

(d) Problems caused by operator design

An operator is a rule in a heuristic search algorithm that extends the search direction and nodes. In heuristic search algorithms, some basic operators may not effectively guide the algorithm, leading to issues such as low efficiency and disruption of testing feedback information. For instance, in the particle swarm algorithm, the uniform crossover operator improves coverage but significantly disrupts testing feedback during the search process. This results in an increase in evolutionary generations, ultimately reducing the efficiency of test case generation.

Building on the four points mentioned above, improvements can be made to the heuristic search algorithm accordingly.

5. Symbolic execution is a technique that extracts logical paths based on coverage criteria, performs symbolic computation on each path, and generates a system of predicate equations formed by all the branch conditions of the path. Any input data (test data) that satisfies these equations will drive the execution of the path. However, due to the complexity of solving predicate equation systems (which is theoretically an unsolvable problem) and issues such as undetermined array subscripts, loop iterations, and the number of executions in recursive processes at compile time, symbolic execution is challenging to apply in practical engineering.

6.5.4 Test Case Generation Algorithm Based on Models

Model-based test case generation involves a variety of optimization and search techniques. The algorithms commonly used include random algorithms, graph theory algorithms, gradient descent methods, hill-climbing algorithms, genetic algorithms, particle swarm optimization, ant colony algorithms, simulated annealing, and other methods for generating logical test cases, performing executability analysis, and generating test data.

Graph theory algorithms and gradient descent methods are frequently applied in scenarios where state explosion and optimized test case generation are not major concerns. In heuristic search algorithms, hill-climbing, genetic algorithms, and particle swarm optimization are commonly used. Random algorithms are often employed as benchmark methods in experimental studies.

In research and experimentation, various metrics are employed to evaluate and compare algorithms. These metrics allow for an objective assessment of the performance and effectiveness of the algorithms. Key evaluation metrics include test coverage, algorithm complexity, defect detection rate, number of iterations, test case design and execution time, number of test cases, the ratio of test completion to the number of test cases, the average number of test cases to detect the first failure, the

Table 6.10 Evaluation metrics for algorithms

Category	Evaluation metrics
Algorithm Performance Category	Algorithm complexity, number of iterations, number of test cases, number of first failures found on average, number of redundant test cases
Test Adequacy Category	Test coverage rate, full coverage success rate
Test Effectiveness Category	Defect detection rate
Test Efficiency Category	Time ratio of test case design to execution, ratio of test completion rate to number of test cases

number of redundant test cases, and the full coverage success rate. These evaluation metrics can be broadly categorized into four groups: algorithm performance, test adequacy, test effectiveness, and test efficiency, as shown in Table 6.10.

The performance of each algorithm will be summarized based on the evaluation metrics outlined above.

1. The random algorithm generates test cases randomly based on test requirements within the input domain. It then uses the generated test cases to execute the program and finally checks whether the test objectives are covered.

   The random algorithm is simple, does not require testing feedback, is easy to execute, and is efficient in generating test cases. However, it suffers from a lack of direction in test case generation, resulting in low code coverage and excessive coverage of the same errors. It may also generate a large number of invalid or redundant test cases. Additionally, it does not offer advantages in terms of execution time and resource usage. Random algorithms are commonly used as data sources for optimization algorithms and as comparison benchmarks.
2. Graph Theory Algorithm

   Graph theory algorithms are a set of well-established theoretical algorithms used to solve problems related to graphs, which are mathematical structures made up of nodes and edges. These algorithms are applied in various fields, including computer networking, social networks, and transportation systems. Examples include Dijkstra's algorithm for finding the shortest path in a graph and Kruskal's algorithm for determining the minimum spanning tree.
3. Gradient Descent Algorithm

   Gradient descent is a commonly used optimization algorithm known for its simplicity and speed. However, a major limitation of gradient descent is that, for non-convex functions, it can easily get trapped in a local optimum and fail to find the global optimum. Combining interval reduction techniques with gradient descent can help mitigate the issue of local optimization. This combination is particularly effective in generating test data using the gradient descent algorithm.
4. Hill Climbing Algorithm

   Hill climbing is a local search algorithm that starts from a random solution and iteratively improves the solution by making small modifications. The algorithm works by moving to a neighboring solution that has a better evaluation until it reaches a local optimum. The hill climbing algorithm is often used in optimization problems where the goal is to find the best solution in a given search space.

In general, traditional hill-climbing algorithms use linear programming to avoid traversal by solving the problem step by step in a similar way to climbing a mountain top. This method is particularly easy to understand. However, once a highland effect or local optimum occurs, it is easy to get stuck in a loop of local optima.

The hill-climbing algorithm can start from any point in the solution space and quickly converge to a local optimum. As such, it is particularly effective for solving single-peak problems. However, the hill-climbing algorithm has several drawbacks:

(a) Due to its intrinsic nature, the hill-climbing algorithm is unable to escape from local optima. Once it reaches a local optimum, it halts the search.
(b) In solving multi-peak problems, the hill-climbing algorithm can only randomly discover one of the many local optima. It cannot measure the gap between the current solution and the global optimum, nor can it assess the difference between the local optimum and other local optima. As a result, the generated local optimum lacks clarity regarding its relative position.
(c) When applied to optimization problems, the outcome of the hill-climbing algorithm depends heavily on the choice of the initial starting point, which determines whether the global optimum can be reached. Therefore, it is impossible to establish an upper bound on the required computation time.

5. Genetic Algorithm

   The genetic algorithm is a heuristic method that combines the genetic processes observed in nature with Darwin's principle of survival of the fittest. It optimizes test cases by applying selection, crossover, and mutation operators, ultimately obtaining the most optimized test cases.

   Experimental studies have demonstrated that genetic algorithms outperform random algorithms significantly in terms of iteration count and testing efficiency. Genetic algorithms can be optimized through adjustments in fitness function design and operator selection. Optimized genetic algorithms, such as TGGA, show superior performance compared to random algorithms in metrics such as iteration count, error path coverage, and the number of test cases generated.

   The primary advantage of genetic algorithms over random algorithms is their ability to reduce randomness and improve efficiency.

6. Particle Swarm Optimization

   Particle swarm optimization (PSO) is a simple and robust algorithm that dynamically generates test cases through the interaction between individual cognition and swarm intelligence, offering strong guidance and convergence. PSO outperforms genetic algorithms in testing complex objects. However, like other intelligent algorithms, PSO suffers from premature convergence. Premature convergence in PSO is generally attributed to a loss of diversity in the population, causing most individuals to cluster within a narrow search space. When this search space does not contain the optimal solution, the algorithm becomes trapped in a local optimum.

7. Ant Colony Optimization

   Ant colony optimization is a bio-inspired heuristic algorithm commonly used to optimize other algorithms. It is an enhanced learning system with advantages such as distributed computing, easy integration with other algorithms, and strong robustness. However, due to the relative lack of information in the early stages of the search, the algorithm's search efficiency is reduced. Additionally, the positive feedback mechanism is prone to stagnation and premature convergence, leading to long execution times and susceptibility to stagnation.

8. Simulated Annealing

   Simulated Annealing, inspired by the principle of solid-state annealing, is a probability-based algorithm. It probabilistically accepts inferior solutions, which allows it to escape local optima. Experimental studies have shown that there is no clear superiority among simulated annealing, genetic algorithms, and particle swarm optimization when it comes to generating test cases. The performance of each algorithm varies depending on the program and evaluation metrics. In some cases, simulated annealing may provide better test coverage than genetic algorithms.

6.5.5 *Algorithm Optimization Methods*

Test case generation algorithms can be considered as consisting of two components: system structure and control structure. The system structure defines and designs the relevant concepts within the algorithm, while the control structure refers to the extension methods used by the algorithm, which control the operations of the elements in the system structure. The optimization methods for the algorithm can be broadly categorized into three types: optimizing the algorithm by altering the system structure, optimizing the algorithm by modifying the control structure, and optimizing the algorithm through a combination of various methods.

1. The method of optimizing an algorithm by altering its system structure involves introducing new constraints through the definition and design of relevant concepts, thereby optimizing test case generation. Based on this approach, the concept of input domain partitioning can be incorporated into random algorithms to propose new variants that reduce their overhead. Additionally, the input domain of the program under test can be divided into different subdomains. A search path graph can be constructed based on these subdomains, and test cases can then be generated using the ant colony algorithm with this path graph.
2. A substantial body of research, particularly in the field of search and heuristic search algorithms, has focused on optimizing test case generation by altering the extension and control operations employed by the algorithm. Most of these

approaches involve the design and refinement of the fitness function and operation operators in heuristic search algorithms to achieve optimal performance.

Some researchers introduced a path mutation operator and self-adaptive evaporation coefficients to increase the diversity of ant paths, which results in fewer iterations compared to genetic algorithms. Other researchers designed local search operators to improve branch coverage. Patterns are defined during the operator calculation process, and their combinations are used to enhance fitness values and accuracy. Additionally, a local search strategy is applied to the best particle of each generation, and an AVM algorithm is proposed to balance the global and local search performance, thus improving the efficiency of test case generation.

3. Algorithm optimization through the combination of multiple approaches can be divided into two categories. One approach is the combined optimization of control structure and system structure. Some researchers have defined patterns in genetic algorithms, where the combination of patterns during the operator calculation process enhances both adaptability and accuracy.

   Another approach is to optimize test case generation by integrating multiple algorithms, leveraging the strengths of each to compensate for their weaknesses. For example, a genetic algorithm can be integrated into ant colony optimization to create the GAACO algorithm. The GAACO algorithm addresses the limitations of genetic algorithms, such as poor local search capabilities and premature convergence, while also overcoming the issues in ant colony optimization, such as high execution time and susceptibility to stagnation. GAACO enhances test case generation in terms of randomness, speed, and global convergence. Experimental results demonstrate that the GAACO algorithm outperforms genetic algorithms, ant colony optimization, and random algorithms in terms of iteration count.

6.6 Case Study in Practice

This section presents model checking and model-based testing on a case study. Using the garage door control system as an example, a model checking case study is introduced. This example is drawn from Model-Based Testing: A Software Process Engineer's Approach, which applies various modeling techniques to the garage door control system and generates test cases. In this section, the garage door control system is modeled as a state machine, and the model is then checked.

Additionally, a model-based testing case study is conducted on the temperature control module of launch platform control software. A MARTE model of the software is created, and testing is performed based on this model.

6.6.1 *Model Checking on the Garage Door Controller*

6.6.1.1 Requirement Analysis

The garage door control system consists of several components.

1. A drive motor that provides power to the control system and can rotate in both directions.
2. Garage door tracks with sensors that can detect whether the door is open or closed, positioned at the fully open and fully closed positions of the door.
3. A control device that sends control signals to manage the system.
4. A beam sensor located near the floor, which operates only when the garage door is closing.
5. The garage door

When the garage door is in motion, either closing or opening, and the beam is interrupted (such as by a pet or an obstacle), the door will immediately stop and reverse direction. While the door is in operation, either opening or closing, the control device sends a signal that halts the door's movement. When the door resumes movement, the control signal continues in the same direction as the door was previously moving. Sensors detect when the door reaches either the fully open or fully closed position, at which point the door stops moving.

6.6.1.2 Build the Finite State Machine

When extracting a model from software, it is essential to accurately define the state transition relationships. Otherwise, the verification results will not reliably reflect the software's issues. Below are the steps required to build a model for the garage door control system.

1. Identify states, group them, and divide the concurrent areas.
   From the specification, we can identify three objects: the garage door, the laser beam, and the motor. By analyzing each object, we can derive the states listed in Table 6.11.
2. Identify events and actions

Table 6.11 Object analysis

Garage door	Beam sensor	Motor
Door fully open	Beam sensor not activated	Motor off
Door fully closed	Beam sensor activated	Motor rotating downward
Door stopping closing	Beam not obstructed	Motor rotating upward
Door stopping opening	Beam obstructed	
Door closing		
Door opening		

Events can be considered as inputs, representing external stimuli to the system. Through system analysis, events related to the garage door object can be identified, such as receiving control signals, reaching the upper end of the operating track, and reaching the lower end of the track. For the beam sensor, an event such as the blockage of the laser beam can be identified. Actions can be considered as outputs, representing the system's responses. Through system analysis, the following actions can be identified for the garage door control system: driving the motor downward, driving the motor upward, stopping the motor, and reversing the motor from down to up.

3. Identify the transitions between states

   We can use a state transition table to analyze and define the complex relationships between states, as shown in Table 6.12. Figure 6.12 illustrates the final result of the analysis of state transitions.

4. Consider the concurrent design and determine the inputs and outputs of the transitions

   In a system, objects are not independent of each other, and there is mutual influence between the states of several parallel domains, which necessitates the introduction of synchronous design. Signals can be regarded as another term for actions after the introduction of synchronous design. Signals serve as the output

Table 6.12 State-transition table of the motor

Status	Motor off	Motor downward	Motor upward
Motor Off	×	√	√
Motor Downward	√	×	√
Motor Upward	√	√	×

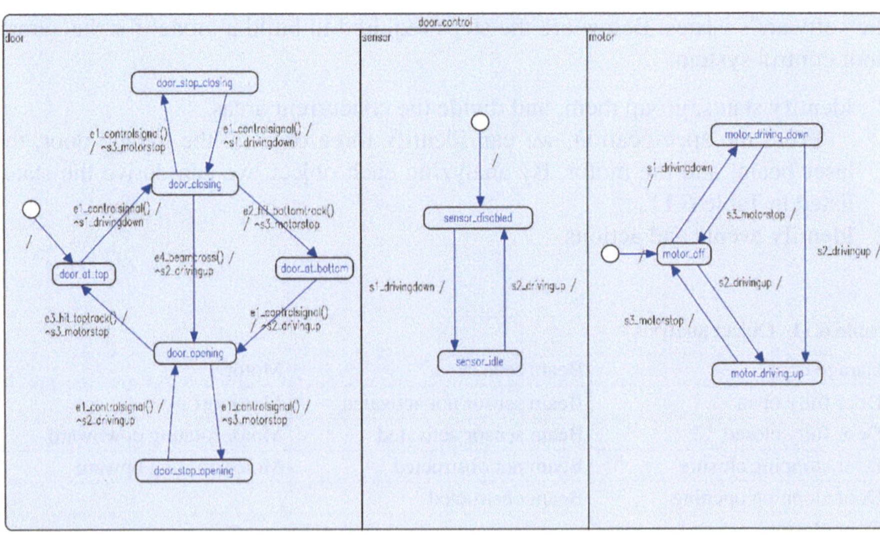

Fig. 6.12 State Machine Diagram

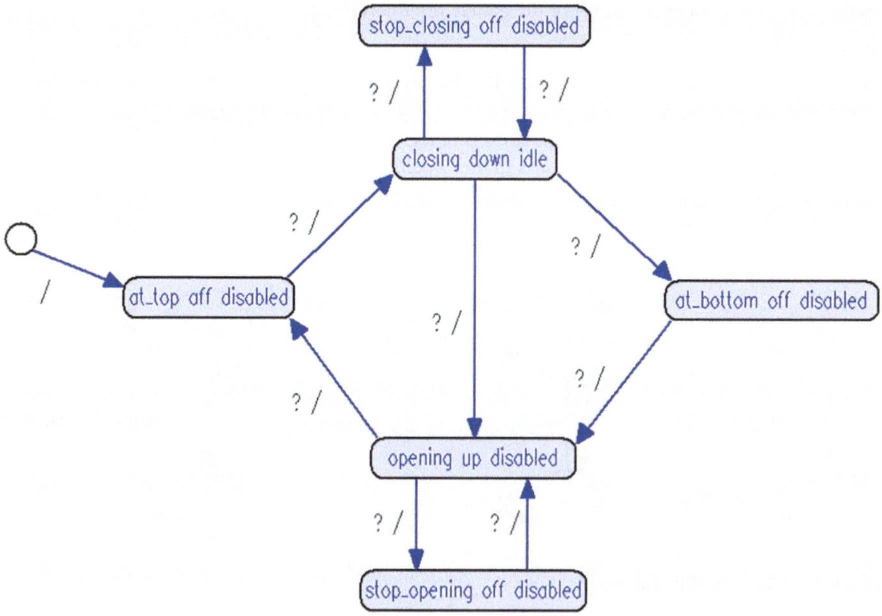

Fig. 6.13 Combined State Machine Diagram

of a transition and simultaneously function as the input of a transition. For instance, when the state machine is in the "door up" state, if the event "receiving control signal" occurs, the state machine will issue a signal to "drive the motor down." At this point, the state machine in the motor domain will transition to the "motor driving down" state, while the state machine in the light beam sensor domain will transition to the "sensor enabled" state. Simultaneously, the "door up" state will transition to the "door closing" state.

The steps mentioned above result in a state machine composed of three concurrent state machines. We will then use the open-source modeling tool visual STATE to create a graphical model, as shown in Fig. 6.12.

To meet the requirements of the model checking algorithm, the concurrent state machines will be merged into a single state machine according to certain rules. The resulting model is shown in Fig. 6.13.

6.6.1.3 Design of Property Specification

Based on the merged state machine of the garage door control system, three variables—namely the states of the garage door, motor, and sensor—are identified as the objects for LTL constraints. Figure 6.13 shows the values of these variables for each state. For example, the initial state "at_top off disabled" indicates that the door

Table 6.13 System specifications of the garage door controller

Property	LTLTL expressions
At any moment, if the motor is going up, the door will eventually reach the top.	G (motor == 'up' -> (F door == 'at_top'))
It is possible that at some moment the door is in the "stop_closing" state and the next moment it is in the "opening" state.	F (door == 'stop_closing' && X door == 'opening')
Under any circumstances, the sensor "disabled" and the motor "down" cannot be true at the same time.	G·!(sensor == 'disabled' && motor == 'down')

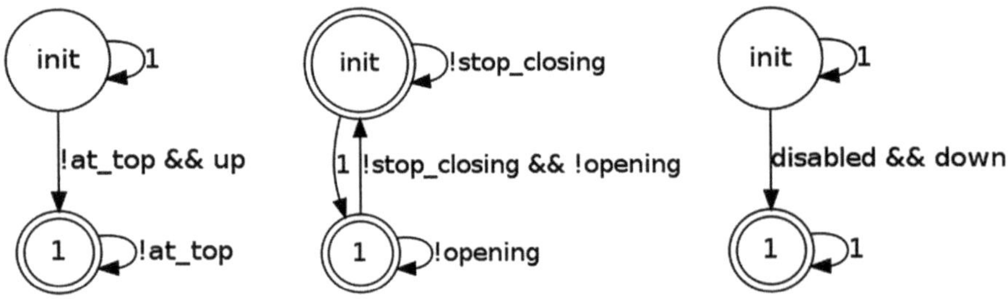

Fig. 6.14 Three Büchi Automata

is at the top, the motor is off, and the sensor is disabled. Table 6.13 presents three selected specifications.

Using the LTL2BA tool, the negations of the three LTL statements are transformed into a Büchi automaton, as shown in Fig. 6.14.

6.6.1.4 Independent Model Checking

First, the model state machine needs to be represented in a format suitable for input into the Python program, specifically as an FSM object. The code is as follows:

```
model = FSM([('at_top','off','disabled'),#The three values
refer to the states of the garage door, motor, and sensor.
 ('closing','down','idle'),
 ('stop_closing','off','disabled'),
 ('at_bottom','off','disabled'),
 ('opening','up','disabled'),
 ('stop_opening','off','disabled')],
 #The trans set is shown below, with each tuple consisting
of (starting point, endpoint).
 [[('at_top','off','disabled'),('closing','down','idle')],

[('closing','down','idle'),('s
top_closing','off','disabled')],
```

```
 [('stop_closing','off','disabled'),('closing','down'
,'idle')],
 [('closing','down','idle'),('at_bottom','off','disabled')],
 [('closing','down','idle'),('opening','up','disabled')],
 [('at_bottom','off','disabled'),('opening','up','di
sabled')],

[('opening','up','disabled'),('st
op_opening','off','disabled')],
 [('stop_opening','off','disabled'),('opening','up','di
sabled')],
 [('opening','up','disabled'),('at_top','off','disabled')]])
```

Next, the property automaton is represented as an AccFSM object. States represented by single-layer circles are non-accepting states, while those represented by double-layer circles are accepting states. Each transition in the property automaton has a precondition that governs whether the transition can be executed. Therefore, the precondition must be included in the trans set. The model checking algorithm can check only one specification at a time. In this case, the first specification is used as an example.

```
buchi_1 = AccFSM([('init'),('1')], #node set
[('1')],# accepted nodes set.[('1')]means ('1') in node set
is accepted.
#The following are the transitions, with the last element of
each transition representing its precondition.
#When describing the preconditions, a Python lambda
statement is used, which functions as a boolean function to
determine whether the precondition is satisfied.
[[('init'),('init'),1],
[('init'),('1'),lambda state:not(state[0]=='at_top') and
state[1]=='up'],
[('1'),('1'),lambda state:not(state[0]=='at_top')]])
```

After defining the two objects, execute the statement modelCheck(getSynProduct(model, buchi_1)). This first computes the synchronous product of the model and buchi_1, then passes the synchronous product to the modelCheck function, which calculates and outputs the emptiness checking results.

The first specification.

The checking result shows that the model does not satisfy the property specification, with the following counterexample:

```
(('at_top', 'off', 'disabled'), 'init')
(('closing', 'down', 'idle'), 'init')
(('at_bottom', 'off', 'disabled'), 'init')
(('opening', 'up', 'disabled'), 'init')
(('stop_opening', 'off', 'disabled'), 'init')
(('opening', 'up', 'disabled'), '1')
The following is the loop section.
(('stop_opening', 'off', 'disabled'), '1')
(('opening', 'up', 'disabled'), '1')
```

The second specification.

Checking result: the model does not satisfy property specification. The counterexample is shown below:

```
(('at_top', 'off', 'disabled'), 'init')
(('closing', 'down', 'idle'), 'init')
(('at_bottom', 'off', 'disabled'), 'init')
(('opening', 'up', 'disabled'), 'init')
(('stop_opening', 'off', 'disabled'), 'init')
(('opening', 'up', 'disabled'), '1')
(('at_top', 'off', 'disabled'), '1')
(('closing', 'down', 'idle'), '1')
The following is the loop section.
(('stop_closing', 'off', 'disabled'), '1')
(('closing', 'down', 'idle'), '1')
```

The third specification

The checking result shows that the model satisfies the property specification.

6.6.1.5 Model Checking Using Tools

All model checking tools are capable of verifying the correctness of system models against specified properties. However, they may describe models and properties in different languages and formats, and may also exhibit varying verification features. Therefore, the first step in using a tool for model checking is to select an appropriate tool. From a requirements perspective, the following key requirements apply to the example presented in this section:

1. The control system software is implemented on an embedded platform using C++.
2. The control system is interrupt-driven, and the interleaved execution of interrupt and non-interrupt handler functions implicitly generates the characteristics of a concurrent system.
3. This section aims to verify the general safety properties of the software, such as array out-of-bounds, deadlock, and infinite loops, as well as to validate user-defined special properties, such as assertions and linear temporal logic (LTL) formulas.
4. The verification tool should output counterexamples and be capable of reproducing the counterexample path to identify the error's cause and resolve the issue.
5. The verification tool should provide options to reduce the scale and compress the state space to limit its generation, ensuring that the time and space consumed during verification remain within feasible limits.
6. The model created should include only the components of the system that are relevant to the properties under verification.
7. The transformation from software to model requires abstraction, so that the model remains relatively small and can be easily traced during execution.

Based on the above requirements, using source code directly as a model tool is not suitable because it does not directly support concurrency, making it difficult to describe interrupt handler functions. Moreover, these tools typically only support assertions and do not support LTL statements. Therefore, tools based on specific modeling languages are considered, as they allow for the abstraction of a more simplified model during the conversion process. Tools that meet these requirements include SPIN and NuSMV. SPIN is chosen here because its modeling language, PROMELA, is closer to C++ source code, and SPIN supports the use of custom LTL statements as property specifications, while NuSMV only supports CTL.

The PROMELA language used by the SPIN tool is not a typical programming language but a modeling language specifically designed for model description. It allows uncertain branching jumps and the concurrent execution of multiple processes in an undetermined order. Processes can be synchronized via global variables or message channels. A PROMELA model mainly includes variable and process declarations. Basic variable types include bit, byte, bool, short, and int, and PROMELA also provides types such as arrays, enumerations, and structures similar to those in C. A process declaration consists of the process name, parameters, local variable declarations, and the process body. The syntax atomic{...} in PROMELA allows multiple statements to be grouped into an atomic sequence, meaning that when a process executes an atomic sequence, other processes cannot interrupt its execution. Atomic sequences can be used for operations that cannot be split or to reduce the complexity of model checking.

The following is a part of the code representing the garage door control system using the PROMELA language. The model contains only one process, door_control. Since state_0 appears at the beginning of the process body, execution starts from the state_0 label, meaning the initial state is state_0. The atomic block sets the values of the variables for this state. The subsequent if statement indicates the possible states to which transitions from this state can lead. It can be seen that the successor state of state_0 is only state_1, whereas state_1 has three possible successor states. Since there is no determined next state, the model checker will explore all three paths, ensuring complete traversal of the state transition paths.

```
......
active proctype door_control(){
state_0:
  atomic{
    lock = 0;
    door_open = 0;
    door_closed = 1;
  }
  if (lock_button == 1){
    goto state_1;
  }
state_0_end:
  skip;
state_1:
  atomic{
    lock = 1;
  }
  if (unlock_button == 1){
    goto state_2;
  }
  if (l ock_button == 1){
    goto state_0;
  }
  if (open_button == 1){
    goto state_3;
  }
state_1_end:
  skip;
state_2:
  atomic{
    unlock = 1;
    lock = 0;
  }
  if (unlock_button == 1){
```

```
      goto state_3;
    }
    if (lock_button == 1){
      goto state_0;
    }
  state_2_end:
    skip;
  state_3:
    atomic{
      door_open = 1;
      door_closed = 0;
    }
    if (close_button == 1){
      goto state_4;
    }
    if (timer_done == 1){
      goto state_5;
    }
  state_3_end:
    skip;
    ...
  }
  ......
```

The SPIN tool lacks a graphical user interface and can only be run from the terminal (e.g., Windows Command Prompt). To make model checking more intuitive and clear, the iSPIN visualization interface, developed using TCL, is used. Unlike the manually implemented program described earlier, SPIN does not require converting LTL statements into state machines. Instead, LTL statements can be directly incorporated into the PROMELA language model. The code is as follows:

```
active proctype door_control(){
......
}
ltl L1 { [](motor==up -> <> (door==at_top)) }
ltl L2 {<>(door==stop_closing && X ( door==opening))}
ltl L3 { [] ! (sensor == disabled && motor==down) }
```

In the code, “ltl” is a PROMELA keyword, signaling that the following text is an LTL statement. “L1/L2/L3” are unique identifiers for each LTL statement. In iSPIN, the statement's name is entered into the parameter box, and the model checking algorithm will use this name to select the corresponding statement for verification.

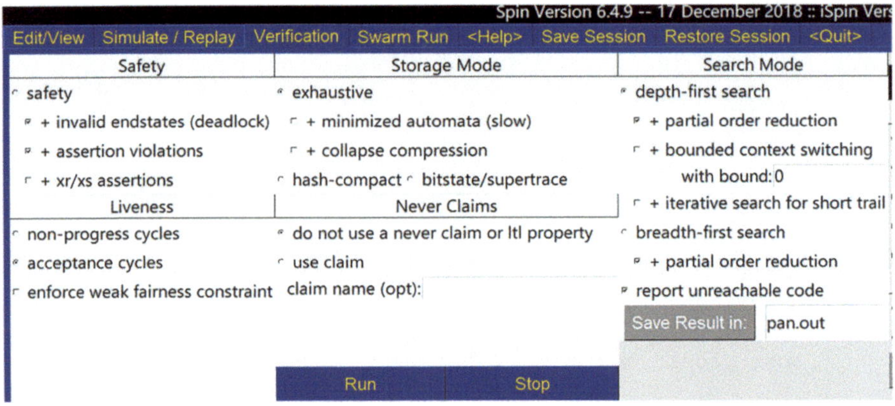

Fig. 6.15 iSPIN Verification Interface

iSPIN is quite simple to use. First, click the "Open" button to load the prepared PROMELA model. Then, select and input the necessary parameters in the Verification interface to begin the verification, as shown in Fig. 6.15. In addition to basic LTL statement verification, SPIN offers various other verification options, such as detecting illegal termination states and progress-free loops. SPIN also employs optimization techniques to enhance verification efficiency, such as state machine compression, bit-hash mapping, and partial order reduction, all of which can be easily configured in iSPIN.

Upon completion of the verification process, the SPIN tool outputs information such as the time and space consumed during the verification, the number of states explored, and whether any properties remain unsatisfied. If the verification fails, the counterexample path can be accessed in the Simulate/Replay interface. By stepping through the path, the values of various variables at each step can be observed. After verifying the three LTL statements, it was found that although the output format of SPIN differs from the independently implemented model checking program, both approaches yielded identical results in determining whether the model satisfied the properties and in tracing the counterexample path. This outcome also aligns with the actual behavior of the garage door control system.

6.6.2 *Model-Based Testing of the Launching Platform Control Software*

6.6.2.1 Software Requirements Specification

The temperature control module of a certain launch platform control software is responsible for deciding whether to activate heating based on the switch signal and the platform's temperature value.

Input: switch signal and the platform's current temperature value.

Table 6.14 System under test and its connected devices

Serial number	System or communication equipment	Function
1	Host Computer (HostComputer)	Send switch signal
2	Control Module (ControlModule)	Temperature control
3	Temperature Sensor (TempSensor)	Collect temperature signal
4	RS-485 Bus Line	Transmit and receive data

Software processing requirements are as follows:

1. When the switch signal is 0, stop the heating.
2. When the switch signal is 1, the temperature analog signal is monitored. If the platform temperature exceeds 60 °C continuously for 10 s, stop the heating.

Table 6.14 lists the system modules and their corresponding functions. The hardware requirements include at least 512 MB of memory, 100 MB of hard disk space, and a high-performance CPU.

We will use MARTE to model the launch platform control software, and testing will be conducted based on the created model.

The selection of MARTE is based on the following considerations: MARTE offers strong modeling capabilities in the embedded systems domain and can be applied to both sides of the V-model in the development process. Moreover, the models described using MARTE are UML-based, ensuring isomorphism in both syntax and semantics. Additionally, open-source tools that support the construction of MARTE models are already available, significantly streamlining the embedded system development process. In this case study, MARTE is well-suited for describing the launch platform control software, building a test model, and conducting testing based on this model.

Testing based on the MARTE model begins with analyzing software requirements, followed by the creation of a static structural model and a dynamic interaction model. Test scenarios are explored from both dynamic and static perspectives, and the test requirements model is derived by analyzing constraints in conjunction with these scenarios. This test requirements model is then further transformed into a test case model, which serves as the foundation for generating test cases.

6.6.2.2 Construct the Test Requirements Model

1. Identify software and hardware object.

   Based on the requirements, identify the software and hardware components, high-level application objects, and their interconnections, and define the fundamental classes and signals, as shown in Fig. 6.16.
2. Use MARTE to create the high-level application model.

Incorporate structural modeling to build the MARTE high-level application object model, as shown in Fig. 6.17.

3. Create the MARTE Object Connection Model

   Establish the object connection model based on the class diagram, high-level application objects, and their interconnections and interface relationships, as shown in Fig. 6.18.

4. Create the Object Interaction Model

   Based on the object connection model and the constraint analysis from the requirements, the object interaction model is constructed using a sequence diagram, as shown in Fig. 6.19.

5. Message-Node Transformation and Construction of a Directed Graph for the Dynamic Interaction Model.

The steps for the message-node transformation are as follows:

1. Analyze the messages executed sequentially. The reception of the signal "wow" is a sequentially executed message, mapped to node 1. The message "setBrakeLabel(wow, temp)" is mapped to nodes 3, 7, and 8.
2. Analyze the messages executed in a loop. The reception of the temperature signal "temp" is a loop-executed message, mapped to nodes 5 and 6.

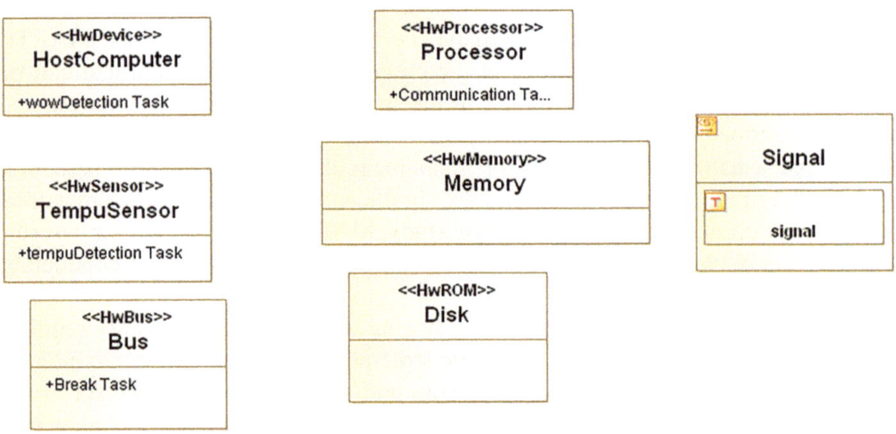

Fig. 6.16 Basic Class Diagram and Signals of the Software

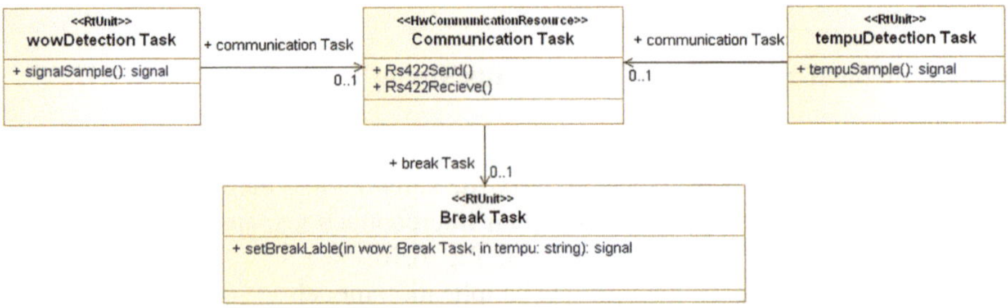

Fig. 6.17 High-Level Application Object Model

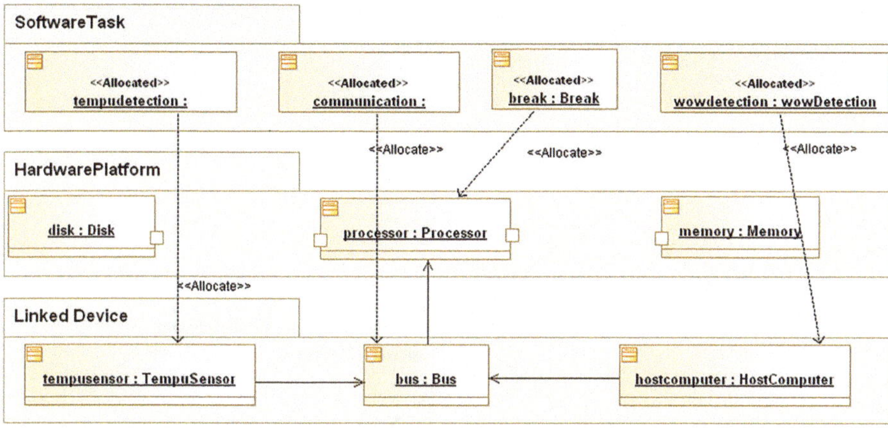

Fig. 6.18 Object Connection Model

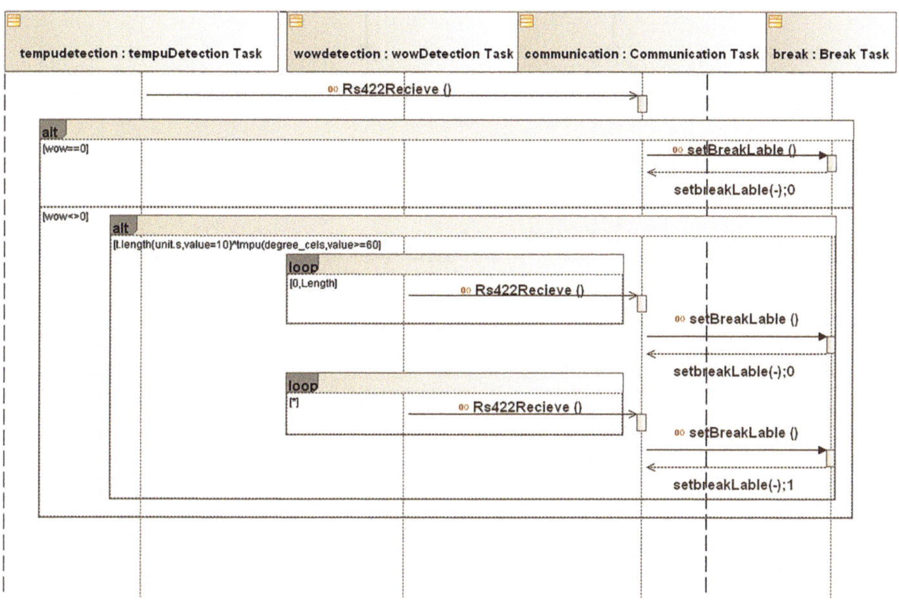

Fig. 6.19 Object Interaction Model

3. Analyze the constraints of combined fragments with branches. Figure 6.20 includes two "alt" combined fragments, with the constraints mapped to pseudo-nodes 2 and 4, respectively.

   Finally, transform the dynamic interaction model into a directed graph, as shown in Fig. 6.20.
4. Create the Adjacency Matrix for the Directed Graph

   Based on the connections between the nodes in the directed graph, the corresponding adjacency matrix A for Fig. 6.21 is established as follows.

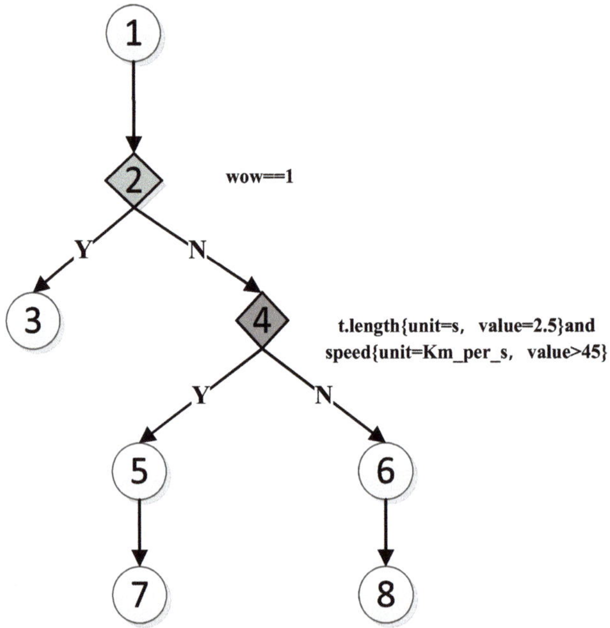

Fig. 6.20 Directed Graph Corresponding to the Sequence Diagram

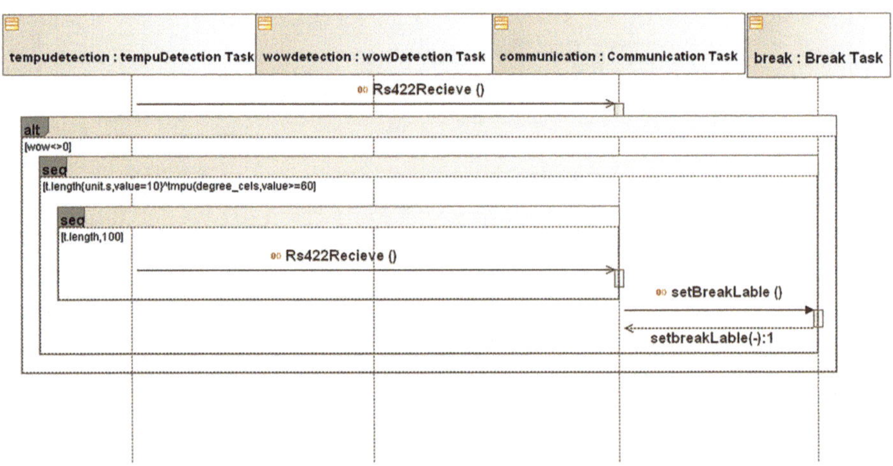

Fig. 6.21 Test Scenario Fragment Model

$$A=\begin{bmatrix}1&2&3&4&5&6&7&8\\0&1&2&2&4&4&5&6\\0&0&1&0&0&0&1&1\end{bmatrix}^{\mathrm{T}}$$

A=

12345678
01224456
00100011T

5. Searching for Paths in the Directed Graph

   Perform a path search on the directed graph represented by adjacency matrix A to identify the following three paths:

   Path 1: $1 \to 2 \to 3$

   Path2: $1 \to 2 \to 4 \to 5 \to 7$

   Path 3: $1 \to 2 \to 4 \to 6 \to 8$

6. Model of Test Scenario

   To traverse the test scenarios in the object interaction model, begin by applying Message-Node transformation to the dynamic interaction model using Object Constraint Language (OCL) to generate a directed graph. Next, traverse the executable paths in the directed graph, and perform the Message-Node transformation again to obtain the test scenarios for the software under test. This process generates the test scenario fragment model, with Fig. 6.21 illustrating the corresponding test scenario fragment model for test scenario 2.

   Analyze the test scenario set based on the test scenario fragment model and perform test scenario modeling. The resulting test scenario model is shown in Fig. 6.22.

7. Generate Test Requirements

   We will analyze the test scenarios and constraints, and based on the test objectives, generate the test requirement model, as shown in Table 6.15.

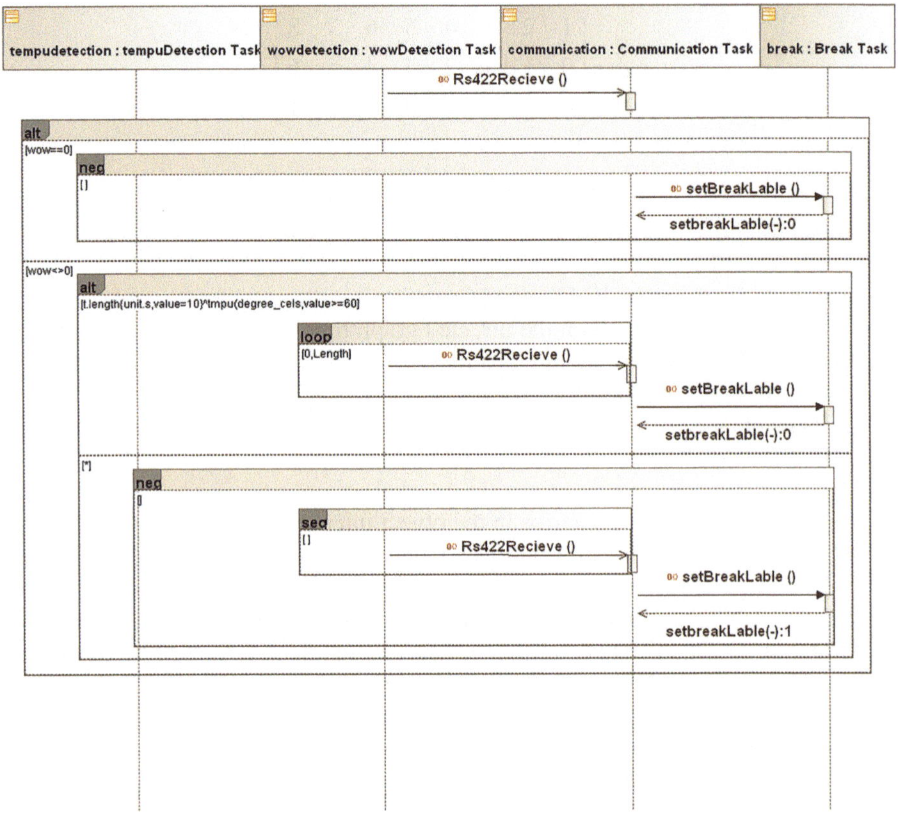

Fig. 6.22 Test Scenario Model

6.6.2.3 Transform into the Test Case Model

For clarity, we will first use the test requirement corresponding to Test Scenario 2 in the test requirement model to illustrate the transformation process, and then explain the remaining test requirements. The test requirement for Test Scenario 2 is as follows:

(wow<>0) and (60=<tempu<=100) and (t.status==continuous) and (t.length==10s) and (Memory.memorysize> =512M and Disk.diskcapacity > =100M).

According to the XML test requirement model conversion algorithm, the test objects in the XML test case model can be directly transformed from the

Table 6.15 Test requirement model

Scenario	Test scenario constraints	Test objective
1	(wow == 0) and (0 < tempu < 60) and (Memory.memorysize >= 512M and Disk.diskcapacity >= 100M)	Heating
2	(wow<>0) and (60 <= tempu <= 100) and (t.status == continuous) and (t. length == 10s) and (Memory.memorysize >= 512M and Disk. diskcapacity >= 100M)	Stop Heating
3	(wow<>0) and not (60 <= tempu <= 100) and (t.status == continuous) and (t.length == 10s) and (Memory.memorysize >= 512M and Disk. diskcapacity >= 100M)	Heating
	(wow<>0) and (60 <= tempu <= 100) and not (t.status == continuous) and (t.length == 10s) and (Memory.memorysize >= 512M and Disk. diskcapacity >= 100M)	Heating
	(wow<>0) and (60 <= tempu <= 100) and (t.status == continuous) and not (t.length == 10s) and (Memory.memorysize >= 512M and Disk. diskcapacity >= 100M)	Heating
	(wow<>0) and not (60 <= tempu <= 100) and not (t.status == continuous) and (t.length == 10s) and (Memory.memorysize >= 512M and Disk. diskcapacity >= 100M)	Heating
	(wow<>0) and not (60 <= tempu <= 100) and (t.status == continuous) and not (t.length == 10s) and (Memory.memorysize >= 512M and controlmodule.diskcapacity >= 100M)	Heating
	(wow<>0) and (60 <= tempu <= 100) and not (t.status == continuous) and not (t.length == 10s) and (Memory.memorysize >= 512M and Disk. diskcapacity >= 100M)	Heating
	(wow<>0) and not (60 <= tempu <= 100) and not (t.status == continuous) and not (t.length == 10s) and (Memory.memorysize >= 512M and controlmodule.diskcapacity >= 100M)	Heating

corresponding tests in the test requirement model. The pre-condition constraint, such as the switch signal "wow=1", remains unchanged, as does the time state. When converting the temperature value to the XML test case model, the value space is "a[60,100,random(60,100),random(0,60)]". The test time is continuous and can be either less than 10s or greater than 10s, with a value space of "a[random(0,10), random[10,n)]". Invariant constraints are directly involved in the transformation. The test objective is determined by comparing the test input values with the test requirements to check for matches.

Based on the defined XML structure, the following XML test case model can be generated:

```
<?xml version="1.0" encoding="ISO-8859-1"?>
<!-- Edited with XML Spy v2007 (http://www.altova.com) -->
<CASE>
<testSequence2>
<case_number>n</case_number>
<test_object>SUT</test_object>
<wow>1</wow>
<time>a1[random (0,10),random[10,n)]</time>
<timeiscontinuous>continuous</timeiscontinuous>
<tempu>b1[60‖100‖random(60,100)‖random(0,60)]</tempu>
<Memory_memorysiz>100M</Memory_memorysiz>
<Disk_diskcapacity>512M</Disk_diskcapacity>
<test_target>c1["heating","stop heating"]</test_target>
</testSequence2>
</CASE>
```

Based on the analysis process described above, we can list the complete XML model of the software under test, as shown in the code below:

```
<?xml version="1.0" encoding="ISO-8859-1"?>
<!-- Edited with XML Spy v2007 (http://www.altova.com) -->
<CASE>
<testSequence1>
<case_number>n</case_number>
<test_object>SUT</test_object>
<wow>0</wow>
<time>a1[random(0,10),random[10,n)]</time>
<timeiscontinuous>continuous</timeiscontinuous>
<tempu>b1[60,100,random(60,100),random(0,60)]</tempu>
<Memory_memorysiz>100M</Memory_memorysiz>
<Disk_diskcapacity>512M</Disk_diskcapacity>
<test_target>c1["heating","stop heating"]</test_target>
</testSequence1>
<testSequence1>
<case_number>n</case_number>
<test_object>SUT</test_object>
<wow>0</wow>
<time>a1[random(0,10),random[10,n)]</time>
<timeiscontinuous>not continuous</timeiscontinuous>
<tempu>b1[60,100,random(60,100),random(0,60)]</tempu>
<Memory_memorysiz>100M</Memory_memorysiz>
<Disk_diskcapacity>512M</Disk_diskcapacity>
<test_target>c1["heating","stop heating"]</test_target>
</testSequence1>
<testSequence2>
<case_number>n</case_number>
```

```
<test_object>SUT</test_object>
<wow>1</wow>
<time>a1[random (0,10),random[10,n)]</time>
<timeiscontinuous>continuous</timeiscontinuous>
<tempu>b1[60||100||random(60,100)||random(0,60)]</tempu>
<Memory_memorysiz>100M</Memory_memorysiz>
<Disk_diskcapacity>512M</Disk_diskcapacity>
<test_target>c1["heating","stop heating"]</test_target>
</testSequence2>
<testSequence3>
<case_number>n</case_number>
<test_object>SUT</test_object>
<wow>1</wow>
<time>a1[random (0,10),random[10,n)]</time>
<timeiscontinuous>not continuous</timeiscontinuous>
<tempu>b1[60||100||random(60,100)||random(0,60)]</tempu>
<Memory_memorysiz>100M</Memory_memorysiz>
<Disk_diskcapacity>512M</Disk_diskcapacity>
<test_target>c1["heating","stop heating"]</test_target>
</testSequence3>
</CASE>
```

The XML generated based on the test requirement for Test Scenario 1, along with the test requirement models for Test Scenarios 2 and 3, contain overlapping information regarding "t.status=continuous." Therefore, the "<test-Sequence2>…</testSequence2>" element in the XML test requirement model reflects the "t.status=continuous" information, while the "<test-Sequence3>…</test-Sequence3>" element reflects the "t.status=not continuous" information.

6.6.2.4 Test Case Generation

Based on the test cases defined in JSP, test case information can be displayed in HTML, as shown in Table 6.16. For each "testSequence" element, 25 test cases are generated, resulting in a total of 100 test cases. These test cases cover all the normal and abnormal inputs derived from the software requirements analysis. To ensure accuracy and completeness, at least one test case for each input (whether corresponding to a normal or an abnormal path) must be included in the test case set. For easy identification, the test case identifier is the same for all test cases corresponding to the same input. A partial set of the test cases is listed in Table 6.15.

Table 6.16 Test case set list

Test case identifier	Quantity	Time	Time continuity	Temperature	Memory/B	Disk capacity/B	Expected target
testSequence1-8	wow = 0	random[10,n)	Continuous	random(0,60)	100M	512M	Stop Heating
testSequence1-5	wow = 0	random[10,n)	Continuous	60	100M	512M	Stop Heating
testSequence1-3	wow = 0	random(0,10)	Continuous	random(60,100)	100M	512M	Stop Heating
testSequence1-1	wow = 0	random(0,10)	Continuous	60	100M	512M	Stop Heating
testSequence1-8	wow = 0	random[10,n)	Continuous	random(0,60)	100M	512M	Stop Heating
testSequence1-7	wow = 0	random[10,n)	Continuous	random(60,100)	100M	512M	Stop Heating
testSequence1-1	wow = 0	random(0,10)	Continuous	60	100M	512M	Stop Heating
testSequence1-4	wow = 0	random(0,10)	Continuous	random(0,60)	100M	512M	Stop Heating
testSequence1-7	wow = 0	random[10,n)	Continuous	random(60,100)	100M	512M	Stop Heating
testSequence1-5	wow = 0	random[10,n)	Continuous	60	100M	512M	Stop Heating
testSequence1-3	wow = 0	random(0,10)	Continuous	random(60,100)	100M	512M	Stop Heating
testSequence1-4	wow = 0	random(0,10)	Continuous	random(0,60)	100M	512M	Stop Heating
testSequence1-1	wow = 0	random(0,10)	Continuous	60	100M	512M	Stop Heating
testSequence1-5	wow = 0	random[10,n)	Continuous	60	100M	512M	Stop Heating
testSequence1-6	wow = 0	random[10,n)	Continuous	100	100M	512M	Stop Heating
testSequence1-7	wow = 0	random[10,n)	Continuous	random(60,100)	100M	512M	Stop Heating
testSequence1-5	wow = 0	random[10,n)	Continuous	60	100M	512M	Stop Heating
testSequence1-4	wow = 0	random(0,10)	Continuous	random(0,60)	100M	512M	Stop Heating
testSequence1-5	wow = 0	random[10,n)	Continuous	60	100M	512M	Stop Heating
testSequence1-1	wow = 0	random(0,10)	Continuous	60	100M	512M	Stop Heating
testSequence1-8	wow = 0	random[10,n)	Continuous	random(0,60)	100M	512M	Stop Heating
testSequence1-7	wow = 0	random[10,n)	Continuous	random(60,100)	100M	512M	Stop Heating
testSequence1-3	wow = 0	random(0,10)	Continuous	random(60,100)	100M	512M	Stop Heating
testSequence1-7	wow = 0	random[10,n)	Continuous	random(60,100)	100M	512M	Stop Heating

(continued)

Test case identifier	Quantity	Time	Time continuity	Temperature	Memory/B	Disk capacity/B	Expected target
testSequence1-5	wow = 0	random[10,n)	Continuous	60	100M	513M	Stop Heating
testSequence1-15	wow = 0	random[10,n)	Discontinuous	60	100M	514M	Stop Heating
testSequence1-14	wow = 0	random(0,10)	Discontinuous	random(0,60)	100M	515M	Stop Heating
testSequence1-15	wow = 0	random[10,n)	Discontinuous	60	100M	516M	Stop Heating
testSequence1-13	wow = 0	random(0,10)	Discontinuous	random(60,100)	100M	517M	Stop Heating
testSequence1-14	wow = 0	random(0,10)	Discontinuous	random(0,60)	100M	518M	Stop Heating
testSequence1-17	wow = 0	random[10,n)	Discontinuous	random(60,100)	100M	519M	Stop Heating
testSequence1-15	wow = 0	random[10,n)	Discontinuous	60	100M	520M	Stop Heating
testSequence1-18	wow = 0	random[10,n)	Discontinuous	random(0,60)	100M	521M	Stop Heating
testSequence1-11	wow = 0	random(0,10)	Discontinuous	60	100M	512M	Stop Heating
testSequence1-11	wow = 0	random(0,10)	Discontinuous	60	100M	512M	Stop Heating
testSequence1-13	wow = 0	random(0,10)	Discontinuous	random(60,100)	100M	512M	Stop Heating
testSequence1-16	wow = 0	random[10,n)	Discontinuous	100	100M	512M	Stop Heating
testSequence1-14	wow = 0	random(0,10)	Discontinuous	random(0,60)	100M	512M	Stop Heating
testSequence1-11	wow = 0	random(0,10)	Discontinuous	60	100M	512M	Stop Heating
testSequence1-13	wow = 0	random(0,10)	Discontinuous	random(60,100)	100M	512M	Stop Heating
testSequence1-11	wow = 0	random(0,10)	Discontinuous	60	100M	512M	Stop Heating
testSequence1-17	wow = 0	random[10,n)	Discontinuous	random(60,100)	100M	512M	Stop Heating
testSequence1-12	wow = 0	random(0,10)	Discontinuous	100	100M	512M	Stop Heating
testSequence1-13	wow = 0	random(0,10)	Discontinuous	random(60,100)	100M	512M	Stop Heating
testSequence1-17	wow = 0	random[10,n)	Discontinuous	random(60,100)	100M	512M	Stop Heating
testSequence1-18	wow = 0	random[10,n)	Discontinuous	random(0,60)	100M	512M	Stop Heating
testSequence1-11	wow = 0	random(0,10)	Discontinuous	60	100M	513M	Stop Heating
testSequence1-16	wow = 0	random[10,n)	Discontinuous	100	100M	514M	Stop Heating
testSequence1-15	wow = 0	random[10,n)	Discontinuous	60	100M	515M	Stop Heating
testSequence1-13	wow = 0	random(0,10)	Discontinuous	random(60,100)	100M	516M	Stop Heating

6.7 Summary

Model-based testing, as an emerging and efficient software testing method, has been widely promoted and applied both domestically and internationally. Complex software, characterized by large system scale, high heterogeneity among system components, increasing network connectivity, and system emergence, presents many challenging issues in software engineering and software verification. Model-based testing, approaching the problem from the perspective of models, utilizes modeling techniques, model verification, test case generation, and other related technologies. It attempts to address the issues inherent in complex software. We believe this approach represents a valuable area of exploration and research.

Chapter 7
Proactive Defect Prevention Based on Human Errors

7.1 Proactive Defect Prevention

7.1.1 The Need for Defect Prevention

As more and more functionalities in computer systems are realized by software, the complexity and size of software have been increasing. Consequently, software faults contribute to many system failures. Numerous approaches have been used to improve software quality and reduce software faults, such as Software Process Improvement, and Software Verification and Validation. According to a report published by the National Institute of Standards and Technology (NIST) in 2002 [1], software engineers spend an average of 70–80% of their time on testing and debugging, with each defect taking about 17.4 h to be found and fixed. This "introducing-testing-fixing" cycle is estimated to consume 40–50% of the development budget.

Early defect prevention becomes an important approach for improving the reliability of software systems within the limited resources. Defect prevention aims to reduce the fault introduction rate by identifying the root causes of failures, serving as a crucial means of improving software quality and reliability. Cognitive errors are the primary cause of software defects [2], as software is essentially a representation of human thought [3, 4]. Therefore, it is important to directly enhance developers' cognitive abilities to prevent human errors. If software developers understand why, how, and when they are prone to making errors, they should be able to prevent them more effectively.

This chapter first provides an overview of the traditional DP process and then introduces a human-centered DP approach, "Defect Prevention based on Human Error Theories" (DPeHE) [5], to improve software developers' cognitive abilities with respect to human error prevention. This approach emphasizes learning from software errors based on their underlying causal mechanisms.

Y. Wang et al., *Reliability Technology for Complex Software Systems*,
https://doi.org/10.1007/978-981-95-3889-8_7

7.1.2 *The Nature of Software and Software Defect*

Programming is a special type of writing, performed by programmers [3]; a computer program is a pure cognitive product that describes its designers' thoughts [4, 6, 7]. Software defects are the manifestations of cognitive errors of software engineers, practitioners and/or of miscommunication between them [6, 7].

Since Huang and Liu [8] proposed the initial framework for advancing software reliability based on the human errors of software practitioners in 2011, noticeable progress has been achieved in this emerging field. Huang et al. [7] proposed a human error taxonomy to identify the root causes of software defects at various stages of software development cycle. The taxonomy was later used as a skeleton for an approach to promote developers' cognitive knowledge and strategies to prevent human errors in various stages of software development [5]. Several taxonomies were proposed for identifying causes of defects in a specific stage of software development. For instance, a human error taxonomy for identifying causes of faults in requirement documents [9]. Mohammadnazar et al. proposed a taxonomy of contextual factors in the software development process, including culture, time zone, product size, number of staff, etc. [10].

Recent advancements from Huang et al. studies [11, 12] show that Error-Prone Scenario (EPS), defined as the situation under which a human error tends to occur, is an excellent approach to capture the human error causal mechanisms behind software defects. The EPS-based approach was able to forecast, at the requirement phase, 75.7% of the defects later committed by all of the programmers [11]. The EPS-based code review has significantly improved True Positives and Sensitivity by about 400% [12].

7.2 Overview of Traditional Defect Prevention

The defect prevention (DP) processes reported in the existing literature are generally similar to the traditional Defect Prevention process [13] or the corresponding Defect Causal Analysis (DCA) process [14]. The conventional DP process, as summarized by Card [14] is represented by the rectangle enclosed by the dotted line in Fig. 7.1.

The typical agenda of the defect prevention (DP) process includes the following major steps [15]: (1) selecting defect samples from a historical database, (2) holding a causal analysis meeting to identify defect causes and develop prevention action proposals, and (3) implementing preventative actions. To implement DP, three resource elements are required: (1) a database and tool for defect sample selection and action tracking, (2) a cause classification method for identifying root causes, and (3) an action team with strong expertise in root cause analysis. DP has proven effective in preventing software defects, particularly from a process improvement perspective [16, 17].

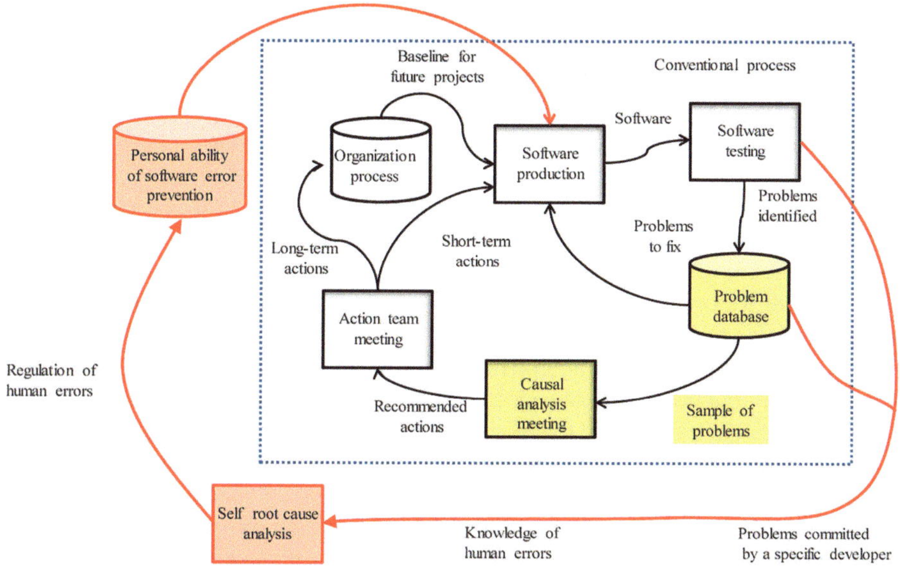

Fig. 7.1 Conventional DP and DPeHE (from [5])

Table 7.1 The summary of existing root cause classifications

Authors	Classifications
Mays et al. [18]	Communication, oversight, education, and transcription
Leszak et al. [19]	Change coordination, lack of domain knowledge, lack of system knowledge, lack of tool knowledge, lack of system knowledge, lack of tools knowledge, lack of process kn., Individual mistake, introduced with other repair, communication problems, missing awareness for need of communication, and not applicable
Schadler [20]	Inadequate training, breakdown of communication and not accounting for all details of task
Card [15]; Kalinowski et al. [21]	Methods, which may be incomplete, ambiguous, wrong, or unenforced; tools and environment, which may be clumsy, unreliable, or defective; people, who may lack adequate training or understanding; and input and requirements, which may be incomplete, ambiguous, or defective

The key to conventional defect prevention is identifying root causes. However, this process heavily depends on expert experience. Root cause taxonomies have been proposed to aid root cause analysis, and these are summarized in Table 7.1. Unfortunately, these taxonomies often seem too abstract to be helpful for organizations with limited experience. Root causes are generally classified into four categories: method, people, tool, and requirement. Detailed causes are analyzed through brainstorming with cause-effect diagrams [14, 21, 22]. While brainstorming is a good method for generating new ideas, it is not as reliable for repeated use. For example, defect causes analyzed by different people may fall into different categories, overlap, or even be omitted. This inconsistency reduces the likelihood of achieving recurring good results across different organizations. It is no accident that Defect Cause Analysis (DCA) is often misunderstood and misapplied in the software industry [14].

Another issue is how to choose defect samples. The common method is to select them according to defect type, typically defined by Orthogonal Defect Classification (ODC) [23]. However, there is a lack of scientific criteria to help people decide which historical defects are heuristically relevant to the current project. This process, once again, heavily relies on the experience of the team members. Consequently, it is difficult for organizations with limited data to implement DP, as they have few defect samples for reference.

These challenges are highlighted in the yellow background in Fig. 7.1. To address them, a structural taxonomy of root causes has been recently proposed [7].

From Fig. 7.1, we observe that the existing framework of defect prevention (DP) offers a systematic approach to preventing defects through process improvement. However, there is a need for further efforts to provide insights into the causal mechanisms of software defects. By making psychological knowledge explicit and offering appropriate strategies, software developers may enhance their awareness and abilities to prevent human errors. Such capabilities could empower developers to actively prevent errors during the programming process for the task at hand.

The Defect Prevention based on Human Error Theories (DPeHE) introduced in this chapter is depicted in red in Fig. 7.1. In contrast to the conventional DP process, DPeHE places greater emphasis on software developers' self-improvement in preventing cognitive errors. Initially, developers gain knowledge about human error mechanisms and prevention strategies through the DPeHE training program. Subsequently, in real software development practice, they utilize this knowledge and provided tools to conduct root cause analysis independently. The defects used in self-checking can be those identified through independent testing or debugging. Through this self-checking experience, developers are expected to enhance their awareness and self-regulation under error scenarios, resulting in improved abilities to prevent human errors.

7.3 Related Cognitive Concepts and Models

7.3.1 Related Cognition Concepts

7.3.1.1 Schema

"A schema is a modifiable information structure that represents generic concepts stored in memory. Schemata represent knowledge that we experience--interrelationships between objects, situations, events, and sequences of events that normally occur. In this sense, schemata are prototypes in memory of frequently experienced situations that individuals use to interpret instances of related knowledge." [24].

7.3.1.2 Working Memory

Working Memory was first proposed by Baddeley and Hitch [25] in 1974. Working memory refers to a brain system that provides temporary storage and manipulation of the information necessary for complex cognitive tasks as language comprehension, learning, and reasoning [26].

Working Memory includes three subcomponents [26]: (1) the central executive, which is assumed to be an attentional-controlling system, is important in skills such as chess playing and is particularly susceptible to the effects of Alzheimer's disease; and two slave systems, namely (2) the visuospatial sketch pad, which manipulates visual images and (3) the phonological loop, which stores and rehearses speech-based information and is necessary for the acquisition of both native and second-language vocabulary.

7.3.1.3 Chunking

Chase and Simon [27] proposed the concept of Chunking for explaining the phenomenon where the memory capacity of experienced performers far exceeds that of novices. It is believed that experts, through training, not only acquire knowledge about the domain but also develop a memory skill that enables them to efficiently utilize the knowledge in the field. This knowledge consists of organized patterns of information (or chunks) stored in long-term memory, which experts can use to encode and retrieve newly presented domain-specific knowledge.

In memory chess tasks, researchers present chess positions to players and then ask them to reproduce them. Studies have found that skilled chess players store a large number of chess patterns, or chunks, in long-term memory. With the help of these chunks, players can represent multiple chess piece positions holistically, rather than memorizing them individually. In contrast, entry-level players, due to the limited patterns stored in long-term memory, use their short-term memory to store various simple patterns. Because of the capacity limit of working memory, novice players can recall only a very limited number of successful chess moves. According to chunking theory, skilled players activate existing information (chunks) in long-term memory by recognizing chess positions, rather than generating new information or establishing new connections in memory.

Chunking theory hypothesizes that the excellent memory performance of professional chess players is mediated by familiar and meaningful patterns of information. The level of players' chess skills interacts with the meaningfulness of the chess positions in memory performance. That is, when chess positions are meaningful real-game scenarios, skilled players' memory performance is significantly better than that of novices. However, when the positions of the pieces are randomly arranged, the memory advantage of skilled players disappears.

Chunking theory can effectively explain the phenomenon of individuals memorizing large amounts of information in complex cognitive activities. As units of information processing, chunks possess characteristics such as dynamism and expansiveness. They can change with changes in individuals' knowledge, experience, and cognitive structure. They can increase the amount of information stored in working memory. Although the number of activated chunks remains around 7 [28], the amount of information within each chunk expands, thereby increasing the total amount of activated information. Therefore, chunking theory can explain why, under the same working memory capacity limit, experts can recall more information than novices. Chess masters can recall chess positions with just one glance at the board because they have more and larger chunks than novices, allowing them to quickly obtain key information and find better moves.

7.3.1.4 Cognitive Load

Cognitive load was proposed by John Sweller [29], refers to the total amount of mental effort or resources required to perform a particular cognitive task. It encompasses the various mental processes involved in understanding, processing, and storing information in working memory. It comprises three types of loads: intrinsic load, extraneous load, and germane load [30].

Intrinsic load is determined by the inherent aspects of a task and arises from the natural complexity of the information that must be processed. However, this complexity varies based on the proficiency levels of individuals. As expertise increases, experts can "chunk" their schema—a large number of interacting elements for one learner may be consolidated into a single element for another learner with better expertise [2, 4, 31]. Therefore, intrinsic load can only be determined with reference to a specific level of expertise.

Extraneous load refers to the mental effort required to deal with factors external to the task that can interfere with learning or task performance. It includes distractions, irrelevant information, or poorly designed instructional materials that increase cognitive demand without contributing to learning.

Germane load is the cognitive load resulting from cognitive activities in working memory. It is associated with intentional learning and extends beyond simple task performance. Germane load occurs when individuals engage in schema construction, involving activities like restructuring problem representations and meta-cognitive processes.

7.3.1.5 Meta-cognition

Meta-cognition refers to the "cognition about cognition" [32]. Meta-cognition includes two essential components [33, 34]: (1) knowledge of cognition, referring to knowledge about cognitive processes and knowledge that can be used to control cognitive processes and, (2) regulation of cognition, as activities regarding self-regulatory mechanisms during an ongoing attempt to learn or solve problems, including planning activities, monitoring activities, and evaluating outcomes.

7.3.2 The Cognitive Features of Software Development

Software development encompasses two main activities: software design and coding. It exhibits the following cognitive features:

1. Ill-defined problems [6, 35, 36]
   The problems to be solved are often difficult to clearly define in advance; typically, only a goal and constraints can be pre-defined. Many aspects of the problem become clear only during the design process, where problem representation intertwines with problem-solving. The solution plan is usually not pre-defined. In real industrial scenarios, software requirements are often ambiguous and/or incomplete at the project's outset. Programmers must gradually discover the missing information, refine the requirements, and establish evaluation criteria to form a comprehensive problem representation.
2. Creativity
   Software development entails numerous problem-solving activities [37]. Even for software systems familiar to programmers, implementing new features may be necessary. Moreover, programmers may frequently need to implement entirely new features in unfamiliar environments. This requirement for creativity renders it impossible to define the problem-solving path in advance.
3. Integration of knowledge from multiple domains. Software development demands expertise from various domains, including domain-specific knowledge, knowledge of software architectures, and proficiency in programming languages.

7.3.3 Integrated Cognitive Model for Software Development

While software developers routinely solve software problems, many are not consciously aware of their thought processes as much of their cognitive activity occurs unconsciously. Before understanding error mechanisms, software developers should first explicitly comprehend the general cognitive processes involved in software development.

In software development, performance encompasses both routine and design activities. Routine activities, such as typing or compiling a program, do not entail problem-solving efforts [38]. However, the primary aspect of software development lies in designing and coding. Coding differs from design in that it involves implementation characteristics, such as defining a variable using a familiar rule. Although coding exhibits problem-solving features, it is considered routine problem solving, whereas software design is viewed as non-routine problem solving [37].

Huang et al. [5] have synthesized these activities into an integrated cognitive model of software development, as illustrated in Fig. 7.2. This model was constructed by integrating important research findings in software psychology, including those of Adelson and Soloway [39] and Détienne [6]. The cognitive performance

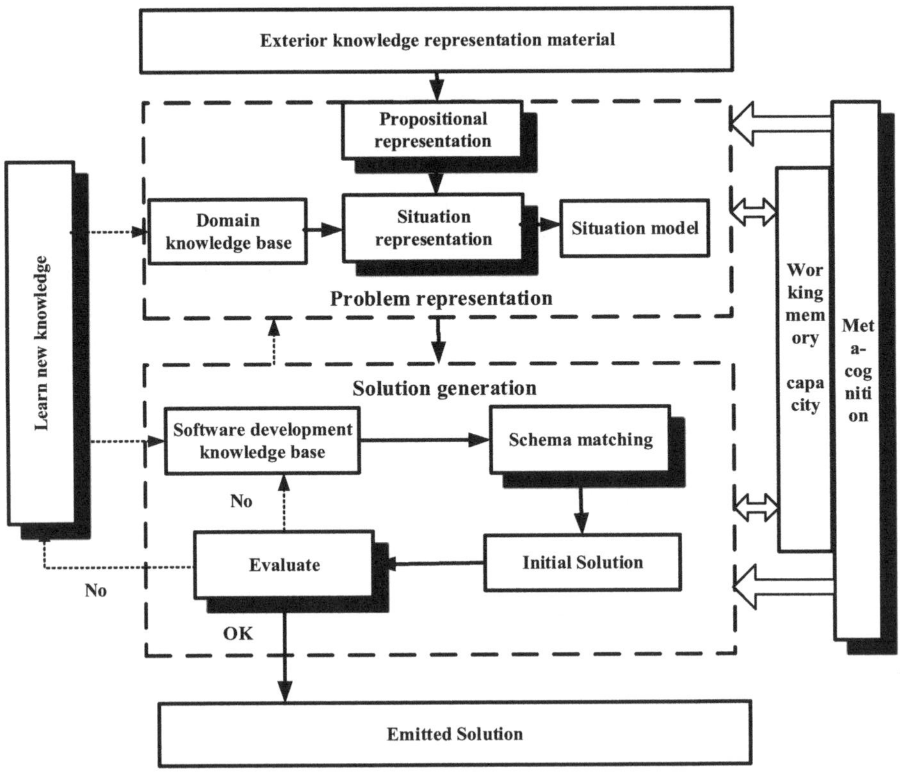

Fig. 7.2 Huang's integrated cognitive model for software development (from [5])

of software development comprises two main stages: problem representation and solution generation.

Problem representation entails understanding the exterior representation material, such as software requirement specifications. Through this process, the information contained in the external document is converted into the developer's mental model. The accuracy of the mental model is significantly influenced by the format of the external representation material and the developer's knowledge base.

Solution generation involves iterative processes of schema retrieval and matching, solution evaluation, and the construction of new schemas through learning. Each of these processes is linked to different error patterns [7]. All these processes are constrained by limited working memory capacity and meta-cognition, which includes monitoring and regulating cognitive processes.

7.4 The Human Error Theories for Studying Software Defects

7.4.1 Rasmussen's Performance Framework [40]

Rasmussen's performance framework has been widely accepted as a fundamental theory in psychology [41]. For instance, J. Reason, one of the most well-acknowledged psychologists in the field of human errors, uses Rasmussen's framework to integrate various human error modes discovered by many psychologists [41]. Rasmussen's performance framework [40] classifies cognitive activities into three levels: skill-based, rule-based and knowledge-based level. Different performance levels have different cognitive characteristics, and are associated with different error modes [41].

Skill-based (SB) performance follows from the statement of an intention, "rolls along" automatically without conscious control. Skill-based activities in programming include typing a text string, compiling a program by pressing a button in the programming environment.

Skill-based errors are the human errors occurring in skill-based performances. Examples of skill-based errors in the psychology domain include such as perceptual confusions, e.g. taking an object that looks like another. In software development, typos and entering a wrong letter which looks similar to the correct one (e.g. taking 0 for o) are typical examples of skill-based errors [11].

Rule-based (RB) performance is applicable for tackling familiar problems. It is typically controlled by stored rules that have been derived from a person's experiences. The mind matches the situation at hand to the preconditions for such stored rules, allowing quick selection of actions. In programming, there are many rule-based performances, such as printing of a string line, and defining a variable in one's familiar programming language.

Rule-based errors are the errors occurring in rule-based performances. A typical rule-based error is "strong-but-now-wrong" error: one tends to use a rule that has been frequently used, but not necessarily matches well with the current situation [41].

Knowledge-based (KB) performance comes into play when individuals face novel situations, and no rules are available from previous experiences. At this level, actions must be planned using an analytical process. Errors at this level can arise from resource limitations and incomplete or incorrect knowledge. In programming, cognitive performances such as constructing the mental model of the system to understand a specific programming task and attempting to figure out a solution for a novel problem are considered knowledge-based performances.

Knowledge-based errors are human errors that occur in knowledge-based performances. Typical knowledge-based errors include "problems with complexity", and "confirmation bias"—the tendency of people to seek evidence that verifies their hypotheses rather than refuting them [41].

7.4.2 Reason's General Error-Modeling System (GEMS) [41]

Reason proposed a General Error-Modeling System, shown in Fig. 7.3.

The Generic Human Error Model System is one of the most representative dynamic cognitive error models. GEMS was based on Rasmussen' SRK model. Skill-based behavior involves auto-pilot responses to external information. At this time, individuals are prone to distractions and may exhibit "lapses" in attention. When abnormal conditions are detected, individuals process this abnormal information and apply existing rules to problem-solving. This is known as rule-based behavior, and the corresponding errors are called rule-based errors. If the occurrence of an abnormal situation is highly complex and unprecedented, individuals must consider the causes of the abnormal situation using their basic knowledge and take appropriate measures. This is the knowledge-based problem-solving process, and the corresponding error type is knowledge-based error. Reason's Generic Human Error Dynamic Model reflects the iterative nature of human cognition and the inevitable transitions between multiple levels' cognitive activities in a task.

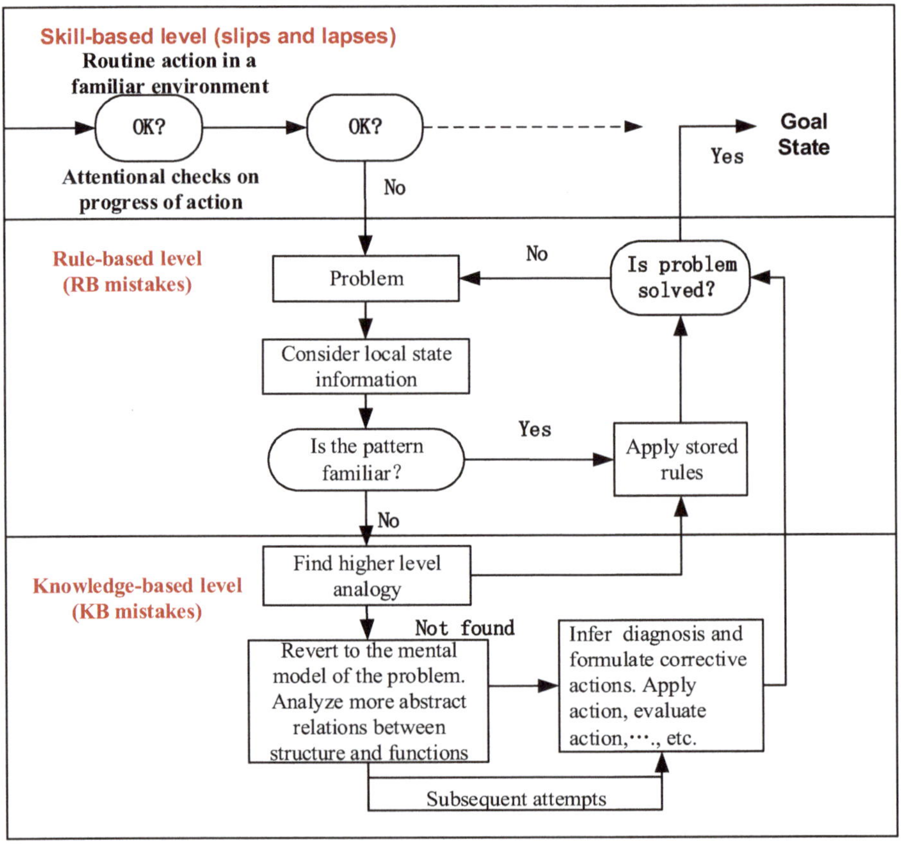

Fig. 7.3 Reason's General Error-Modeling System (cited from [41])

7.4.3 Human Error Modes for Studying Software Engineers

The author developed a pool of human error modes suitable for application in software engineering. These modes are based on widely accepted error patterns in the psychological community, such as those summarized by Reason [41] and the theory of "post-completion errors" by Byrne and Bovair [42].

Four strategies were used to build this pool of human error mechanisms:

Direct Inclusion: Human error modes that are typical, clearly specified in psychological research, and validated by scientists in software engineering were directly included in the pool. For instance, "difficulties with exponential developments" was described in Reason's work [41] as a human error pattern at the knowledge-based performance level. Huang and Strigini's experimental study demonstrated that this error mode indeed manifested as a software defect in programs [11]. Another example is the "post-completion error," first proposed by psychologists Byrne and Bovair [41], Huang [43] designed an experiment embedding the "error-prone scenario" of post-completion error in programming requirements, and the results showed that this scenario indeed triggered many programmers to commit this error, statistically more likely than any random errors.

Reorganization: To retain a systematic relationship between the included error modes. For example, "Countersigns and non-signs," "Rule strength," "General rules," and "Redundancy" are four sub-modes under the mode "misapplication of good rules" in Reason's work [41]. These modes describe different situations where people are prone to applying "strong-but-now-wrong" rules, but they pertain to the same mechanism. Therefore, they are considered not as separate error modes but as scenarios where the error mode of "applying 'strong-but-now-wrong' rules" is prone to occur.

Exclusion: Some human error modes unlikely to occur in software development, and/or difficult for software practitioners to understand, were excluded. Software development tends to be dominated by rule-based and knowledge-based performances. Some errors at the skill-based level are unlikely to occur in software development and/or hard for software practitioners to understand. For instance, the "Overattention" category of errors occurs when an individual commits an error by paying extra attention to the task at hand. Such skill-based errors could occur in daily life or operation-intensive activities, such as playing the piano for an advanced player.

Addition: New human error modes specific to software development were added based on the Integrated Cognitive Model of Software Design. For instance, "lack of knowledge" is a category found by Huang's experimental studies [11], but was not listed in Reason's book.

The final pool of human error modes recommended by the author is described in Table 7.2. It is important to note that these are some human error modes the author considers typical in software engineering, not the full set of human error modes discovered by psychologists.

Table 7.2 Typical human error modes for software engineering

Performance level	Human error mode	Description
Skill-based	Perceptual confusions	People tend to mistakenly use one item for another because these two items share some similarities
	Omissions following interruptions	Original sequence of activity is picked up one or two steps further along after an interruption
Rule-based	Applying the "strong-but-now-wrong" rule	In a context that is similar to past circumstances, people tend to behave in the same way, neglecting the signs of exceptional or novel circumstances. The typical scenarios include: the person encounters a new feature for the first time ("First Exception"); the competing rule is very "strong", in that has been successfully used many times before ("Rule Strength"); and the competing rule is a general rule ("General Rules") [41]. We should expect that programmers tend to prefer rules that have been successfully used in the past. The more frequently and successfully the rule has been used before, the more likely it is to be recalled and used
	Rule encoding deficiencies	Features of a particular situation are either not encoded at all or misrepresented in the conditional component of the rule
Knowledge-based	Lack of knowledge	Software defects or vulnerabilities are introduced when one lacks knowledge, or even does not realize some other knowledge is required. This error mode is liable to appear especially when the problem belongs to an unfamiliar application domain
	Difficulties with exponential developments	Humans tend to underestimate the rate of change, either growth or decline, and tend to construct linear models, when exponential models are required to understand a situation in reality
	Selectivity	Psychologically salient, rather than logically important information is paid attention to
	Biased review	Humans tend to believe that all possible courses of action have been considered, when in fact only a subset have been considered. When programmers generate test cases, they may fail to take all conditions into consideration, e.g. exception and boundary conditions
	Confirmation bias	Humans tend to verify ones' assumptions rather than refute them
Cross-level	Post-completion error	If the ultimate goal is decomposed into sub-goals, a sub-goal is likely to be omitted under the following conditions: the sub-goal is not a necessary condition for the achievement of its super-ordinate goal, and the sub-goal is to be carried out at the end of the task

7.5 The DPeHE Approach

7.5.1 DPeHE Framework

DPeHE provides a process to understand these mechanisms, learn from historic errors in a systematic way, and improve developers' awareness and ability to preventing human errors.

The framework of DPeHE is based on meta-cognitive theories. Meta-cognition refers to the "cognition about cognition" [32]. Meta-cognition includes two essential components [33, 34]: (1) knowledge of cognition, referring to knowledge about cognitive processes and knowledge that can be used to control cognitive processes and, (2) regulation of cognition, as activities regarding self-regulatory mechanisms during an ongoing attempt to learn or solve problems, including planning activities, monitoring activities, and evaluating outcomes.

Meta-cognition helps people to perform cognitive tasks more effectively. It is a salient feature of high- performance individuals in learning, problem solving, and engineering design [44–47]. There is a great deal of research demonstrating that those who are flexible and perseverant in problem solving, and who consciously apply their intellectual skills, are those who possess well-developed meta-cognitive abilities [44]. Since meta-cognition involves monitoring and regulating cognitive activities, it is also assumed to be closely related to cognitive failures [48, 49].

Researchers use meta-cognition training to improve cognitive performance, e.g., learning ability [50, 51], math problem solving [48, 52–54] and engineering design [47, 55]. Meta-cognitive knowledge and regulation can be improved using a variety of instructional strategies. For example, Schoenfeld [54] elaborated a regulatory strategy consisting of heuristics to be used in mathematical problem solving. He divided problem solving into three stages: analysis, exploration, and verification. For example, strategies used in problem analysis include drawing a diagram, examining special cases of the problem, and trying to simplify the problem. Schraw [56] proposed a strategy evaluation matrix (SEM) in classroom settings, which includes strategies of skimming, slowing down, activating prior knowledge, mental integration and diagrams.

Based on the above evidence, it is reasonable to believe that providing meta-cognitive training for error prevention can be a helpful way of improving the software developers' ability to prevent errors, since it is widely accepted that proper meta-cognitive training can significantly improve problem solving performance [48, 53, 55, 57, 58]. Meta-cognition theory provides us with a reliable DPeHE framework from a theoretical viewpoint. However, only after directing the knowledge and regulations at the specific purpose (human error prevention) and the specific context (software development) does the approach become practical. In the DPeHE, the meta-cognitive knowledge concerns the cognitive process of software problem solving, such as why software developers commit errors, what kinds of errors they tend to commit under certain scenarios, and what strategies they can use. DPeHE regulation enables developers to become aware of error-prone scenarios

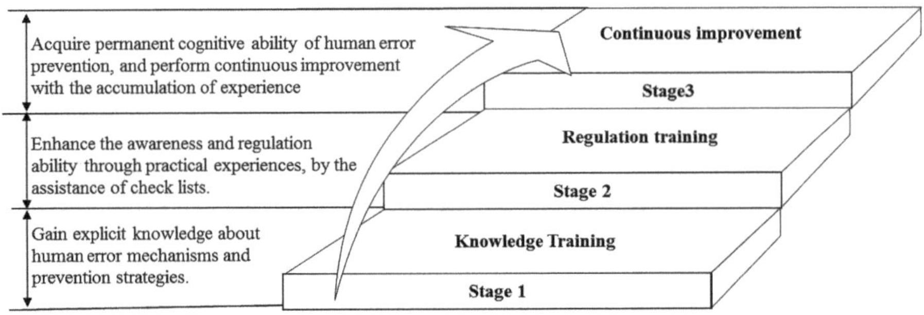

Fig. 7.4 The framework of DPeHE

when solving software design problems, to employ strategies to avoid committing errors, and to make appropriate evaluations of the outcomes.

The DPeHE framework is shown in Fig. 7.4. The DPeHE consists of three stages. The first stage is called "knowledge training". Software developers are trained to know why humans commit errors, what kinds of errors they might commit during programming, under what scenarios humans tend to commit errors, and how to prevent human errors during programming.

The second stage is called "regulation training". Regulation training is the process to help software developers promote the awareness and self-regulation of human error prevention. Two DPeHE checklists are distributed to software developers. Before and during the course of programming tasks, software developers are asked to plan and monitor their problem solving processes assisted by the problem solving regulation list. After defects are found by debugging or testing, developers are required to identify causes for each defect they introduced, assisted by the root cause identification checklist. This stage continues until developers reflect that they can be consciously aware of some symptoms under error-prone situations.

At the final stage, software developers have acquired the ability to self-regulate under error-prone situations. They become consciously aware of error-prone situations and use strategies to prevent errors. Developers are encouraged to improve such abilities continuously with the accumulation of experience.

7.5.2 Human Error Knowledge Training

Software development is a complex and flexible cognitive activity. Systematic knowledge about human errors is the basis to prevent errors; moreover, the amount of such knowledge deposited in programmers' long-term memory is an important determinant of their meta-cognitive monitoring ability [59]. Such *systematic* knowledge should contain not only error modes, but also the underlying process that governs human thoughts and actions. Understanding of the cognitive mechanisms involved in software development can be helpful for comprehending error mechanisms, as "correct performance and systematic errors are two sides of the same coin" [41].

Therefore, the knowledge training of DPeHE contains two aspects: cognitive model of the software design and human errors. The cognitive model of software design (described in Sect. 7.3.3), which provides the cognitive mechanism that governs human errors, is the other side of the "coin". Knowledge about human errors are designed in the frame of "why, when, what and how": why and when software developers tend to commit errors, what patterns of errors they might commit under different scenarios, and "how" to prevent errors by using heuristics strategies.

7.5.2.1 Why Human Commits Errors: human Error Mechanism

Errors are the manifestations of brain bottlenecks [41]. The first bottleneck is the capacity limitation of working memory, which leads to errors in "the attentional mode"; the other bottleneck lies in the human knowledge base, which leads to errors in "the schematic control mode" [41].

The Attentional Mode

Errors in the attentional mode originate from the limited capacity of working memory. Working memory is the system that actively holds information in the mind to do verbal and non-verbal tasks such as reasoning and comprehension, and to make it available for further information processing. Working memory is characterized as limited in terms of capacity and the time to sustain information.

The limitation of working memory confers the important benefits of two cognitive mechanisms: "selectivity" of inputs and "automatic processing" of information. In the real world, information is extremely abundant. With limited attention resource, humans only pay attention to several features of information. In such selective processes, if one has allocated attention to wrong features or ignored some important features, errors occur. Meanwhile, humans tend to "save" attention in familiar situations, leaving the task in automatic processing. In the course of such automatic processing, if one has omitted proper checks or performed mistimed checks, errors may occur.

Another group of error modes is also originated from working memory limitation. If one's working memory is overloaded that the available attention is not sufficient for the task on hand, errors are prone to occur.

"Working memory overload" is a cognitive state, which appears when cognitive load exceeds the individual's working memory capacity. A diagram (Fig. 7.5) is extracted from Schnotz and Kurschner's cognitive load theory [30] to visually demonstrate the main mechanism of working memory overload. Working memory load consists of three kinds of loads: intrinsic load, extraneous load and germane load [30]. *Intrinsic load* is imposed by the intrinsic aspects of a task. It is due to the natural complexity of the information that must be processed. However, complexity varies with subjects' proficiency levels. With the practice level enhanced, experts can "chunk" their schema [2, 4, 31]—a large number of interacting elements for one

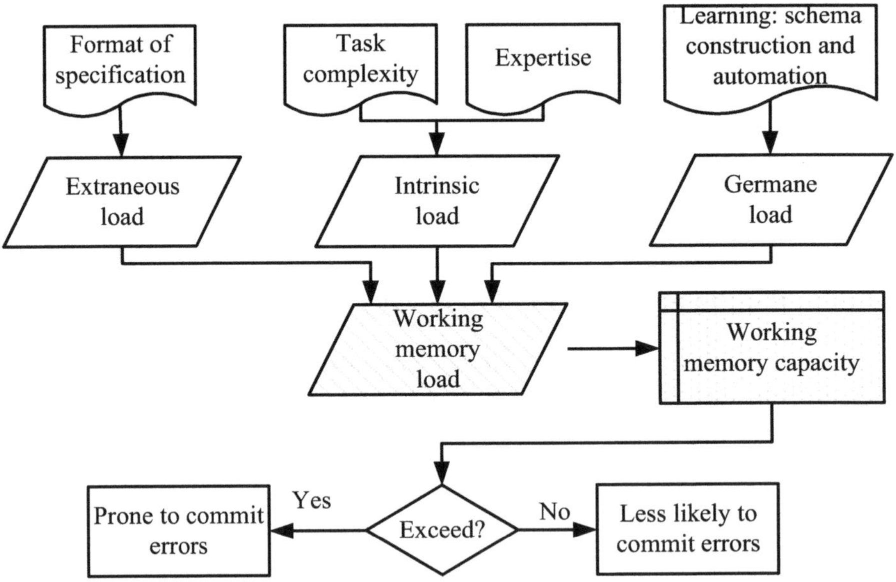

Fig. 7.5 Factors leading to working memory overload (from [5])

learner may only be a single element for another learner with better expertise. Accordingly, intrinsic load can be determined only with reference to a particular level of expertise. *Extraneous load* is caused entirely by the format of the problem representation material, e.g., requirement specifications. *Germane load* is cognitive load caused by cognitive activities in working memory. It aims at intentional learning and goes beyond simple task performance, which is caused when the subject is engaged in schemata construction. Activities such as restructuring of problem representations and meta-cognitive processes are typical cognitive activities that increase germane load.

In summary, working memory overload is a mental state when working memory load exceeds the limitation of working memory capacity. Under this situation, people are liable to commit errors. Error modes called "informational overload", "workspace limitations" and "problems with complexity" by Reason [41] are all associated with working memory overload.

The Schematic Control Mode

The generic cognitive structure underlying all aspects of human knowledge and skill is schema, which was described as "an active organization of past reactions, or of past experiences, which is supposed to be operating in any well-adapted organic response" [41]. Through using schemata, most everyday situations do not require effortful processing—automatic processing is all that is required. People can quickly organize new perceptions into schemata and act effectively without effort. Under most circumstances, these "default" behaviors are sufficient; however, under exceptional or novel circumstances, they may lead to errors. Reason calls this "the schematic control mode".

In the "schematic control mode", one tends to ignore countersigns indicating that the previous rules are inapplicable to the current situation, especially on the first occasion when the individual encounters a significant exception to a general rule. Humans tend to use those rules that have been successfully and frequently used in past. When it comes to solution construction, one tends to fit his/her previous experience. This tendency manifests itself as "illusory correlation" and "halo effects" [41]. When it comes to solution checking, one tends to seek evidence to verify his/her hypotheses rather than to refute the hypotheses (e.g., "confirmation bias" and "overconfidence"). Furthermore, humans tend to believe that all possible courses of an action have been checked, when in fact very few have been considered ("biased review") [41].

7.5.2.2 What and When Errors Are Committed: error Mode Base

Though human errors appear in different "guises" in different contexts, they take a limited number of underlying modes [41]. Understanding such recurring error modes is essential to improving the cognitive ability of error prevention. Therefore, we built an error mode base that contains common error modes committed by software developers.

Being the basis for DPeHE training, the error modes included in the base must be very reliable. Therefore, the error modes included in DPeHE are those that have been widely validated and accepted, e.g., classical modes summarized by Reason [41], "post completion error" [42], lack of knowledge [7], and problem representation errors [2, 60, 61]. Furthermore, we adapted, tailored, and reorganized these modes to make a balance between usability, consistency and completeness. For instance, "perceptual confusions" proposed by Reason (1990) was not included in the base as we think this error mode rarely occurs in software development. "Countersigns and nonsigns" and "rule strength" [41] share the same underlying mechanism, "apply strong but now wrong rules" mode, so they are integrated. Error modes such as "repetitions" [41] are excluded as they are hard to understand by engineers.

Furthermore, to enable the error modes to be practically used, explanations, scenarios and programming error examples are provided with each error mode, shown in Table 7.2. These scenarios and examples demonstrate how the general psychological error modes manifest themselves as programming errors. Knowing such scenarios is essential to build conscious awareness of error-prone situations. The examples were obtained from the controlled experiments [2, 62] or historical data in the Chinese aviation industry [63]. They are recurrent error patterns committed by different people in the same ways and governed by the same human error mechanisms.

7.5.2.3 How to Prevent Error: Strategy Heuristics

Based on the knowledge about the cognitive mechanisms of problem solving and error committing, we then provide strategies to guide software developers to prevent errors. Strategy heuristics have proven to be an effective and idiomatical way to enhance self-regulation in education domain [54, 64].

In meta-cognitive training, heuristics are proposed as powerful general strategies to support learner in selecting the right strategy to solve a problem. General heuristics can be applied in a variety of problems; they are usually content-free and thus apply across many different situations. Researchers agree that students who know about the different kinds of heuristics will be more likely to use them when confronting different classroom tasks [54, 65, 66].

Schoenfeld [54] and Schraw [56] have done a lot of work on meta-cognitive regulatory strategies in problem solving education. Schoenfeld [54] elaborated a regulatory strategy teaching method consisting of heuristics to be used in the stages of analysis, exploration and verification. Strategies used in problem analysis include *drawing a* diagram, *examining special cases of the problem*, and *simplifying the problem*. Strategies used in exploration include *replacing conditions by equivalent ones*, *decomposing the problem in sub-goals*, and *looking for a related problem where the solution method may be useful with a view to solving the new problem*. Strategies used in verification contain checking whether all pertinent data are used, and checking whether the solution can be obtained differently.

Schraw [56] has proposed a strategy evaluation matrix (SEM) in classroom settings, which includes strategies of *skim*, *slow down*, *activate prior knowledge*, *mental integration* and *diagrams*. SEMs was found to be useful by teachers who used them [56]. One strength of SEMs is that they promote strategy use (i.e., a cognitive skill), which is known to significantly improve performance. A second strength is that SEMs promote explicit meta-cognitive awareness, even among younger students. A third strength is that SEMs encourage students to actively construct knowledge about how, when, and where to use strategies.

Integrating the above strategies [54, 56], the error mechanisms described in Table 7.2 with the problem solving model described in Sect. 3.2.1, a list of strategy heuristics are designed, shown in Table 7.3. For instance, the strategy "slow down" is borrowed from Schraw [56]; while "searching countersigns" is proposed based on mechanisms of the targeted error mode: "apply 'strong but now wrong' rules" described in Table 7.2. To prevent this type of errors, one needs to delicately allocate attention to searching the signs that may contradict to the retrieved strong schema.

The strategies are arranged according the problem-solving stages and under a systematic framework of "when to use, how to use, and why to use". We divide the problem-solving process into three stages, problem representation, solution generation and solution evaluation. "When to use" describes the error-prone situations under which the strategy applies. The classification on problem-solving stages combined with the error-prone situations aim to promote one's *awareness* of error prevention. "How to use", which describes the ways to use the strategies, is proposed for improving one's *self-regulation* under error-prone situations. "Why to use" explains the researchers' motivation underlying the proposed strategies—error modes that the strategies are used to prevent.

Table 7.3 Strategy heuristics for error prevention

Stage of problem solving	Strategy	When to use (Error-prone situations)	How to use	Why to use (Targeted error modes)
Problem representation	Slow down	When information seems especially important	Stop, read, and think about information	Enhance focus of one's attention
	Retrospectively reason	When solution is generated automatically and quickly	Ask oneself how the solution was produced, what schema was used in previous experiences	Enhance one's attention in rule-based performance, preventing "strong but now wrong" errors
	Searching countersigns	The task in hand seems extremely familiar	Check if some information indicates the condition in hand is different from previous experiences	Focus one's attention on exceptional conditions, preventing "strong but now wrong" errors
Solution generation	Decomposition	The problem is complex	Decompose the problem into sub-problems and work on them in the top-down hierarchical way	Prevent errors caused by working memory overload and problems with complexity
	Mental integration	Learn complex information or require deeper understanding of a problem	Relate main ideas. Use these to construct a theme or conclusion	Prevent errors caused by lack of knowledge
	Making notes or draw diagrams	When there is a lot of interrelated information	Identify main ideas, connect them, and list supporting details under main ideas, connecting supporting details	Extend working memory capacity, preventing errors caused by working memory overload and problems with complexity

(continued)

Table 7.3 (continued)

Stage of problem solving	Strategy	When to use (Error-prone situations)	How to use	Why to use (Targeted error modes)
Solution evaluation	Hierarchically tracking	The problem is complex and is decomposed into hierarchical sub-goals	Check if all the goals have been achieved in the hierarchical way	Prevent sub-goal omission errors and errors caused by inattention
	Examine special cases	The problem is complex, or interactive with many other functions	Examine if some special cases or boundary conditions are lack of consideration	Prevent errors caused by lack of knowledge and biased review
	Falsification	Perform unit test by oneself	Try to search evidences to reject the solution. Design falsification test cases to check the solution	Prevent errors caused by confirmation bias
	Exchanging review	One thinks the solution is completed and there is no any other problem detected by himself or herself	Find someone else to check the program, conduct an exchanged review with a partner, or independent test	Prevent errors caused by biased review and confirmation bias

7.5.3 Human Error Regulation Training

Knowing error mechanisms and prevention strategies does not mean software developers will use them in practice. Therefore, simply providing knowledge without experience or vice versa does not seem to be sufficient for the development of cognitive control. Only when a software developer is consciously aware of the error-prone situations and applies the strategies he/she learned in real programming practice, has the error-prevention ability been obtained.

Self-questioning is an effective way to promote cognitive regulation for self-directed learners and it is widely used in various domains [67]. For example, Schraw [56] proposed a regulatory checklist (RC), which provides a heuristic with three main categories: planning, monitoring, and evaluating. Questions like "what is my goal?" at the planning stage, "am I reaching my goals?" in monitoring and "what didn't work" are proposed to learners. Such self-questioning strategies can guide the learner's performance before, during, and after task performance, thus improve his/her self-awareness and control over thinking and thereby improve performance.

We designed two regulation checklists to promote the ability to preventing errors. One checklist called problem solving regulation checklist (PSRC) is used before and in the course of software development. The other checklist, root cause identification checklist (RCIC), is used after defects are found in debugging or testing.

7.5.3.1 Problem-Solving Regulation Checklist (PSRC)

The awareness of the problem-solving process is essential to preventing errors. Different stages of problem solving involve different mental activities, along with different error modes. The problem regulation checklist is designed to help developers monitor their problem-solving process and promote awareness under error-prone scenarios. The checklist can be used before and during the course of the developers' task, shown in Table 7.4.

Each scenario is relevant to a specific error mode, shown in the right column of Table 7.4. The PSRC and the strategy heuristics in Table 7.3 are closely related: error prevention is logically part of problem solving. Each error mode in Table 7.4 can be tracked to corresponding strategy heuristics in Table 7.3. PSRC is focused on arousing an individual's awareness of error-prone scenarios during his or her problem solving in a real task. Whereas strategy heuristics focus on what strategies can be used when the individual is aware of the error-prone scenarios. The strategy heuristics can be presented to individuals in the knowledge training report beforehand.

Table 7.4 Sample of problem-solving regulation checklist

Problem solving stage	Scenarios (Triggering factors) questionnaire self-awareness regulatory	Relevant error modes
Problem representation	– Does the task involve disciplinary domains? – Have you ever solved such problem before? – Have you ever received related training? – Do you feel hard to find solution?	Lack of knowledge
	– Is there any ambiguous item in the specification document? – Do you find something inconsistent with what you think before? – Are there some items you are not sure about the meaning? – Is there any implicit specification?	Problem representation error
	– Is there any information that you did not pay attention to? – Is there any special case you should consider?	Selectivity

(continued)

Table 7.4 (continued)

Problem solving stage	Scenarios (Triggering factors) questionnaire self-awareness regulatory	Relevant error modes
Solution generating	– Is this the first occasion you encounters a significant exception to a general rule? – Is there any countersign indicating that the more general rule is inapplicable? – Is there any countersign omitted as the information is too abundance? – Are the rules you use to solve this problem have been used successfully and many times before (any exceptions this time)?	Apply "strong but now wrong" rules
	– Is there any properties of the problem space are not encoded? – Are there any properties of the problem space encoded inaccurately? – Are there any domain-specific exceptional rules?	Rule encoding deficiencies
	– Is the problem seems complex? – Are there many interfaces you have to handle? – Do you need to construct any exponential models (e.g., in the form of $y = ab^x$) instead of linear models (e.g., in the form of $y = a + bx$)? – Do you feel your brain is working under high load?	Problems with complexity and difficulty with exponential development
Solution evaluation	– Have you tried to refute your solution? – Are there any other possible solutions? – How may the mistakes reveal themselves if your solution is wrong?	Confirmation bias
	– Any boundary conditions are out of your consideration? – Any extreme conditions you need to consider?	Biased review
	– Is there any sub-goal to be carried out at the end of the task but not a necessary condition for the achievement of its corresponding super-ordinate goal? – Are there any sub-goals omitted?	Post-completion error
All the process	– Interrupted by someone or something else, have you checked breakpoint? – Have you been engaged in the same activity for a long time, and is there anything needed to pay attention to? – Have you been engaged in two or more tasks, and is there any interference occurred?	Omissions following interruptions
	– Are there any characters/words look like each other? – Are there any similar function/object names? – Are there any similar directories?	Perception confusions

7.5.3.2 Root Cause Identification Checklist (RCIC)

The evaluation of the outcomes is also important for the development of meta-cognitive regulation [33]. Learners who observe and evaluate their performance accurately may react appropriately by keeping and/or changing their study strategies to achieve the goal of, for example, maximizing their grade in a course. Thus, it is desirable to engage learners in activities that will help them assess themselves explicitly [51].

For software development, generally, not all errors can be prevented during the problem-solving processes. Some errors may be introduced into the program and reveal themselves as defects, which may be found in debugging or testing. It is desirable to engage developers to assess and explain explicitly why these defects are introduced and what types of errors they have committed. Through such self-checking process, they can enhance their awareness of error committing mechanism and prevention heuristics.

In order to assist developers in identifying their defect causes, a retrospective checklist designed is provided in Table 7.2. In addition to human error modes included in the table, the users may consider adding other factors leading to software defects such as process and tools, into the checklist.

7.6 Summary

Only by thoroughly understanding the mechanisms behind the occurrence of an event can we truly prevent and control it. Human error is a key factor in software failures, yet traditional defect prevention techniques have not delved deep into the realm of human error mechanisms. Cognitive errors of programmers are the primary factors leading to software defects. The awareness and regulatory abilities of programmers in error-prone situations are fundamental to preventing software defects. This chapter proposes the concept of proactive fault prevention based on human error mechanisms.

This chapter proposes preventing software defects by enhancing programmers' metacognitive abilities in terms of human error (DPeHE). A DPeHE process model based on metacognitive training framework has been constructed. DPeHE training consists of two main stages: the first stage involves metacognitive knowledge training on human error, and the second stage involves metacognitive regulation experiences on human error. Metacognitive knowledge training includes knowledge on cognitive processes in program design, human error mechanisms, and measures for human error prevention. Metacognitive regulation experiences are conducted through reflection during problem-solving processes and self-evaluation after discovering faults.

Compared to traditional fault prevention techniques, DPeHE targets the blind spots that traditional fault prevention techniques cannot reach—programmers' cognitive errors. Moreover, traditional fault prevention techniques are usually implemented by only a few companies, while DPeHE is accessible to all companies regardless of their level of process maturity, whether at CMM5 or initial levels. DPeHE can also be used for self-improvement of programmers' cognitive abilities in defect prevention. DPeHE provides the software industry with a new perspective for reducing fault introduction rates.

References

1. P.J. Hopcroft, G. Broadfoot, Cutting errors in software design. Information Professional **1**, 30–35 (2004)
2. F. Huang, B. Liu, Y. Song, S. Keyal, The links between human error diversity and software diversity: Implications for fault diversity seeking. Sci. Comput. Program. **89**(Part C), 350–373 (2014)
3. G.M. Weinberg, *The Psychology of Computer Programming* (VNR Nostrand Reinhold Company, London, 1971)
4. F. Huang, B. Liu, Y. Wang, Review of software psychology. Comput. Sci. **40**, 1–7 (2013)
5. F. Huang, B. Liu, Software defect prevention based on human error theories. Chin. J. Aeronaut. **30**, 1054–1070 (2017)
6. F. Détienne, *Software Design—Cognitive Aspects* (Springer, New York, 2002)
7. F. Huang, B. Liu, B. Huang, A Taxonomy System to Identify Human Error Causes for Software Defects, in *The 18th International Conference on Reliability and Quality in Design* (Boston, USA, 2012), pp. 44–49
8. F. Huang, B. Liu, Systematically Improving Software Reliability: Considering Human Errors of Software Practitioners, in *23rd Psychology of Programming Interest Group Annual Conference (PPIG 2011)* (York, UK, 2011)
9. V. Anu, W. Hu, J.C. Carver, G.S. Walia, G. Bradshaw, Development of a human error taxonomy for software requirements: a systematic literature review. Inf. Softw. Technol. **103**, 112–124 (2018)
10. H. Mohammadnazar, M. Pulkkinen, H. Ghanbari, A root cause analysis method for preventing erratic behavior in software development: PEBA. Reliabil. Eng. Syst. Saf. **191**, 106565 (2019)
11. F. Huang, L. Strigini, HEDF: A method for early forecasting software defects based on human error mechanisms. IEEE Access **11**, 3626–3652 (2023)
12. F. Huang, B. Zhao, H. Madeira, A New Code Review Method Based on Human Errors, in *2022 IEEE 22nd International Conference on Software Quality, Reliability and Security (QRS)* (2022), pp. 321–332
13. C.L. Jones, A process-integrated approach to defect prevention. IBM Syst. J. **24**, 150–167 (1985)
14. D.N. Card, Myths and Strategies of Defect Causal Analysis, in *Proceedings of the Twenty-Fourth Annual Pacific Northwest Software Quality Conference* (2006), pp. 469–474
15. D. Card, Learning from our mistakes with defect causal analysis. IEEE Softw. **15**, 56–63 (1998)
16. M. Kalinowski, D.N. Card, G.H. Travassos, Evidence-based guidelines to defect causal analysis. IEEE Software **29**, 16–18 (2012)
17. R.G. Mays, Applications of defect prevention in software development. IEEE J. Selected Areasin Commun. **8**, 164–168 (1990)
18. R.G. Mays, C.L. Jones, G.J. Holloway, D.P. Studinski, Experiences with defect prevention. IBM Syst. J. **29**, 4–32 (1990)
19. M. Leszak, D.E. Perry, D. Stoll, A Case Study in Root Cause Defect Analysis, in *Proceedings of the 22nd International Conference on Software Engineering* (Limerick, Ireland, 2000), pp. 428–437
20. M. Schadler. Casual Analysis and Resolution [Online]. https://opal.msoe.edu/sdl/cmmi_archive/2008-2009/Causal%20Analysis%20and%20Resolution/CausalAnalysis_schadlerm.pdf
21. M. Kalinowski, G.H. Travassos, D.N. Card, Towards a Defect Prevention Based Process Improvement Approach, in *34th Euromicro Conference Software Engineering and Advanced Applications* (2008), pp. 199–206
22. J.H. van Moll, J.C. Jacobs, B. Freimut, J.J.M. Trienekens, The Importance of Life Cycle Modeling to Defect Detection and Prevention, in *Proceedings of the 10th International Workshop on Software Technology and Engineering Practice* (2002), pp. 144–155

23. R. Chillarege, in *Orthogonal Defect Classification*, ed. M.R. Lyu. Handbook of Software Reliability Engineering (McGraw-Hill and IEEE Computer Society Press, 1996), pp. 359–400
24. R. Glaser, Education and thinking: The role of knowledge. Am. Psychol. **39**, 93–104 (1984)
25. A.D. Baddeley, G.J. Hitch, *Working Memory* (Academic, New York, 1974)
26. A. Baddeley, Working memory. Science **255**, 556–559 (1992)
27. W.G. Chase, H.A. Simon, Perception in chess. Cogn. Psychol. **4**, 55–81 (1973)
28. G.A. Miller, The magical number seven, plus or minus two: Some limits on our capacity for processing information. Psychol. Rev. **63**, 81–97 (1956)
29. J. Sweller, Cognitive load during problem solving, effects on learning. Cogn. Sci. **12**, 257–285 (1988)
30. W. Schnotz, C. Kurschner, A reconsideration of cognitive load theory. Educ. Psychol. Rev. **19**, 469–508 (2007)
31. F. Détienne, Expert Programming Knowledge: a Schema-Based Approach, in *Psychology of Programming*, ed. by J.-M. Hoc, T.R.G. Green, R. Samurcay, D. Gilmore, (Academic, 1990), pp. 205–222
32. J.H. Flavell, Metacognitive aspects of problem solving, in: *The nature of Intelligence*, vol. 12, pp. 231–235, 1976
33. A.L. Brown, Metacognition, executive control, self-regulation and other more mysterious mechanisms, in *Metacognition, Motivation, and Understanding*, (Lawrence Erlbaum Associates, New Jersey, 1987), pp. 65–116
34. G. Schraw, D. Moshman, Metacognitive theories. Educ. Psychol. Rev. **7**, 351–371 (1995)
35. R. Guindon, Designing the design process: Exploiting opportunistic thoughts. Hum. Comp. Interact. **5**, 305–344 (1990)
36. L.J. Ball, J.S.B.T. Evans, I. Dennis, Cognitive processes in engineering design: a longitudinal study. Ergonomics **37**, 1753–1786 (1994)
37. W. Visser, Dynamic Aspects of Design Cognition: Elements for a Cognitive Model of Design, INRIA, France, Research Report 2004
38. A.J. Ko, B.A. Myers, A framework and methodology for studying the causes of software errors in programming systems. J. Vis. Lang. Comput. **16**, 41–84 (2005)
39. B. Adelson, E. Soloway, A model of software design. Int. J. Intell. Syst. **1**, 195–213 (1986)
40. J. Rasmussen, Skills, rules, and knowledge; signals, signs, and symbols, and other distinctions in human performance models. IEEE Trans. Syst. Man Cybern. **13**, 257–266 (1983)
41. J. Reason, *Human Error* (Cambridge University Press, Cambridge, 1990)
42. M.D. Byrne, S. Bovair, A working memory model of a common procedural error. Cogn. Sci. **21**, 31–61 (1997)
43. F. Huang, Post-completion Error in Software Development, in *The 9th International Workshop on Cooperative and Human Aspects of Software Engineering, ICSE* 2016, (Austin, 2016), pp. 108–113
44. L.H. Swanson, Influence of metacognitive knowledge and aptitude on problem solving. J. Educ. Psychol. **82**, 306–314 (1990)
45. S. Bryant, Rating expertise in collaborative software development, in *Proceedings of the 17th Workshop on the Psychology of Programming Interest Group* (2005), pp. 19–29
46. B. Berardi-Coletta, L.S. Buyer, R.L. Dominowski, E.R. Rellinger, Metacognition and problem solving: A process-oriented approach. J. Exp. Psychol.: Learn. Memory Cogn. **21**, 205–223 (1995)
47. R. Hargrove, Assessing the long-term impact of a metacognitive approach to creative skill development. Int. J. Technol. Des. Educ. **23**, 489–517 (2013)
48. V. Pennequin, O. Sorel, I. Nanty, R. Fontaine, Metacognition and low achievement in mathematics: The effect of training in the use of metacognitive skills to solve mathematical word problems. Think. Reason. **16**, 198–220 (2010)
49. A. Efklides, G.D. Sideridis, Assessing cognitive failures. Eur. J. Psychol. Assess. **25**, 69–72 (2009)
50. M.V.J. Veenman, B.H.A.M. Van Hout-Wolters, P. Afflerbach, Metacognition and learning: conceptual and methodological considerations. Metacogn. Learn. **1**, 3–14 (2006)

51. C.A. Gama, *Integrating Metacognition Instruction in Interactive Learning Environments*, Doctor of Philosophy, University of Sussex (2004)
52. S.K. Teong, The effect of metacognitive training on mathematical word-problem solving. J. Comput. Assist. Learn. **19**, 46–55 (2003)
53. B. Kramarski, Z.R. Mevarech, Enhancing mathematical reasoning in the classroom: The effects of cooperative learning and metacognitive training. Am. Educ. Res. J. **40**, 281–310 (2003)
54. A.H. Schoenfeld, Learning to think mathematically: Problem solving, metacognition, and sense-making in mathematics, in *Handbook for Research on Mathematics Teaching and Learning*, ed. by D. Grouws, (Macmillan, New York, 1992), pp. 334–370
55. O. Lawanto, Metacognition changes during an engineering design project, in *Frontiers in Education Conference, 2009. FIE '09. 39th IEEE*, 2009, pp. 1–5
56. G. Schraw, Promoting general metacognitive awareness. Instr. Sci. **26**, 113–125 (1998)
57. M. Chinnappan, M.J. Lawson, The effects of training in the use of executive strategies in problem solving. Learn. Instr. **6**, 1–17 (1996)
58. E. Kapa, Transfer from structured to open-ended problem solving in a computerized metacognitive environment. Learn. Instr. **17**, 688–707 (2007)
59. T.O. Nelson, L. Narens, Metamemory: A theoretical framework and new findings. Psychol. Learn. Motiv. **26**, 125–141 (1990)
60. A. Tang, M.F. Lau, Software architecture review by association. J. Syst. Softw. **88**, 87–101 (2014)
61. B. Meyer, On formalism in specifications. IEEE Softw. **2**, 6–26 (1985)
62. F. Huang, B. Liu, Study on the correlations between program metrics and defect rate by a controlled experiment. J. Softw. Eng. **7**, 114–120 (2013)
63. F. Huang, B. Liu, S. Wang, Q. Li, The impact of software process consistency on residual defects. J. Softw. Evol. Process **27**, 625–646 (2015)
64. M.V.J. Veenman, P. Wilhelm, J.J. Beishuizen, The relation between intellectual and metacognitive skills from a developmental perspective. Learn. Instr. **14**, 89–109 (2004)
65. A. Whimbey, J. Lochhead, R. Narode, *Problem solving & comprehension* (Routledge, 2013)
66. J.D. Bransford, A.L. Brown, R.R. Cocking, *How People Learn: Brain, Mind, Experience, and School* (National Academy Press, 1999)
67. X.D. Lin, J.D. Lehman, Supporting learning of variable control in a computer-based biology environment: Effects of prompting college students to reflect on their own thinking. J. Res. Sci. Teach. **36**, 837–858 (1999)

MIX
Papier aus verantwortungsvollen Quellen
Paper from responsible sources
FSC® C105338

If you have any concerns about our products,
you can contact us on
ProductSafety@springernature.com

In case Publisher is established outside the EU,
the EU authorized representative is:
Springer Nature Customer Service Center GmbH
Europaplatz 3, 69115 Heidelberg, Germany

Printed by Libri Plureos GmbH
in Hamburg, Germany